Viscoelastic Behavior of Rubbery Materials

Viscoelastic Behavior of Rubbery Materials

C. M. Roland
Polymer Physics Section, Naval Research Laboratory
Washington DC

OXFORD
UNIVERSITY PRESS

OXFORD
UNIVERSITY PRESS

Great Clarendon Street, Oxford OX2 6DP

Oxford University Press is a department of the University of Oxford.
It furthers the University's objective of excellence in research, scholarship,
and education by publishing worldwide in

Oxford New York

Auckland Cape Town Dar es Salaam Hong Kong Karachi
Kuala Lumpur Madrid Melbourne Mexico City Nairobi
New Delhi Shanghai Taipei Toronto

With offices in

Argentina Austria Brazil Chile Czech Republic France Greece
Guatemala Hungary Italy Japan Poland Portugal Singapore
South Korea Switzerland Thailand Turkey Ukraine Vietnam

Published in the United States
by Oxford University Press Inc., New York

First published 2011

British Library Cataloguing in Publication Data
Data available

Library of Congress Cataloging in Publication Data
Data available

Typeset by SPI Publisher Services, Pondicherry, India
Printed in Great Britain
on acid-free paper by
CPI Antony Rowe, Chippenham, Wiltshire

ISBN 978–0–19–957157–4

1 3 5 7 9 10 8 6 4 2

--- *Dedicated to Carol, my wife and best friend* ---

Preface

The preface is where an author gives his excuse for writing a book. My motivation in part was the fear that the things I've learned or discovered were exceeding my ability to recall them. This book describes the relaxation dynamics of rubbery materials, with the objective of providing a molecular basis for many physical properties. As the term comprises any amorphous, flexible macromolecule above its glass-transition temperature, rubber includes a broad class of substances, with a richness of behavior rivaled by few materials. The focus is mainly on the phenomenology, emphasizing anomalies and aspects that are incompletely understood and thus productive avenues for future research. Rubber is especially interesting because it has unique properties. It can exist in a state of equilibrium, unlike glassy or semicrystalline plastics, thermosetting resins, fibers, etc. These polymers have path-dependent morphologies and process-specific properties, which frustrate scientific inquiry, notwithstanding their practical utility. Among all materials only rubber exhibits high elasticity – the ability to recover from very large deformations. This property underlies most applications of elastomers and gave rise to its own field of study (Chapter 4). Despite these singular characteristics, rubber is arguably the prototype for relaxation in soft matter. By copolymerizing different monomers, an enormous variety of chemical structures are available that, along with the ease of avoiding crystallization, makes rubber ideal for the study of the glass transition (Chapters 2 and 6), a major unsolved problem in condensed-matter physics. In the glassy state or when vitrification is imminent, polymers cannot easily be distinguished from molecular liquids, and the correspondence of many phenomena makes distinctions between molecular and polymeric liquids artificial. Accordingly, the scope of this book is not limited to polymer science, with the discussion often extending to small-molecule compounds, including simple liquids and liquid crystals.

Science is a collaborative effort and I am indebted to many colleagues, in particular Riccardo Casalini, Peter Mott, Kia Ngai, Marian Paluch, Don Plazek, Chris Robertson, and Pat Santangelo. I thank C. Alba-Simionesco, G. Caetano-Anolles, S. Capaccioli, J. Colmenero, J. Czub, J.R. Dutcher, M.D. Ediger, S. Etienne, P. Gong, T. Hogan, C. Holland, R. Inoue, W.G. Knauss, H.-N. Lee, D. Leporini, Y. Miyamoto, K.L. Ngai, L. Nikiel, M. Paluch, A. Petelin, C. Riedel, E. Rossler, C.G. Robertson, K. Schmidt-Rohr, S. Simon, A.P. Sokolov, P. Strehlow, O. Tsui, S. Urban, D.C. Venerus, and B. Vogt for providing data and images, and A.V. Roland, A.P. Saab, D.E. Sheehan, and especially R.J. Colton for careful reading of the manuscript. The financial support of the Office of Naval Research is gratefully acknowledged.

Contents

1
Introduction

Molecular motions in fluids are of long-standing interest to physicists, chemists, materials engineers, and biologists since the dynamics bear a direct relationship to the physical and rheological properties underlying many applications. A first-principles approach to understanding material properties would address how the chemical structure determines the forces between molecules, and in turn how these forces govern the molecular motions. Fluctuation-dissipation theory then provides a connection between theoretical models and experimental observables. The latter usually involve the decay of stress, polarization, or other measurable quantities, which are the macroscopic manifestation of the molecular diffusions, reorientations, and conformational transitions returning the perturbed system to equilibrium.

The relaxation[1] of even simple van der Waals fluids remains a continuing field of study. Analysis of liquid motions is simplest for non-interacting species, and solutions were obtained in the early twentieth century for translational Brownian diffusion[2] [1] and rotational diffusion [2] in dilute solution. For viscous liquids of sufficient density that molecules exert a reciprocal influence, mutual interactions can dominate the behavior – neighboring species must make small adjustments in position in order for a given molecule to change configuration. The behavior deviates from thermally excited transitions between potential wells, and theoretical efforts to describe the many-body dynamics remain largely at the model-building stage. It is in the study of such complex correlated dynamics that polymers in general, but rubber in particular, can be most useful – the former because of their diverse properties, unique phenomena, and facile supercooling;[3] the latter because only rubber among polymers exists in a state of structural and mechanical equilibrium. For this reason, rubber is the most fundamentally interesting polymeric material. Its unique properties have also led to wide application in industry, the military, and for consumer products (world consumption of natural and synthetic rubber exceeded 23 million tons in 2008 [3]), which in turn has spurred further research.

[1] The term "relaxation" is used to describe both the experimental data (e.g., relaxation spectra) and the underlying motions (e.g., segmental relaxation times).

[2] Thermally driven random movement of molecules is known as Brownian motion, after the 1827 Scottish botanist Robert Brown, who used a microscope to observe the irregular paths followed by pollen particles suspended in water. Initially ascribed to the particles being alive, the correct mechanism – molecular collisions – was identified by Einstein in 1905. Living cells also exhibit migration, but the motion is well beyond that due to thermal fluctuations.

[3] Traditionally "supercooling" referred to quenching below the melting point to a metastable liquid state; however, many polymers are incapable of crystallization due to irregularity of their chemical repeat units and thus have no melting temperature. Nevertheless, reducing their temperature towards the glass transition is still regarded as supercooling.

1.1 Viscoelasticity and high elasticity in polymers

The distinctive feature of polymers is the enormous size of the constituent molecules ("macromolecules"), which can be two orders of magnitude larger than the distance between segments. This is quite different from ordinary liquids, which, unless near a critical point or the glass transition, have only one characteristic length scale, the intermolecular separation, equal to a few Angstroms. It is the size and consequent large aspect ratio of chain molecules that give polymer melts[4] (that is, amorphous polymers above the glass-transition temperature T_g) their special properties, in particular high elasticity and viscoelasticity.

High elasticity (rubber elasticity) refers to the ability of flexible-chain polymers to recover from large strains, a property unique to chain molecules. Experiments have shown that high molecular weight, uncrosslinked polymers exhibit recoverable strains (stretched length divided by length after recovery) of as much as 10 [4,5]. Entanglements give rise to a transient network that responds to applied forces through conformational changes of the polymer backbone (e.g., for extensional strains *gauche* rotamers convert to the more extended *trans* form). The rotations of backbone segments are highly correlated (intramolecular cooperativity) to avoid large displacements of the chain. Upon removal of the external force, the material recovers some of its original dimensions since the conformational transitions are reversible. High elastic strains are attained in polymeric liquids only if the strain rate is large relative to the molecular response time; flow, which changes the relative position of the molecules, leads to unrecovered strain ("set"). A network of chemically bonded chains, however, retains it shape, with any permanent set the result of defects, chain scission, or mechanically labile network junctions. As implied by the name, permanent set is distinct from recoverable strain that may be retarded due to the viscoelastic nature of polymers. This behavior is discussed in Chapter 4.

A measure of the hypothetical maximum extension of a polymer chain is the ratio of its fully extended length to its mean length. From eqn. (1.22) below, this quantity is proportional to the square root of the number of backbone bonds. For a typical polymer having a molecular weight $\sim 10^5$ g/mol ($\sim 10^3$ repeat units), this extension is on the order of 100. Of course, reversible strains this large are unattainable for real networks, which are limited by defects and finite extensibility.

Viscoelasticity describes a time-varying reaction to a transient perturbation in an unchanging material; equivalently, the perturbation could be dynamic, resulting in a frequency-dependent response. The requirement is that the response changes with time even though the material does not. Viscoelasticity includes mechanical properties such as stress relaxation and the dynamic modulus, as well as experimental quantities involving related molecular motions, such as dielectric polarization and the quasi-elastic scattering of light, X-rays, or neutrons. The term itself implies a second, equivalent definition: energy is both dissipated and stored during deformation of a viscoelastic material; that is, the material response has viscous and elastic components.

Non-polymeric materials exhibit viscoelastic effects, a prominent example being supercooled small-molecule liquids [6,7]. Viscoelasticity is manifested in metals as creep under

[4] "Melt" is commonly used for high molecular weight, amorphous polymers above T_g, with the term liquid reserved for unentangled polymers; this distinction reflects the fact that entanglements confer solid-like properties over a limited frequency range. However, the terminology is confusing insofar as "melt" implies the loss of an ordered structure, such as found in crystals. Since near T_g the viscoelastic behaviors of molecular liquids and polymer melts are indistinguishable, the terms are used interchangeably herein.

sustained loading [8] and becomes significant at temperatures above about 0.5 times their melting point [9]. This high-temperature creep is due to changes in the microstructure of the metal, but even well below the melting point metals are not entirely elastic. For example, long term creep in the Cu-Be wires of a torsional balance introduced small errors into measurements of the gravitational constant [10]. Strictly speaking, the behavior of metals is not truly viscoelastic, since the changing response to a load is usually caused by changes in the material itself, e.g., its dislocation structure or grain size.

Polymers are the prototypical viscoelastic materials, exhibiting large rate dependences at almost all temperatures. The normal forces accompanying shear flow of polymers is another consequence of their viscoelasticity (although normal stresses are not unique to viscoelastic fluids). Viscoelasticity is a consequence of the diverse range of molecular motions in polymers, which means that for any applied stress or strain, some relaxation modes are moving on the timescale of the perturbation and thereby dissipating energy. This gives rise to a peak in the out-of-phase, or loss, component of the dynamic response, at a frequency equal to the frequency of the underlying molecular motions. Other modes, moving relatively fast, respond either elastically or, for the entire molecule in the absence of crosslinking, as flow. Some molecular motions may have very long time constants, so as to be "frozen" or unresponsive, on the timescale of the perturbation; an example would be conformational transitions of the polymer backbone below T_g. In rheology two parameters are used to describe the competition between the timescale of the imposed perturbation and the time characterizing the material response. The Deborah number[5] is the ratio of the response time to the experimental timescale. A large Deborah number means the fluid responds in a more solid-like fashion. The related Weissenberg number[6] is the ratio of the material relaxation time to the process time during flow, and is used to characterize the amount of molecular orientation induced by the flow.

In Figure 1.1 the in-phase and out-of-phase components of the dynamic modulus are shown for high molecular weight 1,4-polyisoprene (PI) subjected to a periodic shear strain [11]. The energy stored per cycle is proportional to the storage modulus, G', and the energy dissipated to the loss modulus, G''. Energy storage and loss are both expected whenever relaxation occurs, although the two quantities are not independent. According to the Kramers–Kronig relation [12], based on the principle of causality (the time order of cause and effect)

$$\int_0^\infty G'(\omega)\frac{\sin\omega t}{\omega}d\omega = \int_0^\infty G''(\omega)\frac{\cos\omega t}{\omega}d\omega = \frac{\pi}{2}G(t) \tag{1.1}$$

where $G(t)$ is the stress relaxation modulus measured in a transient experiment and ω is the circular frequency, equal to 2π times ν, the frequency in Hertz. Although valid for all linear responses, eqn (1.1) is difficult to apply because it requires knowledge of the modulus function over the entire frequency range. There are various approximations to the Kramers–Kronig relation [13], for example

$$G''(\omega) \approx \frac{\pi}{2}\frac{dG'(\omega)}{d\ln\omega} \tag{1.2}$$

[5] The term comes from the biblical quote by the prophetess Deborah *"The mountains melted from before the Lord"* (Judges 5:5, KJV).

[6] Named for the early twentieth century Austrian rheologist Karl Weissenberg.

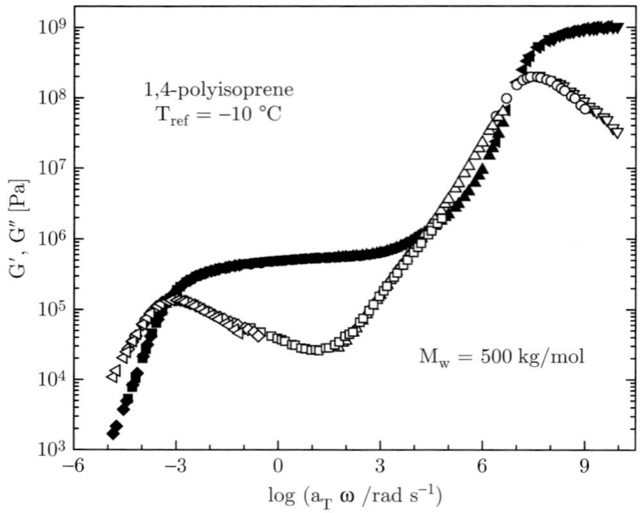

Figure 1.1 Dynamic storage and loss modulus of polyisoprene ($M_w = 500$ kg/mol). The symbols denote the different measurement temperatures, with the isotherms shifted to obtain the master curve at the indicated reference temperature. The spectra exhibit both aspects of viscoelasticity – the modulus varies with frequency and the response has both elastic and dissipative components. Data from ref. [11].

The corresponding equation for the dielectric loss is sometimes used to circumvent the masking of loss peaks by ionic conductivity, since the latter does not contribute to the in-phase response [14].

The frequency of molecular motions is characterized by a time constant or relaxation time. The classical definition is the time required for the perturbation to be reduced by a factor of e ($\sim$2.72). This implies exponential decay (or Debye relaxation in the frequency domain), which is valid for relaxation that is homogeneous and proceeds at a rate proportional to the amount remaining to relax. For example, after imposition of a strain, the stress, $\sigma(t)$, relaxes according to

$$\frac{d\sigma}{dt} = -k_0\sigma(t) \tag{1.3}$$

$$\sigma(t) = \sigma_0 \exp(-k_0 t) \tag{1.4}$$

where σ_0 is the initial stress and k_0 the rate constant (equal to the reciprocal of the relaxation time). Exponential relaxation rarely describes the intermolecularly cooperative motions in dense fluids. Another characteristic timescale is the period required for complete cessation of the decay; however, this is usually too long to be practical and relies on judgment of when equilibrium has been reestablished. (Since variables still fluctuate while in thermal equilibrium, an indicator that equilibrium has been attained is when the change of a variable with time becomes equal to its rate of change due only to equilibrium fluctuations [15].)

An absorption peak in the loss modulus (Figure 1.1) or dielectric loss, reflecting the conversion of mechanical or electrical energy to heat, can be used to define a model-independent, most probable relaxation time as the reciprocal of the peak frequency, $\tau_{max} = \omega^{-1}$. If the peak in the spectrum is symmetric, τ_{max} is also the mean relaxation time. However,

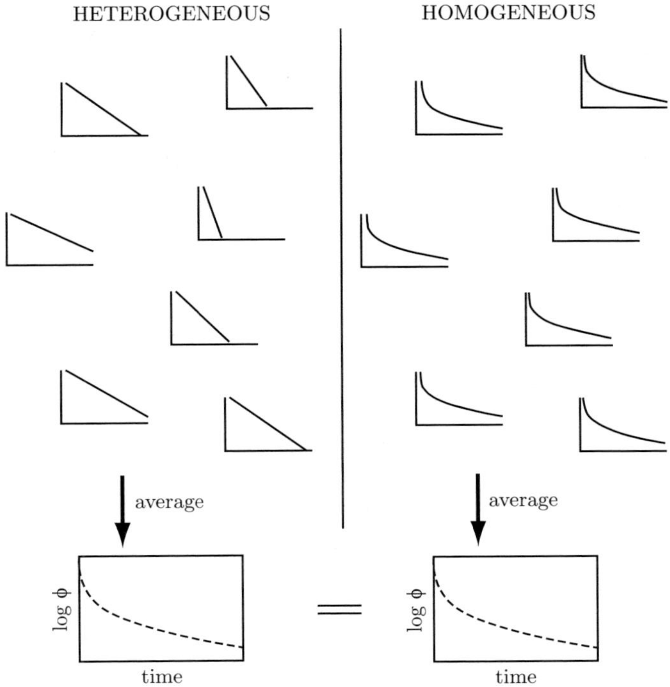

Figure 1.2 Heterogeneous and homogeneous dynamics leading to non-exponential relaxation. The former corresponds to a distribution of exponential relaxation times and the latter to inherently non-exponential relaxation function. The spatially distributed responses are averaged to yield the observed decay function, which is indistinguishable for the two scenarios. Reproduced from ref. [17] with permission.

relaxation functions are often asymmetrically broadened. The asymmetry can arise due to heterogeneous relaxation, in which the response is a weighted summation of simple exponential decays

$$G(t) \sim \int_{-\infty}^{\infty} H(\tau) e^{-t/\tau} d(\ln\tau) \tag{1.5}$$

with $H(\tau)$ the distribution function. Alternatively, the relaxing species may just decay in an inherently non-exponential fashion, in which an empirical equation or a model is used to describe the peak shape. An $H(\tau)$ can always be found that will fit the data, with the mean relaxation time given by the weighted sum

$$\langle\tau\rangle = \frac{\int_0^\infty \tau H(\tau)}{\int_0^\infty H(\tau)} d\tau \tag{1.6}$$

Since molecular motions in polymers encompass such a broad frequency range, a logarithmic timescale is more useful; that is, the distribution is expressed in terms of $\ln\tau$ ($d\ln\tau = \tau^{-1} d\tau$) [16]. Heterogeneous and homogeneous relaxation scenarios are depicted in Figure 1.2 [17].

A common form of asymmetry of relaxation peaks is characteristic skewing toward higher frequencies ("stretched exponential decay") that conforms in the time domain to the Kohlrausch function[7] [18,19]

$$G(t) = G_0 \exp[-(t/\tau_K)^{\beta_K}] \tag{1.7}$$

where G_0 is a constant and $0 < \beta_K \leq 1$. This equation is obtained by making the rate constant in eqn (1.3) time dependent: $k_0(t) \sim t^{\beta_K - 1}$. The one-sided Fourier transform of eqn (1.7) gives the Kohlrausch function in the frequency domain and, while written in terms of the mechanical modulus, there are corresponding equations for dielectric relaxation, quasi-elastic scattering, etc. If eqn (1.7) arises from a superposition of exponential relaxation functions

$$G(t) = \int_0^\infty H(\tau) \exp(-t/\tau) d\tau \tag{1.8}$$

the distribution of relaxation times is given by [20]

$$H_K(\tau) = -\frac{1}{\pi\tau} \sum_{k=0}^{\infty} \frac{(-1)^k}{k!} \sin(\pi\beta k)\Gamma(\beta k + 1)\tau^{\beta k} \tag{1.9}$$

where Γ is the gamma function.[8] This distribution function includes several decades of relaxation times shorter than the main (peak) relaxation time. Equation (1.9) must be evaluated numerically, although a good approximation is

$$H_K(\tau) \cong \left[\frac{\beta_K}{2\pi(1-\beta_K)\tau^2}\right]^{\frac{1}{2}} \left[\frac{\beta_K \tau}{\tau_K}\right]^{\frac{\beta_K}{2(1-\beta_K)}} \exp\left[-(1-\beta_K)\left(\frac{\beta_K \tau}{\tau_K}\right)^{\frac{\beta_K}{(1-\beta_K)}}\right] \tag{1.10}$$

The Kohlrausch relaxation time τ_K is related to the mean value as

$$\langle\tau\rangle = \frac{\tau_K}{\beta_K}\Gamma(\beta_K^{-1}) \tag{1.11}$$

Fitting the higher-frequency dispersion in Figure 1.1 to eqn (1.7) gives $\beta_K = 0.52$, with the full width at half-maximum (FWHM) = 2.11 decades. The lower limit for a dispersion is the breadth of a Debye peak (β_K is unity), FWHM = 1.144 decades. The various relaxation times for the segmental peak in Figure 1.1 are $\tau_K = 0.25$ μs, $\tau_{max} = 0.33$ μs, and $\langle\tau\rangle = 0.46$ μs; this rank ordering of magnitudes

$$\tau_K \leq \tau_{\max} \leq \langle\tau\rangle$$

is general and not specific to any particular value of β_K.

Exponential decay usually describes thermally excited single molecule transitions, expected to conform to an Arrhenius temperature dependence

$$\tau = \tau_\infty \exp(-E_a/RT) \tag{1.12}$$

[7] This function is also named after Williams and Watts, who derived it independently [19].

[8] The gamma function $\Gamma(x) = \int_0^\infty e^{-u}u^{x-1}du$. If x is a positive integer, $\Gamma(x) = (x-1)!$

with a constant activation enthalpy[9] E_a (R is the gas constant). A heuristic argument, referred to as the compensation rule, requires that the prefactor, τ_∞, in eqn (1.12) be related to E_a according to [21]

$$\tau_\infty \propto \exp(-E_a) + constant \tag{1.13}$$

If correct, a distribution of exponential decays would be invalid for many observed relaxation phenomena. However, non-exponential relaxation is rarely associated with Arrhenius behavior. If all molecular motions contributing to the response have the same temperature dependence, temperature alters only the timescale of the response and the material is thermorheologically simple; that is, the behavior conforms to the time–temperature superposition principle. This aspect of viscoelastic behavior is addressed in detail in Chapter 6.

1.2 Modes of motion

There are various length scales in polymers, in addition to the intersegment or intermolecular distance in common with molecular liquids: (i) the Rouse length scale ($\sim$ couple of nm), which refers to the shortest distance over which the separation of chain segments has a Gaussian distribution; (ii) the entanglement distance (*ca.* 3 to 8 nm), referring to the effective distance a chain moves transversely to its contour before experiencing topological constraints due to the uncrossability of the chains. It is these entanglement constraints that confer high elasticity to uncrosslinked rubber; and (iii) the molecular size *per se* ($\leq$ 50 nm). Since the conformational arrangement of polymer chains exhibits a dynamic equilibrium and the lengths of different chains may not be the same (molecular weight polydispersity), the macromolecular size is expressly defined by moments. For linear chains the mean square end-to-end distance, $\langle r^2 \rangle$, is an obvious measure of size, while for branched or cyclic chains the distribution of segments around the mass center is more meaningful; thus, the mean square radius of gyration is proportional to the mean square distance between each of the n backbone atoms.

$$\langle R_g^2 \rangle = \frac{1}{2n^2} \sum (r_i - r_j)^2 \tag{1.14}$$

For freely jointed chains or for real linear polymers under Theta conditions,[10] $\langle r^2 \rangle = 6R_g^2$. The molecular size of polymers has an especially significant effect on their rheological properties. When crosslinked, the primary chain length becomes irrelevant, aside from a small influence on network properties (e.g., strength and hysteresis) affected by chain ends.

These length scales refer to the structure of the material. However, in condensed matter including polymers, molecules and segments rearrange at different rates, giving rise to heterogeneous dynamics and thus dynamic length scales. Some of the dynamic properties unique to polymers are manifested in the glassy or semicrystalline states, but they are most apparent in amorphous polymers above their glass-transition temperature. It is these conditions that define a "rubber," which refers to an amorphous, flexible chain, high polymer[11] above T_g. (An elastomer is a rubber that has been chemically crosslinked.) If the polymer backbone lacks

[9] In the chemical literature E_a is referred to as an activation energy but enthalpy is the more correct term for an isobaric process, given the potential for PV work.

[10] Chains in solution under Theta conditions have the same configuration as in the equilibrium melt state.

[11] The term "high polymer" implies a molecular weight sufficiently large that the chain ends have no effect on properties.

flexibility, it cannot undergo the conformational transitions required for high elasticity. A metric of chain flexibility is the persistence length, l_p, describing the distance over which bond orientations become uncorrelated. l_p is equal to the area under the correlation function describing the orientational correlation of backbone bonds

$$C(r) = \langle \tilde{u}(0) \cdot \tilde{u}(r) \rangle \quad (1.15)$$

where $\tilde{u}$ is a unit vector parallel to the bond, r runs along the polymer contour, and the brackets imply the scalar product averaged over all chains. For a completely flexible polymer l_p equals one bond length, and for a completely rigid rod it is equal to the chain length. Another way to distinguish flexible chains from rigid-rod polymers is by their crystallization behavior [22–24]. Without the ability to conformationally rearrange, rigid rods maximize their packing by aligning in parallel fashion to form extended-chain crystals. To minimize the loss of orientational entropy, however, chains with sufficient flexibility will back-fold during crystallization, and this absence of extended chain crystals can serve as a defining characteristic of flexible-chain polymers. Based on a very approximate lattice model, Flory concluded that 0.63 was the fraction of bent (*gauche*) bonds necessary for a chain to crystallize with back-folding and thus be considered a flexible macromolecule [22]. High-speed spinning and related technologies are routes to achieve extended-chain crystallization of flexible-chain polymers by reducing the transient concentration of *gauche* conformers during processing [25,26,27]. Extended-chain crystals form stronger, more dimensionally stable fibers because the back-folds are defects. This Flory approach to chain flexibility leads to another way to contrast flexible and rigid chains – from the size of the chain relative to the size of a completely flexible chain having the same number of repeat units. This characteristic ratio, C_n, is defined in eqn (1.24) below.

Another classification of dynamic properties is between those that are unaffected by chain length, and thus common to materials composed of small molecules and macromolecules, and molecular motions that are unique to polymers. The former are the more local modes, including the glass transition, since one cannot distinguish a polymer from a molecular liquid using only measurements around T_g. As discussed below the dependence of T_g on molecular weight is due to the presence of chain ends; this dependence is not a direct consequence of the long-chain character of the molecule. The chain ends in a high polymer are too sparse to affect T_g and other properties.

The broad range of length scales in polymers leads to a diversity of their dynamics: vibrations, the Boson peak, the nearly constant loss, secondary relaxations including both side group motions and the Johari–Goldstein process, structural relaxation (the glass transition), the Rouses modes, and the entangled dynamics. Only the last two are unique to polymers. In Figure 1.3 various dynamics are evident in the dielectric loss spectrum of 1,4-polybutadiene (PBD) [28]. Temperature is used as the abscissa in order that the processes, which encompass more than 12 decades of frequency, can be viewed together. In general we expect slower dynamic processes to be progressively less sensitive to chemical structure, since motion over longer length scales averages out the structural details.

1.2.1 Vibrations

Beginning around 10^{-14} s and sensibly independent of temperature, the fastest motions of atoms (i.e. neglecting electronic excitations) are vibrations, which change the instantaneous bond angles and lengths and occur even in the glassy state. These are not specific to polymers

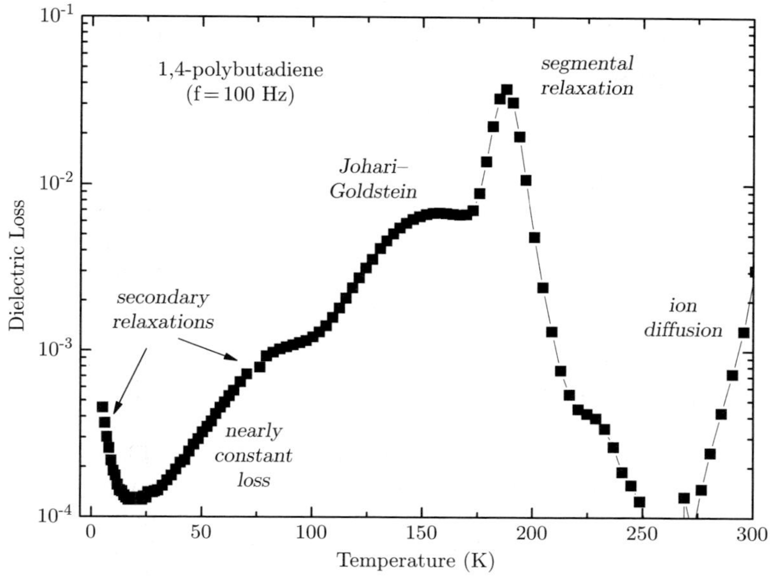

Figure 1.3 Dielectric loss of 1,4-polybutadiene measured as a function of temperature at 100 Hz. The different relaxation mechanisms appear as peaks, while the nearly constant loss and ionic conductivity are manifested as changes in slope. Data from ref. [28]

but do depend on the atoms and chemical structure; thus, vibrational spectroscopy is useful for identification of a material, including the polymer species and any additives. Related to these oscillations of individual moieties are the phonons. A universal feature of glassy and crystalline solids, phonon motion consists of lattice vibrations; i.e. collective modes in the form of extended plane waves. The term itself comes from the fact that long-wavelength phonons underlie the propagation of sound in solids. Phonon excitations transpire at very high frequencies, *ca.* 10^{12} Hz. The earliest model of phonons was due to Einstein, who assumed all vibrational frequencies to be equally probable [29]. This leads to incorrect predictions for the heat capacity of solids. A more accurate description of phonon motion is the Debye model, according to which elastic solids have a density of states $g(\omega)$ (number of states per unit frequency) proportional to ω^2 for ω less than a characteristic value known as the Debye frequency, $\omega_D \sim 10^{12}$ Hz. For frequencies greater than ω_D, $g(\omega) = 0$ [29,30]. Notwithstanding its simplicity the Debye model remains the starting point for treating low-temperature lattice vibrations. However, for disordered solids such as polymeric glasses, scattering experiments reveal an additional vibrational contribution, in excess of the Debye prediction. This excess $g(\omega)$ is seen as a maximum in $g(\omega)\omega^2$ at $\omega \sim 10^{12}$ Hz [31]. This so-called Boson peak, not observed in crystals, has a shape that is independent of chemical structure [32]. An example is shown in Figure 1.4 [33]. The origin of the Boson peak is much investigated, with the phenomenon of general interest because of the possibility that the Boson intensity correlates with properties of the liquid state including the glass transition [34–36].

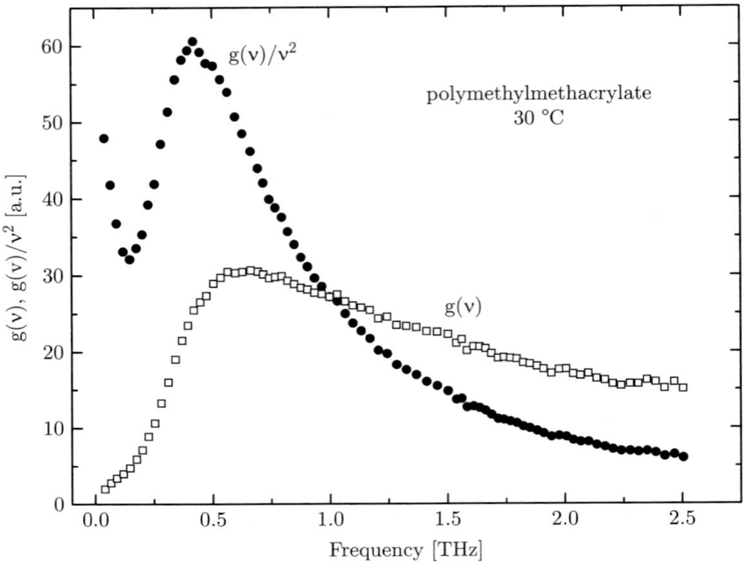

Figure 1.4 Vibrational density of states from the anti-Stokes Raman spectrum of PMMA, as measured (squares) and after division by ν^2 to make apparent the excess over the Debye theory. The spectrum is invariant to physical aging or thermal history. Data from ref. [33].

1.2.2 Nearly constant loss

Spectroscopic measurements on polymers such as PI [37], PBD [28,38], polyisobutylene (PIB) [39], polystyrene (PS) [39], and polymethylmethacrylate (PMMA) [40], as well as ions [41–45] and small molecules [46–49], reveal that the dynamics over a range of frequencies can be described by a slowly decreasing function of frequency, varying as $\omega^{-\lambda}$ where λ is a positive number close to zero. This is illustrated in light-scattering spectra for PIB both below and above T_g (Figure 1.5 [39]) and in dielectric spectroscopy on two PBDs, an oligomer and a higher molecular weight sample (Figure 1.6 [38]). Given the very weak frequency dependence, the process is referred to as the nearly constant loss (NCL). The frequency regime of the NCL is much higher than structural relaxation, so the response represents motion of species (ions, molecules, or chain segments) within the near-neighbor "cage." This cage confinement is responsible for the weak energy dissipation connoted by the term NCL.

Since the local liquid structure loosens with increasing temperature, the magnitude of the NCL increases with T (Figure 1.7); however, at higher T its temporal extent is reduced by encroachment on the low-frequency side from structural relaxation. For this reason the NCL is only observed at low temperatures and high frequencies. Thus, while the NCL should be ubiquitous, its detection is limited to conditions where there is no overlap with the more intense vibrational and reorientational contributions.

1.2.3 Secondary relaxations

Secondary relaxations are any of the myriad processes occurring at higher rates (or observed at lower temperatures) than the structural relaxation underlying the glass transition. The

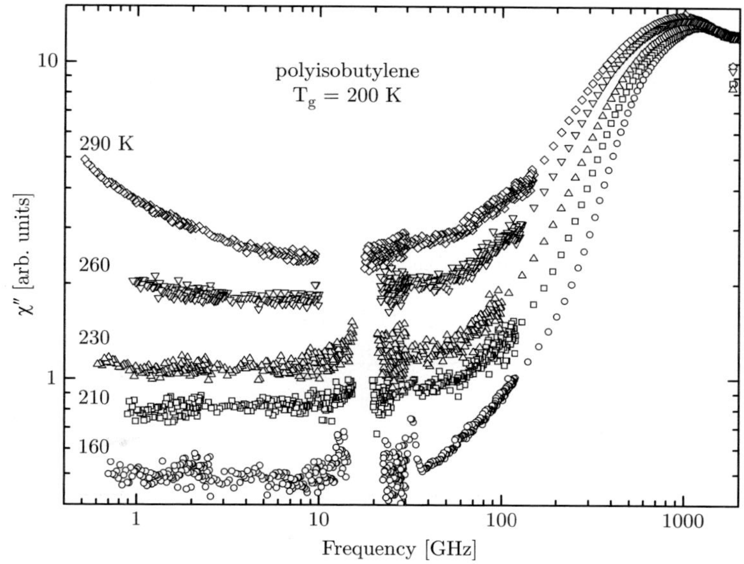

Figure 1.5 Depolarized light scattering susceptibility spectra for PIB, showing a flat, weakly T-dependent response around 1 to 30 GHz referred to as the NCL. The vibrational contribution to the spectra is evident at higher frequencies. Data from ref. [39].

most facile means to observe secondary relaxation peaks is dielectric spectroscopy because of its sensitivity and broad frequency range. Secondary relaxations are named using successive Greek letters beginning with β; the glass transition is the α-process. In Figure 1.8 dielectric spectra are shown for polycyclohexylmethacrylate, which has a complex structure and therefore multiple secondary processes. The relaxation times (inverse of peak frequency) corresponding to each process are shown in an Arrhenius plot in Figure 1.9 [50]. Secondary relaxations, found in small molecules as well as polymers, fall into two classes: the Johari–Goldstein (JG) relaxation and higher-frequency motions. The latter are more trivial, in the sense of being specific to a given material and generally having a limited effect on macroscopic, physical properties. The higher-frequency secondary relaxations are due to motion of some, but not all, atoms in the molecule or repeat unit. Examples include motion of a pendant group, the chain ends, particular backbone atoms, or motion in the vicinity of crystal defects. Table 1.1 lists representative secondary relaxations for several polymers [51–59]. Secondary relaxations can usually only be resolved unambiguously from the primary α-process below T_g and therefore are measured in the glassy state, where temperature variations only change the thermal energy but not the structure of the material. For this reason secondary relaxations exhibit an Arrhenius temperature dependence (eqn (1.12)). Activation energies are listed in Table 1.1; E_a is on the order of 10 k_BT (k_B is Boltzmann's constant).

Since below T_g a material is (by definition) out of equilibrium, properties of a glass such as volume, enthalpy, and entropy all decrease toward their equilibrium values at a rate that decreases with the extent of supercooling. This phenomenon, known as physical aging, also causes changes in characteristics of secondary relaxations; moreover, the time evolution of properties of the β-process in glasses reflects the structural aging dynamics. This means that

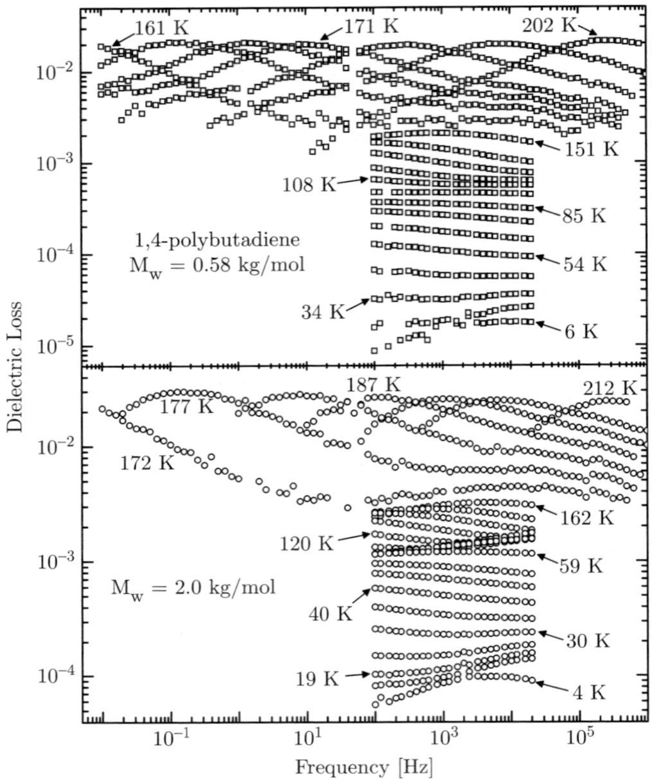

Figure 1.6 Dielectric loss for an oligomeric (upper) and a higher molecular weight (lower) PBD. At the higher temperatures the local segmental peak (α-relaxation) falls within the experimental frequency range, with the Johari–Goldstein secondary process on the high frequency side. At lower temperatures the NCL becomes evident. The measurable span of frequencies is limited at lower temperatures due to the low loss and consequent need to measure using a high-resolution bridge, which has a frequency range of only 2.3 decades. Data from ref. [38].

the time constant for these changes provides a measure of the α-relaxation time, which below T_g is too long for direct measurements. This is illustrated in Figure 1.10, showing α-relaxation times for three materials directly measured above T_g and deduced for the glassy state from changes of the secondary relaxation [60–62].

The JG relaxation is a special type of secondary motion, present in all amorphous or glass-forming materials [63,64]. It is the slowest secondary process and thus can be referred to as the β-relaxation. However, since the JG relaxation is sometimes masked by a strong glass-transition peak, the next higher-frequency secondary process is sometimes inadvertently labeled the β-peak. For small molecules the JG relaxation corresponds to rotation of the entire molecule, although the angle of the reorientation is small, enabling the motion to proceed independently from that of other molecules. The JG relaxation was first discovered in dielectric measurements on rigid molecules [65], whose secondary dynamics cannot involve intramolecular degrees of freedom; all atoms must participate in the process. For polymers

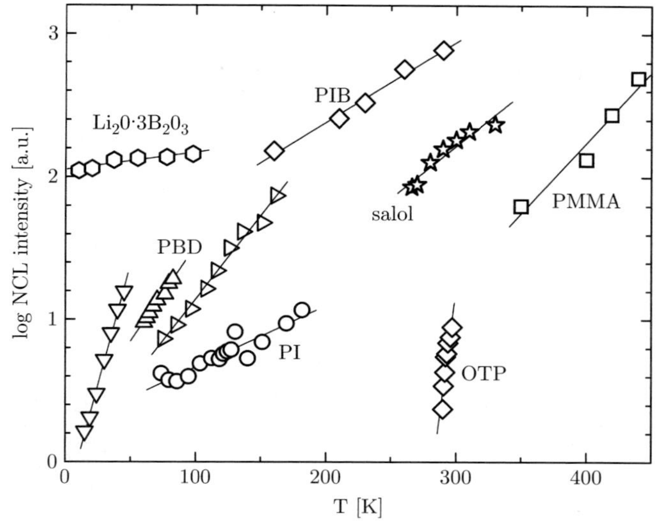

Figure 1.7 Temperature variation of the nearly constant loss in the depolarized light scattering susceptibility of PIB (diamonds, [39]) and PMMA (squares, [40]) and in the dielectric loss of PI (circles, [37]) and PBD (triangles, $M_w = 560$ kg/mol, [28]; inverted triangles, $M_w = 2.0$ kg/mol [38]; right triangles, $M_w = 0.58$ kg/mol [38]). For comparison conductivity of an alkali triborate glass (hexagons [41]) and optical Kerr effect intensities for *o*-terphenyl (pentagons) and phenyl salicylate (stars) [46] in the NCL regime are also shown. Lines are linear fits to the semilogarithmic plots.

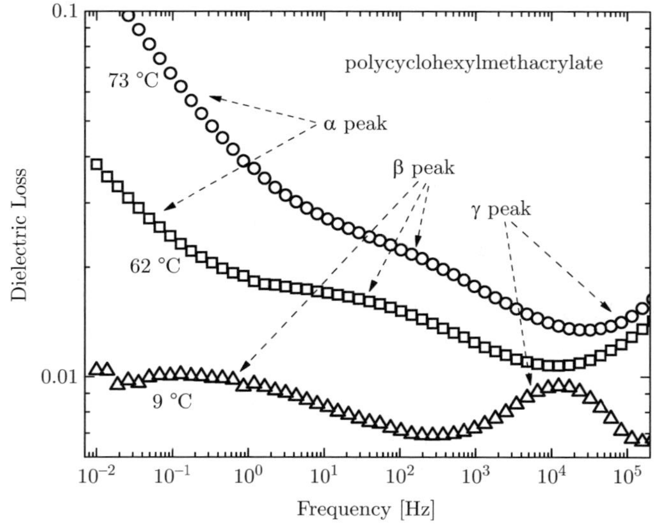

Figure 1.8 Dielectric loss measured for PCHMA [50] at temperatures for which the weak β-peak falls in the measured range of frequencies. The high-frequency side of the strong α-peak appears at higher temperatures. The γ-peak can be seen at frequencies beyond $\sim 10^4$ Hz.

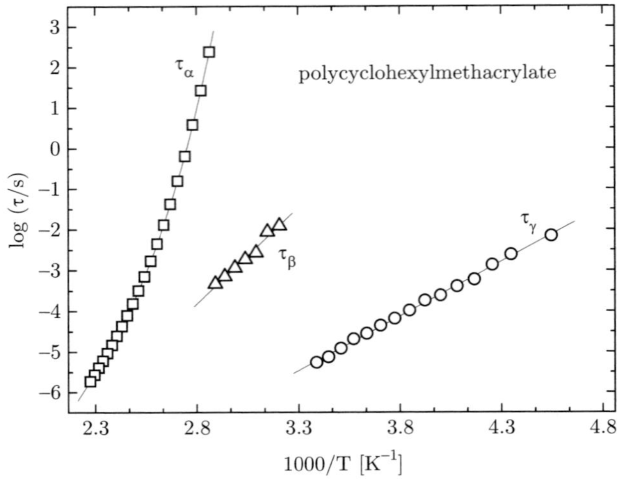

Figure 1.9 Relaxation times of PCHMA determined from the frequency of the maxima in the dielectric loss for the local segmental and two secondary relaxations [50]. The respective temperature sensitivities of the τ increase with increasing timescale of process, becoming non-Arrhenius for the segmental dynamics.

the dynamics of the JG process can be more complicated, but nevertheless a JG relaxation is usually present as the slowest secondary relaxation and one involving the entire repeat unit, not only pendant groups or specific moieties. As an example of a JG relaxation in a polymer, multidimensional NMR experiments on PMMA reveal that the β-relaxation involves 180° flips of the pendant carboxyl group; however, these are coupled to restricted rocking motion of the chain backbone in order to avoid steric interference from other segments (Figure 1.11) [66,67]. The amplitude of this rocking motion increases with temperature, eventually becoming the conformational transitions of the backbone that underlie the segmental dynamics and T_g.

Table 1.1 Non JG secondary relaxations in selected polymers.

polymer	motion	E_a (kJ/mol)	ref.
polyethylmethacrylate	pendant alkyl group motion	7.1	51
polycarbonate	carbonate motion	26	52
	phenyl torsional motion	48	
epoxy polymer	epoxide end group motion	28	53
polyvinyl carbazole	polar side group motion	29	54
nylon	methylene segments rotation	31	55
polystyrene	phenyl group rotation	38	56
styrene–acrylonitrile copolymer	cyano group motion	39	57
polycyclohexylmethacrylate	cyclohexyl flipping	45	58
polyoxymethylene	backbone twisting about crystal defects	80	59

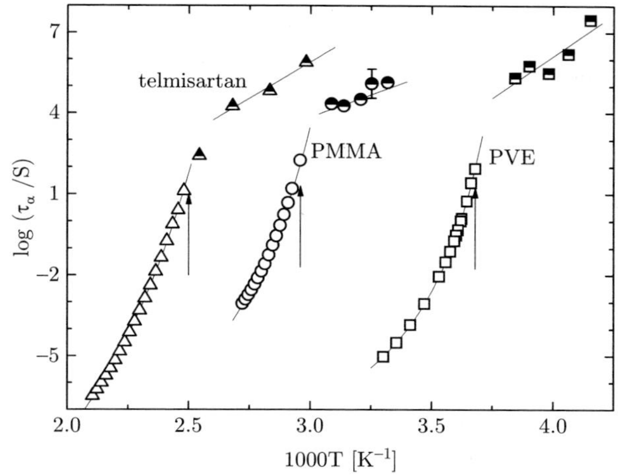

Figure 1.10 α-relaxation times measured above T_g (open symbols) and the values of τ_α deduced from the kinetics of changes in the secondary relaxation due to physical aging (half-filled symbols) for the pharmaceutical telmisartan (triangles) [60], polymethylmethacrylate (circles) [61], and polyvinylethylene (squares) [62]. The departures from Arrhenius behavior above T_g (indicated by the arrows) are absent in the glassy state.

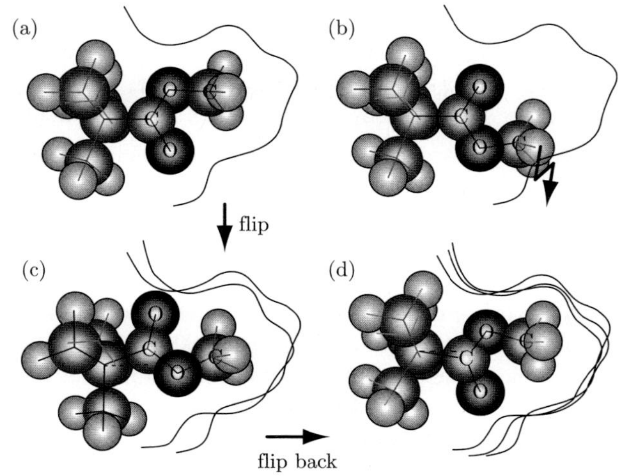

Figure 1.11 Depiction of the motion of the side group in PMMA (a) Initial orientation of the pendant group. (b) Rotation of side-group without backbone motion is sterically hindered by neighboring segments. (c) Twisting about backbone allows the side group to be accommodated by the local empty space; this in turn causes some readjustment of neighboring segments. (d) Counterrotation of the side group returns it approximately to its original orientation but with some alterations due to the changes in the orientations of the backbone and the neighboring segments. [66] (*reproduced with permission*)

Thus, the secondary relaxation in PMMA detected mainly as a side-group motion involves all the atoms and is therefore a JG process.

At higher temperatures and longer times JG relaxations evolve into the primary structural process [68], and for this reason the properties of JG relaxations are correlated with those of the segmental dynamics. The JG relaxation mimics [69] or even serves as the precursor [70] to the glass transition. Nuclear magnetic resonance (NMR) is the most unambiguous method to distinguish whether a secondary process is a JG motion, although this can sometimes be inferred from the greater sensitivity to pressure or volume changes in comparison to motions involving only a portion of the molecule or repeat unit. The properties of JG relaxations are considered in more detail in Chapter 2 dealing with the segmental dynamics of polymers.

1.2.4 Glass transition

The glass transition refers to structural relaxation occurring in the liquid as it vitrifies. T_g is operationally defined as the pressure-dependent temperature at which the local segmental dynamics of polymers (or correspondingly, the reorientation of small molecules or the translation of colloidal particles) transpires on a timescale significantly longer than the experimental time.[12] The local and intermediate range structural order of polymers, as revealed by X-ray and neutron diffraction, changes negligibly as the material is taken from the melt to the glassy state; it is for this reason that amorphous polymer glasses can be regarded as "frozen" liquids. Segmental relaxation of polymers involves skeletal bond conformational transitions, with motion over large length scales avoided by cooperative rotations of neighboring units along the chain (intramolecular cooperativity) [71,72].

Although structural relaxation *per se* is not unique to polymers, interference from crystallization is less of a problem with polymers, as alluded to above, since crystallization is often slow (or even non-existent in the case of copolymers or atactic homopolymers, which have non-uniform repeat units). Molecular liquids have to be quenched to avoid crystallization; hence, the term "supercooled" for amorphous liquids below their melting temperature. This facile vitrification makes polymers well suited for the study of the glass transition, which is a major unsolved problem of condensed-matter physics.

The glass transition is a property of polymers, along with their strength and modulus, that is influenced by the concentration of chain ends [73–75]. If the molecular weight, M, is very high, the concentration of chain ends is negligible and molecular weight effects are absent. The value of M necessary to attain this high-polymer limiting behavior depends on the material. A general rule of thumb is that chain ends exert a measurable effect only for M below *ca.* 10^5 g/mol. However, especially flexible chains, such as those with an oxygen atom in the backbone, retain an insensitivity to molecular weight down to much shorter chain lengths. Note that the high-polymer limiting behavior refers only to the effect of chain ends, but not to chain length *per se*. The slow dynamics of polymers such as the viscosity and frequency range of the rubbery plateau depend non-linearly on M, but this is a size effect and not due to any dependence on the chain-end concentration.

There have been many attempts to model the effect of M on T_g, primarily by invoking an addition from the chain ends to the free volume or configurational entropy. Simulations show that the chain ends are associated with an excess volume compared to the inner chain segments (Figure 1.12 [76]). The difficulty is that the control parameter (density, thermal

[12] This timescale can be estimated as the ratio of T_g to the rate at which temperature is varied.

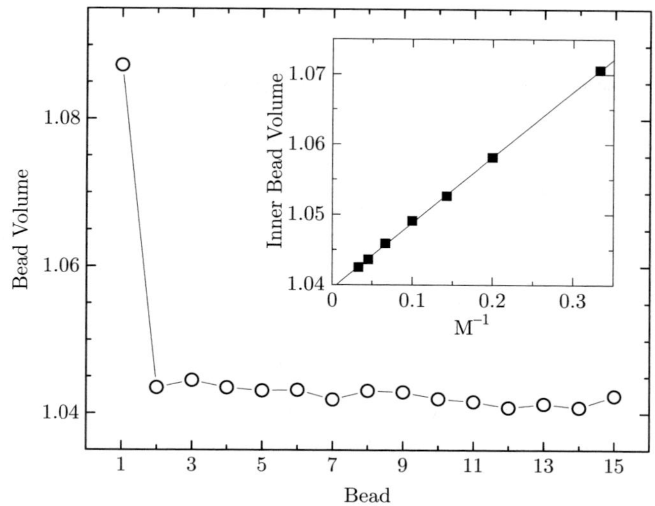

Figure 1.12 Average volume (both occupied and free) per repeat unit as a function of the position of the bead along the polymer chain. The occupied volume of a bead equals unity. The inset shows the variation of the average volume of an inner bead with inverse of the degree of polymerization. The line is the linear fit. Data from ref. [76].

energy, configurational entropy) underlying the segmental dynamics and thus T_g remains an open question, although much progress has been made (see Chapter 2). An early effort due to Fox and Flory [77] assumed that T_g corresponds to a state of constant free volume, leading to an expression for the variation of T_g with the number average molecular weight, M_n, of linear polymers

$$T_g = T_{g,\infty} - k_{FF}/M_n \tag{1.16}$$

where M_n is the mean

$$M_n = \frac{\int_0^\infty \hat{n}(M) M dM}{\int_0^\infty \hat{n}(M) dM} \tag{1.17}$$

$\hat{n}$ is the number of chains of weight M. In eqn (1.16) $T_{g,\infty}$ is the glass-transition temperature in the absence of chain ends and k_{FF} is a material constant whose value depends on the identity of the end groups [78,79]. An empirical expression more accurate for low molecular weights is [80]

$$T_g^{-1} = T_{g,\infty}^{-1} + K_{UK}/M_n \tag{1.18}$$

where the constant k_{UK} serves the same function as k_{FF}. This equation is seen in Figure 1.13 [81] to describe the T_g of various polystyrenes, including mixtures with linear or ring PS. Alternative expressions not wedded to free volume ideas have been derived [82,83], but other than possibly offering insights into the phenomenon, these have no advantage in describing experimental $T_g(M_n)$ data for linear polymers. Equation (1.16) has been applied to star-branched polymers by multiplying the last term by the junction functionality [83] and to dendrimers by accounting for the generation number [84]. An implication from studies of branched polymers is that the concentration of chain ends governs the glass-transition temperature. However, this is at odds

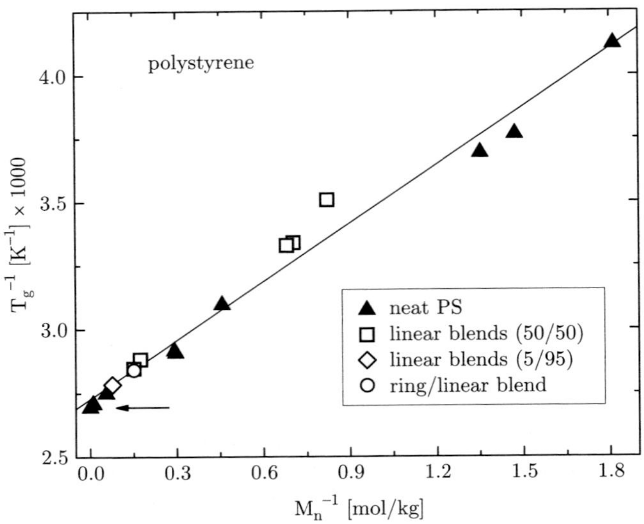

Figure 1.13 Molecular weight dependence of the glass-transition temperature of linear PS plotted in the form of eqn (1.18) for the neat polymer, its blends with 5 and 50% by weight of low M_n (= 739 g/mol) linear PS, and blends with 50% cyclic PS (M_n = 4.4 kg/mol). For the latter 100 kg/mol was used in calculating M_n of the blend, since the ring structure has no chain ends. The arrow indicates the T_g measured for this cyclic PS. Data from ref. [81].

with the invariance of T_g to the degree of branching for PBD 4-, 8- and 12-arm stars, in which T_g depends only on M_n [85]. It may be that in highly branched architectures the freedom conferred by chain ends is countervailed by the constraining effect of the junction points. Other properties are also sensitive to chain ends. For example, since chain ends are not load bearing, they reduce the modulus and strength of a rubber network. Such properties can be described generally by an equation having the form of eqn (1.16).[13]

Since T_g is defined as the temperature at which the material response (i.e. the local segmental dynamics) becomes significantly slower than the experimental measurement rate (Deborah number » 1), the glass transition is rate dependent. This means rapidly strained rubber can behave as a glass; that is, the rubbery state may not persist at high strain rates. As discussed in Chapter 2, strain-induced glass transitions in rubber underlie some applications of elastomers. (The other phase change induced by the deformation of rubber, strain-induced crystallization, can also be affected by strain rate; for example, strain crystallization of natural rubber is curtailed at higher strain rates [86–88])

In addition to the static length scales referred to above, the glass transition is associated with a dynamic length scale reflecting the heterogeneous nature of the reorientational motions. This length scale can be defined as a correlation length, or distance over which the segmental dynamics (or reorientation of small molecules) are mutually correlated. Entropy models of the glass transition embody a related concept, the minimum distance for segmental

[13] Flory [73,74] proposed an equation of this form for the effect of molecular weight on the tensile strength of rubber. Ironically, he applied the analysis to data for strain-crystallizing materials, which complicates an interpretation in terms of the chain-end concentration.

motions to transpire without being accommodated by motion of other segments; this leads to the idea of cooperatively rearranging regions [89,90]. The length scale of the segmental dynamics grows as T_g is approached, becoming macroscopic in the glassy state (Section 2.3.2). Experimental results indicate this length scale in molecular liquids and polymers to be a few nm for temperatures not too far above T_g [91–93]. The corresponding reorientation angle is on the order of 10 degrees [94].

1.2.5 Unentangled chain dynamics

Motions over length scales larger than the correlation length for segmental relaxation transpire at longer times. Those involving only *ca.* 10 backbone bonds may occur as a result of, or even simultaneously with, the local segmental dynamics; that is, the glass transition comprises conformational transitions that permit relaxation to proceed beyond the locally relaxing segments themselves. These longer modes are designated "sub-Rouse modes" [95,96] to convey that the segments involved, while larger than the couple of conformers involved in local segmental relaxation, contain fewer repeat units than the shortest of the Gaussian submolecules described by the Rouse model. The sub-Rouse motion is not evident in measurements probing only the local modes, such as most dielectric relaxation and quasi-elastic photon and neutron-scattering spectroscopies. However, since almost all modes contribute to the stress, the sub-Rouse dynamics can be detected in favorable cases by mechanical experiments. To be apparent as a distinct spectral feature, however, requires that the sub-Rouse modes are not subsumed by the intense segmental process.

Polyisobutylene has many unusual properties, including a similar temperature variation of its segmental and Rouse dynamics [97,98], due to the weak T-dependence of the local segmental motions (see Chapter 6). The result is that the two viscoelastic mechanisms are widely separated in time (more so than for most polymers) in typical mechanical measurements, enabling resolution of the sub-Rouse modes. This is illustrated in Figure 1.14 showing the loss tangent (tan δ, defined as the ratio of the out-of-phase and in-phase components of the dynamic modulus) for PIB at various temperatures [99]. Note that the abscissa is actual frequency; these are not master curves. The unusually broad, structured nature of the peak is ascribed to the prominence of the sub-Rouse modes in PIB.

If a flexible chain has at least 20 to 50 backbone bonds, correlations between bonds are lost, and the dynamics can be described by the Rouse model [100]. The polymer molecule is modeled as a freely jointed chain undergoing random-flight statistics. The end-to-end distribution function $P(r)$ is based on the inverse Langevin function, which can be expressed as a series expansion of the form [101]

$$P(r) \sim \exp\left\{-n_K\left[\frac{3}{2}\left(\frac{r}{n_K l_K}\right)^2 + \frac{9}{20}\left(\frac{r}{n_K l_K}\right)^4 + \frac{99}{350}\left(\frac{r}{n_K l_K}\right)^6 + \dots\right]\right\} \tag{1.19}$$

where n_K is the number of steps in the random walk and l_K is the step length. For moderate extensions this series can be truncated after the first term to give the Gaussian approximation

$$P(r) = \left(\frac{3}{2\pi n_K l_K^2}\right)^{3/2} \exp\left(-\frac{3r^2}{2n_K l_K^2}\right) \tag{1.20}$$

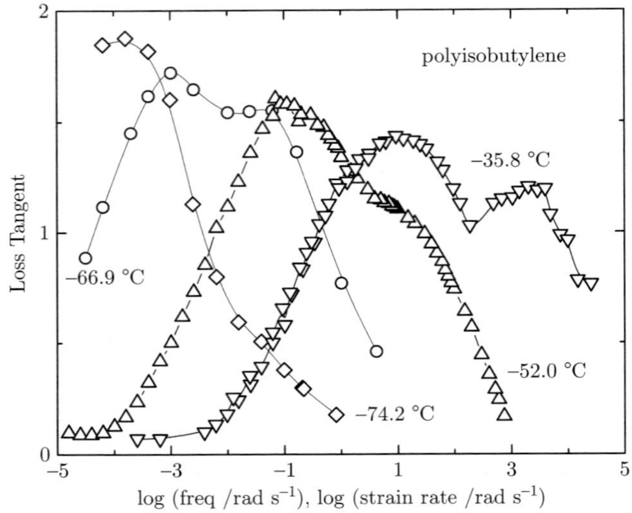

Figure 1.14 Loss tangent of polyisobutylene measured by dynamic mechanical spectroscopy and calculated from the recoverable creep compliance. Note the abscissa is actual frequency; these are not master curves. Data from ref. [99].

The prefactor in eqn (1.20) results from normalization of the distribution function

$$\int_0^\infty P(r)4\pi r^2\,dr = 1 \tag{1.21}$$

Gaussian distributions are common because of the statistical property that the arithmetic mean (normalized sum) calculated for any independent random variable approaches a Gaussian distribution, independent of the original distribution of the variable.[14] Deviations from eqn (1.20) are small over the usual range of interest; moreover, the Gaussian approximation is exact for $\langle r^2\rangle$, since it involves only averages (that is, the Central Limit Theorem applies). The absence of correlations between successive steps leads to a simple relation

$$\langle r^2\rangle = n_K l_K^2 \tag{1.22}$$

A macromolecule, or a section of it, conforming to eqn (1.20) is called a "Gaussian chain," with n_K the number of backbone bonds of length l_K.

Of course, real chains are not freely jointed but are confined to fixed torsional angles involving conformational states of different energy. The connection to random-flight statistics is made by redefining n_K and l_K of the equivalent "Kuhn" chain in terms of the number, n, and length, l, of actual backbone bonds. There are two equivalences: (i) the fully extended dimension

$$n_K l_K = nl\sin(\theta/2) \tag{1.23}$$

[14] This property, known as the Central Limit Theorem, expresses the fact that a Gaussian distribution maximizes the entropy [102].

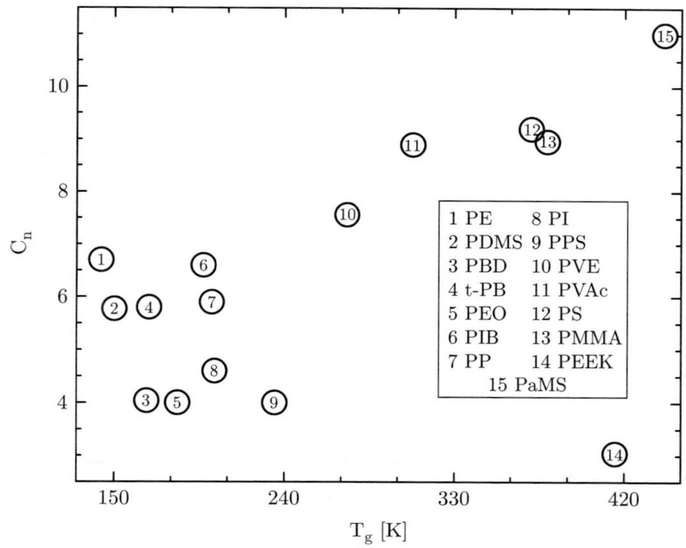

Figure 1.15 Characteristic ratio of various polymers plotted *vs.* their glass-transition temperature. Both properties are affected by the flexibility of the chain. Data from refs. [104 and 103].

where θ is the bond angle (= 109.5° for carbon–carbon single bonds), and (ii) the mean square end-to-end distance

$$\langle r^2 \rangle = C_n n l^2 \tag{1.24}$$

By definition then, l_K is twice the persistence length, l_p. In eqn (1.24) C_n is the characteristic ratio, a measure of chain stiffness. This parameter increases with molecular weight to a constant value at large n, C_∞. Since the glass-transition temperature also increases with stiffness of the polymer backbone, there is a rough correlation between C_∞ and T_g, as illustrated in Figure 1.15 (some of the scatter of these data [103,104] represents differences in the measurement conditions).

The force, f_e, required to maintain a deformation of the chain can be determined from the Helmholtz energy

$$A = E - TS \tag{1.25}$$

where E and S represent the internal energy and entropy, respectively. The thermodynamic relation for an isothermal, isochoric change in chain dimension is [103]

$$f_e = \left.\frac{\partial A}{\partial r}\right|_{T,V} = \left.\frac{dE}{dr}\right|_V - T\left.\frac{dS}{dr}\right|_T \tag{1.26}$$

Using the definition of the entropy[15]

$$S = k_B \ln P(r) \tag{1.27}$$

[15] Equation (1.27) appears at the top of the tombstone marking Ludwig Boltzmann's grave in Vienna, Austria.

eqn (1.20) yields

$$f_e = \left.\frac{dE}{dr}\right|_{T,V} + \frac{3k_B T}{\langle r^2 \rangle} r \tag{1.28}$$

in which r represents the distension of the chain from its equilibrium length. If the chain configurations all have the same conformational energy, the elastic force on a chain is purely entropic and directly proportional to the temperature

$$f_e = \frac{3k_B T}{\langle r^2 \rangle} r \tag{1.29}$$

This force increases linearly with the end-to-end distance; thus, Gaussian chains can be regarded as entropy springs conforming to Hooke's law. There are two obvious deviations for real chains: The underlying random-walk model cannot describe the finite extensibility of polymer chains, so eqn (1.29) is limited to small or moderate strains. And conformations have different energies, so that the internal energy of a real chain changes with conformation ($\left.\frac{dE}{dr}\right|_T \neq 0$), with experiments indicating this energetic contribution is typically 10–20% of the total strain energy [101]. Further discussion of the elastic properties of chains is deferred to Chapter 4.

In the Rouse model the motion of a chain is resisted by drag from the surroundings due to interactions with neighboring segments. These complex interactions can be approximated by an average friction coefficient ζ, which can be identified with the friction coefficient governing local segmental relaxation. This friction deforms the chain, giving rise to a counter force specified by eqn (1.29). Neglecting inertia (since the chain is immersed in a viscous medium and the frequencies are not too high) the equation of motion is

$$\zeta \frac{dr}{dt} + f_e = 0 \tag{1.30}$$

The solution to eqn (1.30) has the form [100]

$$r = r_0 \cos\left(\frac{p\pi n}{\hat{N}}\right) \exp(-t/\tau_p) \tag{1.31}$$

where $\hat{N}$ is the number of Gaussian submolecules comprising the chain. Thus, the complex chain dynamics are decomposed into a series of independent normal modes, corresponding to $p = 1, 2, 3, \ldots$, with $p = 0$ representing center of mass motion (uniform translation). The amplitude is r_0 and the relaxation time of the pth mode is given by

$$\tau_p = \frac{\langle r^2 \rangle \zeta}{12 k_B T \sin^2\left(\frac{\pi p}{2n}\right)} \tag{1.32}$$

Emphasizing the dominant contribution from the longer modes ($n \gg p$), the sine is replaced by its argument to yield

$$\tau_p = \frac{\langle r^2 \rangle \zeta n^2}{3 k_B T \pi^2 p^2} \tag{1.33}$$

Each mode conforms to eqn (1.20) with the shortest mode encompassing one Gaussian submolecule. Equation (1.33) shows that the relaxation time of a Rouse chain varies as n^2. The diffusion coefficient obtained from the Einstein relation

$$D_R = \frac{kT}{\zeta_{chain}} = \frac{kT}{n\zeta} \sim n^{-1} \tag{1.34}$$

varies inversely with chain length, since the friction coefficient for the chain, ζ_{chain}, is the sum over the entire chain of the local friction coefficient of each segment and thus proportional to n. From this description of the motion of the submolecules, the forces on each subchain can be summed to obtain an expression for the modulus

$$G(t) = ck_BT\sum_{p=1}^{n} \exp(-t/\tau_p) \tag{1.35}$$

for c chains per unit volume. This equation has the same form as the modulus expression for n parallel Maxwell elements (a spring in series with a dashpot). However, it is important to recognize that the form of eqn (1.35) comes from a mode analysis. Any smooth closed curve can be represented by a Fourier sum; the Gaussian submolecules are an artifice used to deduce the response of a chain to a perturbation. The additivity of the contributions from each mode does not imply, for example, that different components of stress are additive in any general way. This topic is examined in more detail in Chapter 5.

Originally, the Rouse model was derived for chains moving independently in dilute solution. Since no assumptions are made about the nature of the solvent, Bueche [105] pointed out that such a model can be applied to neat polymers, with neighboring segments serving the role of the solvent. The only restriction is that the effect of the surrounding chains is limited to influencing the magnitude of the friction coefficient, ζ; that is, there is no entanglement effect. Additionally, the absence of excluded volume and hydrodynamic interactions in the Rouse model (see Chapter 3), both of which are important for the dynamics of polymers in dilute solution, enhances the applicability of the Rouse model to concentrated solutions and melts of low molecular weight polymers.

The dynamic moduli corresponding to the Rouse relaxation modulus in eqn (1.35)) are given by the Fourier transforms

$$G'(\omega) = \omega\int_0^\infty G(t)\sin\omega t dt \tag{1.36}$$

$$= ck_BT\sum_{p=1}^{\infty}\frac{\omega^2\tau_p^2}{1+\omega^2\tau_p^2} \tag{1.37}$$

for the storage modulus and

$$G''(\omega) = \omega\int_0^\infty G(t)\cos\omega t dt \tag{1.38}$$

$$= ck_BT\sum_{p=1}^{\infty}\frac{\omega\tau_p}{1+\omega^2\tau_p^2} \tag{1.39}$$

for the loss modulus. At low enough frequencies the limiting behavior $G'(\omega) \sim \omega^2$ and $G''(\omega) \sim \omega$ is obtained. Over an intermediate frequency range, prior to very short times where the model breaks down, the prediction is that the modulus varies according to $\omega^{1/2}$. The quadratic dependence on molecular weight in eqn (1.33) is borne out approximately for the longest ($p = 1$) Rouse mode by experiments on unentangled polymers (see Chapter 3). Although the model

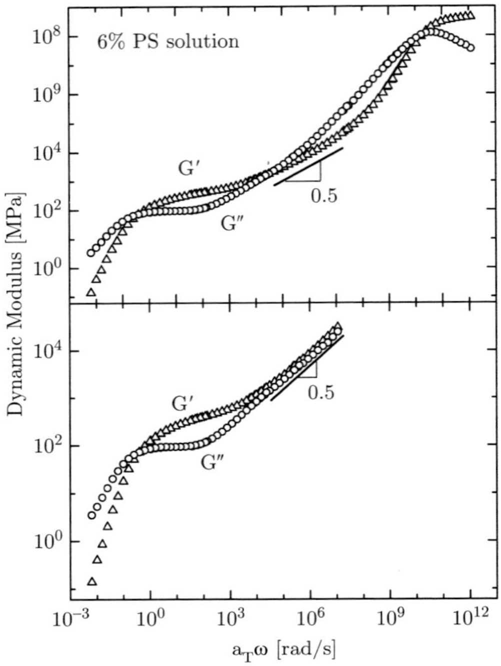

Figure 1.16. Dynamic storage and loss modulus *vs* reduced frequency for 6% PS (5.48×10^6 D) in diethyl phthalate solutions. The lower curve is after subtraction of the solvent and local segmental contributions to the dynamic moduli. Data from ref. [106].

does not address the effect of entanglements, the Rouse description remains valid for high molecular weight melts at times before the onset of entanglements effects. This is illustrated in Figure 1.16 [106], wherein the ω^2 behavior in eqn (1.37) is observed for frequencies higher than the entanglement plateau in measurements on a concentrated solution of high molecular polystyrene. This power-law behavior continues well into the transition zone in this example because the contribution from the segmental dynamics has been subtracted from the spectrum [106]. With its basis in Gaussian chain statistics, the Rouse model applies to flexible polymers. There are indications that stiff chains, having persistence lengths larger than the distance between entanglements, lack Rouse motion at any timescale [107,108].

The Rouse model describes the response of an ensemble of chains. Quantities defined by equations such as (1.20) describe a spherical probability distribution because the conformations are pre-averaged over all chain orientation. However, the shape of chain molecules (the length, width, and depth of the chain conformation), or of Gaussian paths in general, is not spherical. The segment distribution is more extended along one axis than the other two, giving rise to a shape like an oval bar of soap; that is, close to prolate but with unequal minor axes [109–111]. With increasing molecular weight, the departure of the chains from spherical symmetry is reduced; nevertheless, the high molecular weight limiting shape is still highly non-spherical. This asymmetry can be expressed as the ratios of the radius of gyration measured along each of the three principle axes, which for linear chains (the most asymmetric polymers) equal 1:1.7:3.6 [112–114]. It should be emphasized that this non-spherical segment distribution describes the average configuration of a chain molecule; in the absence of orientation the ensemble average values are $\langle R_x^2 \rangle = \langle R_y^2 \rangle = \langle R_z^2 \rangle = \frac{1}{3} \langle R_g^2 \rangle$. The elongated shape of polymer chains influences events transpiring over timescales smaller than the terminal relaxation time. This is the reason why during viscous flow the stress field can orient chains without causing their stretching.

1.2.6 Dynamics of high polymers

Most applications of polymeric materials employ high molecular weight polymers, which have rheological properties distinctly different from unentangled chains. Given its practical importance, the flow of polymeric materials has been an area of intense research over the past half-century. The uncrossablity of polymer chains leads to the idea of topological constraints to explain the rheology. The effects of these constraints are brought out clearly in plots of various dynamic properties as a function of molecular weight. However, the chains of a given material are not all the same length, requiring the use of averages to describe the distribution of molecular weights. These include the arithmetic mean defined in eqn (1.17), the weight average molecular weight:

$$M_w = \frac{\int_0^\infty \hat{n}\,(M) M^2 dM}{\int_0^\infty \hat{n}\,(M) M dM} \tag{1.40}$$

and the so-call z-average:

$$M_z = \frac{\int_0^\infty \hat{n}\,(M) M^3 dM}{\int_0^\infty \hat{n}\,(M) M^2 dM} \tag{1.41}$$

Other measures of molecular weight include the peak value, M_p, which corresponds to the maximum in the distribution, and M_v, determined from the intrinsic viscosity, $[\eta]$, using the Mark–Houwink relation [115]

$$[\eta] = \hat{K} M_v^{\hat{a}} \tag{1.42}$$

where $\hat{K}$ and $\hat{a}$ are empirical, polymer-specific parameters. The size of the molecular can also be expressed in terms of the degree of polymerization, N, which is the corresponding M divided by the molecular weight of the repeat unit, m_0. The breadth of the chain length distribution is described by the polydispersity index

$$P_i = M_w / M_n \tag{1.43}$$

This index is related to the variance, σ_V^2, or averaged squared difference of the chain lengths from their mean, according to

$$\sigma_V^2 = (P_i - 1) M_n^2 \tag{1.44}$$

The different M averages are denoted in Figure 1.17 showing the molecular weight distribution of a styrene–butadiene copolymer, polymerized using an organolithium initiator. This is an example of an anionic "living" polymerization [116], which can yield fairly monodisperse materials, having a molecular weight distribution approaching the Poisson distribution, for which

$$P_i = m_0 / M_n \tag{1.45}$$

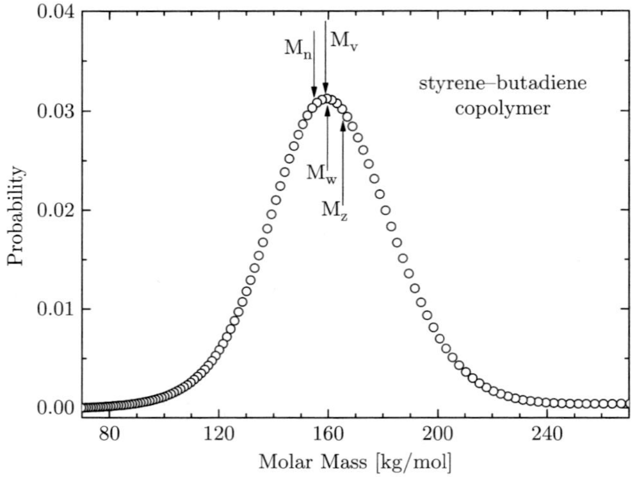

Figure 1.17 Molecular weight distribution measured by gel permeation chromatography for an SBR (19.5% styrene) obtained by anionic polymerization using an organolithium initiator. The polydispersity index = 1.03; for a Poisson distribution with this same M_n, $P = 1.004$. (*Courtesy of Terry Hogan of the Bridgestone Americas Center for Research and Technology.*)

where m_0 is the repeat unit molecular weight. Monodisperse polymers are required for fundamental studies because of the non-linear dependence of the rheological properties on molecular weight, as described below.

For polydisperse polymers the rheological properties are generally expected to be a function of the weight average molecular weight [117]; thus, the zero-shear viscosity, η_0, the length of the rubbery plateau in the relaxation and storage moduli, and the terminal relaxation time, τ_η, are assumed to be proportional to M_w^d where the exponent $d \geq 1$. However, the actual relationship is more complicated. For example, it has been found for model blends of monodisperse polymers that using M_w significantly overestimates the melt viscosity [118–120].

When the chains are long enough to entangle, a rubbery plateau becomes evident in the dynamic mechanical spectra (Figure 1.1). This plateau behavior continues until the chains have sufficient time to relax via disentanglement, with the reciprocal of the characteristic frequency of this relaxation defining a terminal relaxation time, τ_η. Since lateral motions are suppressed, the chain follows an anisotropic path along its own coarse-grained contour [121]. The Doi–Edwards tube model [122] provides a theoretical framework for this reptation concept and, with various refinements, has become predominant in the field of polymer rheology. The tube model has two species-dependent parameters, the Rouse friction coefficient, ζ, and a parameter characterizing the concentration of entanglements.

The snake-like motion of the chain is Brownian diffusion in a tube, with the equation for Fickian diffusion

$$\tau = \langle x^2 \rangle / D \tag{1.46}$$

used to obtain predictions for the chain-length dependences of the terminal relaxation time (i.e. the time for a chain to escape its tube) and the self-diffusion constant. With x ($\propto n$) representing the contour length of the chain and D_R ($\propto n^{-1}$) used for the diffusion coefficient (the chain dynamics inside the tube are unaffected by entanglements), the terminal relaxation time is predicted to vary as $\tau_\eta \sim n^3 \sim M_w^3$. The zero-shear rate viscosity is the integral of the stress relaxation modulus,

$$\eta_0 = \int_0^\infty G(t)dt = \int_0^\infty tG(t)d\ln t \tag{1.47}$$

so η_0 has the same cubic dependence on molecular weight. The macroscopic diffusion coefficient for entangled chains is also obtained from eqn (1.46). During a time on the order of τ_η, the chain travels a distance roughly equal to its size, so $\langle x^2 \rangle$ (now the macroscopic diffusion distance not the length of the tube) is proportional to n (eqn (1.24)) and thus $D_{rep} \sim n^{-2} \sim M_w^{-2}$ according to the tube model.

There exists an enormous amount of experimental data allowing these tube model predictions to be tested [123–126]. The overriding conclusion is that the actual dependence is stronger than predicted for the relaxation time and viscosity; experimentally $\tau_\eta \sim \eta_0 \sim n^{3.4-3.7}$, meaning the tube model underestimates the effect of chain length on the terminal dynamics. In Figure 1.18 viscosity data [127–130] are shown for three polymers having molecular weights exceeding one million g/mol. Since these polymers, amorphous polyethylene (PE), PBD, and PI, have small entanglement molecular weights, M_e, these materials attain exceptional degrees

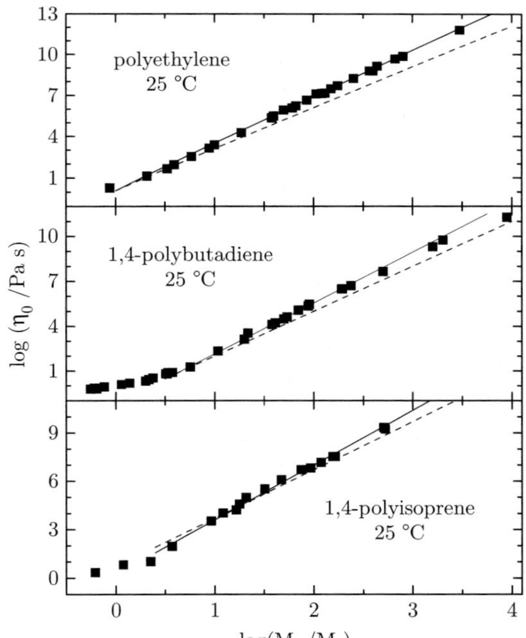

Figure 1.18 Zero-shear rate viscosities of polyethylene [127,128], 1,4-polybutadiene [129], and 1,4-polyisoprene [130] as a function of the number of entanglements per chain, using $M_e = 1.20$, 1.85, and 6.19 kg/mol for PE, PBD, and PI respectively. η_0 have been shifted to 25 °C using reported shift factors. The solid line, conforming well to the data, has a slope of 3.4 and the dashed line has a slope equal to 3.0, representing the tube-model prediction.

Table 1.2 Molecular weight dependence of diffusion coefficients ($D \sim M^{-U}$) for entangled polymers [ref. 126 and references therein].

polymer	U	
	self-diffusion	tracer diffusion
polystyrene	2.3 ±0.3	2.0 ±0.1
polyethylene	2.2 ±0.2	2.0 ±0.1
polydimethylsiloxane	2.2 ±0.2	—
polyethylene oxide	2.2 ±0.2	—
1,4-polybutadiene	> 2.0	—
1,4-polyisoprene	—	1.9

of chain entanglement; nevertheless, the molecular weight dependence in all cases is significantly stronger then M^3. The variation of self-diffusion coefficients with molecular weight is also at odds with the tube model. Experimentally $D_{self} \sim n^{-2.2 \pm 0.2}$ (Table 1.2), rather than a quadratic dependence.

The usual explanation for the deviations in molecular weight dependences is that other relaxation mechanisms contribute to the entangled chain dynamics, thereby diminishing the topological confinements. The tube itself, being composed of neighboring chains, relaxes over the timescale of the reptation time, leading to: *constraint release*, movement of chain segments comprising the tube out of the path of the given chain, and *contour length fluctuations*, movement of the ends of the tube. There is a belief that at very high M "pure reptation" might be obtained as the other relaxation mechanisms become inoperative. For example, if the data in Figure 1.18 are replotted using $M_w^3\eta_0$ as the ordinate variable, there is arguably a suggestion of some flattening of the curve for the higher M_w, consistent with approach to a cubic power dependence on M_w. From the experimental side it is not easy to obtain measurements on polymers with a thousand or more entanglements per chain, since the relaxation times are so long. If solvents are employed for sample preparation, it is difficult to ensure that an equilibrium degree of entanglement has been attained in the material prior to testing. And to reduce the duration of the measurements, temperatures must be used at which chemical degradation can be significant.

Moreover, it is not clear how pure reptation can ever be achieved with homogeneous melts, since the relaxation times of the entangled chain and of the chains comprising the tube scale equivalently with molecular weight. However, for diffusion of probe chains dissolved in a matrix of higher M polymer, constraint release and contour length fluctuations of the tube are suppressed over the timescale of the probe chain motion. This implies that the diffusion coefficient of probe chains should exhibit the expected n^{-2} behavior, an expectation borne out by experimental data (Table 1.2).

1.3 Fluctuations and linear response theory

Small fluctuations of properties from their average value, arising from thermally driven motion of the constituent molecules, are inherent to materials. These microscopic motions allow a system to transition from a metastable state to the thermodynamically stable one. This effect is apparent during the inflation of rubber balloons. Figure 1.19 shows the

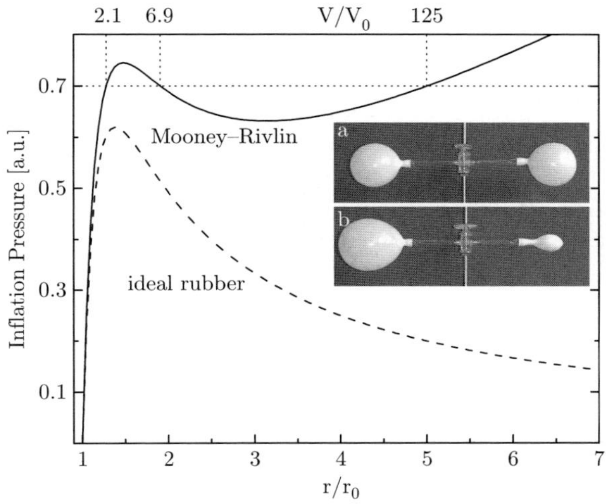

Figure 1.19 Pressure during isothermal inflation of a thin spherical elastomer calculated assuming linear behavior (eqn (1.29); dashed line) and a more realistic response (eqn. (4.9) of Chapter 4; solid line). The intersection with the horizontal dotted line denotes three possible equilibrium states for $P = 0.7$. The photographs (*courtesy of Peter Strehlow of the Technische Universität Berlin*) depict two identical balloons inflated to the same pressure. (a) Initially their volumes are equal. (b) After opening a valve to allow interchange of the contained air, there is a spontaneous change to unequal volumes, with the final pressure unchanged.

pressure–volume curve of a balloon (dashed line) calculated using the ideal chain model (eqn (1.29)), which is valid only for small strains. Introducing non-linearity (see Chapter 4) into the elastic response gives the inflation behavior observed in real balloons (solid curve). As the volume increases there is a steep rise in pressure, followed by a decreasing P until very high volumes are attained (this non-monotonic response is apparent when using one's lungs to inflate a balloon). Note that the variation of the force on the rubber with inflation volume is a combined function of LaPlace's law,[16] the diminishing cross-sectional area, and the form of the force–extension relation for the biaxially strained elastomer. Beyond the maximum in P, the expansion is unstable and non-uniform.

To see the effect of equilibrium fluctuations on inflated balloons, consider the two identical balloons in Figure 1.19, initially at the same pressure P (= 0.7 in arbitrary units) and volume ($V/V_0 = 6.9$) but separated by a closed tube [131]. Upon opening a valve in the connecting tube, the pressures will remain the same and equal for the balloons; however, this pressure balance exists for three different inflation volumes. For moderate pressures and depending on the details of the elastic properties of the material, the balloons spontaneously transition to smaller and larger volumes respectively (e.g., $V/V_0 = 2.1$ and 125). This behavior, depicted in the photographs in the inset to Figure 1.19, is non-intuitive but well understood (and possible

[16] According to LaPlace's law, the surface tension in the balloon at fixed inflation pressure is proportional to its radius of curvature.

only for inflation strains sufficient for deviations from eqn (1.29) [132,133]). The question is, how did the balloon ending at the lower $V/V_0 = 2.1$ in Figure 19 traverse the intermediate volumes associated with higher pressures? The answer is by fluctuations in the pressure. Analogous to chemical reactants overcoming an activation barrier to form lower-energy products, the final inflation conditions of the two balloons can be achieved because of random fluctuations inherent to an equilibrium system.

The same fluctuations prevailing at equilibrium enable a material to return to equilibrium following a small external perturbation. Fluctuation-dissipation theory (FDT) relates the fluctuations to how the system responds to (dissipates) external forces. According to FDT, if the material is driven not too far from equilibrium, it is sufficient to know the behavior of the dissipating and perturbing systems in the absence of any coupling in order to predict their interaction; that is, decay of a spontaneous thermal fluctuation cannot be distinguished from decay of fluctuations induced by small external forces. The fluctuating properties can be described through time correlation functions, defined as

$$C(t) = \langle A(0) \cdot B(t) \rangle \tag{1.48}$$

where $A(t)$ and $B(t)$ are microscopic quantities whose values depend on the coordinates and momenta of the constituent molecules. An autocorrelation function quantifies the correlation of a property with its own prior value

$$C(t) = \langle A(0) \cdot A(t) \rangle \tag{1.49}$$

The ensemble average varies over time due to the stochastic fluctuations. Since the system is stationary at equilibrium, average quantities are time independent and the initial time is artificial; that is, $\langle A(0)A(t) \rangle = \langle A(u)A(t+u) \rangle$ and $\langle A(0)[dA(t)/dt] \rangle = d\langle A(u)A(t+u) \rangle dt$, with t specifying the duration between successive evaluations of A. For an ergodic system this average over the system is equivalent to the time average

$$C(t) = \lim_{\hat{t} \to \infty} \hat{t}^{-1} \int_0^{\hat{t}} A(u)A(t+u)du \tag{1.50}$$

Normalization of $A(t)$ by its initial value, $\langle A^2 \rangle$, and subtraction of the long-time asymptotic, uncorrelated value, $\langle A \rangle^2$, makes $C(t)$ equal unity initially and zero eventually. $C(t)$ is calculated by solving the equations of motion describing the coordinates and momenta over time, and averaging over all phase space. A model with suitable assumptions is used to simplify this general N-body problem.

FDT connects the fluctuations described by $C(t)$ to a susceptibility characterizing fluctuations induced experimentally (the latter more accessible than spontaneous fluctuations). The initial statement of FDT was the Nyquist theorem [134] describing Johnson electrical noise [135]. This is the white noise in a conductor due to thermal motion of the electrons, manifested as voltage fluctuations having a mean square potential proportional to the resistance. The resistance, or impedance, is the response measured when an external voltage is applied. More generally experiments yield a susceptibility function describing the change of an observable induced by an infinitesimal perturbation ς

$$\chi(t) = \lim_{\varsigma \to 0} \frac{\langle A(t) \rangle_\zeta \langle A \rangle_0}{\varsigma} \tag{1.51}$$

According to FDT $\chi(t)$ is related to $C(t)$ as [136,137],

$$\chi(t) = \frac{C(t)}{k_B T} \tag{1.52}$$

The temperature factor accounts for the fact that fluctuations are thermally driven. The Fourier transform of $C(t)$ yields the spectral density and therefore the dynamic susceptibility, $\chi(\omega)$. An example of the FDT is the Einstein relation, eqn (1.34), relating the Brownian motion of particles to viscous dissipation. $A(t)$ and $B(t)$ in eqns (1.48) and (1.49) depend of course on the nature of the experiment. In a transient mechanical experiment, A represents the local stress and the stress relaxation modulus is

$$G(t) = \frac{V}{k_B T} C(t) \tag{1.53}$$

In a dielectric experiment the time-dependent fluctuations of the polarization are measured and $A(t)$ is identified with electric dipoles on the molecules or segments. In most polymers this dipole moment is oriented transverse to the chain, so that the dielectric experiment probes the local segmental dynamics. For those few polymers having a dipole moment parallel to the backbone (which means their repeat unit structure lacks a symmetric center), dielectric relaxation measures the autocorrelation of the chain's end-to-end vector. Note that although the dipoles orient in the presence of the applied field, the interaction energy is much less than the available thermal energy; this is a characteristic of the linear response regime.

Quasi-elastic scattering techniques can be used to probe fluctuations of the polarizability (light scattering), density (X-rays), and nuclei (neutrons) of the molecules, where in favorable cases the motion can be identified with a particular axis or moiety in the molecule. The scattered intensity varies with the wave vector, $q = \frac{4\pi}{\lambda}\sin(\theta/2)$, where θ is the scattering angle and λ the wavelength. This angle dependence of the intensity means that scattering experiments yield information about the variation of the molecular motions with length scale, information not available from mechanical or dielectric relaxation spectroscopies. The characteristic length scales must be on the order of the wavelength, which distinguishes light, X-ray, and neutron measurements. Scattering experiments are able to probe correlations between different particles. For example, the intermediate scattering function, $F(q,t)$, measured by quasi-elastic X-ray and neutron experiments, is related to the correlation function according to

$$C(r,t) = C_s(r,t) + C_c(r,t) = \frac{1}{8\pi^3}\int \exp(iqr)F(q,t)dq \tag{1.54}$$

in which $C_s(r,t)$ and $C_c(r,t)$ described the self- and cross-correlations respectively. Other examples of experimentally accessible correlation functions are for nuclear magnetic dipoles, probed by NMR measurements, and chromophores attached to polymers chains, responsible for fluorescence decay and holographic grating techniques.

Two restrictions in the application of FDT are that the perturbed system begins in a state of equilibrium and that the external perturbation does not drive the material beyond values of $A(t)$ attainable by equilibrium fluctuations. The latter is the linear response requirement that determines the amplitude limit for the perturbation. For an initially non-equilibrium system the correlation and response functions deviate from FDT; nevertheless, the theory can still be useful. For example, in some situations the departure from equilibrium can be quantified by defining an effective temperature in analogy with eqn (1.52) [138–140]

$$T_f = k_B^{-1}\left[\frac{C(t)}{\chi(t)}\right] \tag{1.55}$$

The quantity in brackets is called the fluctuation-dissipation ratio, which approaches unity close to equilibrium. T_f is the temperature at which the observed non-equilibrium properties would have their equilibrium values. For glasses, which are prototypical non-equilibrium materials, T_f is known as the fictive temperature. During physical aging of a glass, the structure moves towards equilibrium and thus T_f approaches the actual temperature.

From the measured susceptibility FDT enables the relevant time correlation function to be obtained. Its interpretation, however, relies on computation of $C(t)$ from a model of the dynamic process probed by the experiment. For the case of reorientational motions, Debye [2] proposed a simple model wherein the liquid molecules change orientation through a series of small angular jumps due to unbalanced torques from the fluctuating surroundings (random "collisions"). The motion is thus a random walk in angular coordinate space, with the time between changes in orientation long compared to the time of a reorientation; i.e. the rotational correlation time is much longer than the correlation time for the angular momentum. Motion of this nature can be described by a diffusion equation whose solution is exponential decay (i.e. the form of a first-order rate equation such as eqn (1.4)). This in turn is the form FDT predicts for the experimental relaxation function. However, there are two obvious cases where the assumptions underlying the Debye model break down. If unbalanced forces are sparse and infrequent, large changes in orientation can occur by free rotation (inertially driven) interrupted by periodic perturbations; such motion cannot be described as a random walk. This situation prevails in low-density fluids, with the relevant $C(t)$ depending on the effect assumed for the perturbations. For example, intermolecular collisions may randomize the instantaneous orientation of the molecule or scramble its angular momentum. This type of motion is not relevant to polymers.

With increasing density in supercooled liquids and polymers, there is a crossover from continuous small-amplitude Brownian motion to discontinuous, large-amplitude hopping as movements become highly correlated – a given segment's change in position or orientation requires motion of its neighbors. The Debye model's mean-field description of a fluctuating environment is inadequate, and measured correlation functions deviate from simple exponential decay. Relaxation in such complex correlated systems often conforms to the Kohlrausch equation (eqn (1.7)). Although it is used as an empirical fitting function, this form can be arrived at in various ways: from models based on free volume [141], hierarchal constraints [142], defect diffusion [143], defect distances [144], random free energy [145], or molecular weight polydispersity [146]; by employing a particular distribution of exponential decay functions [147]; or from specific ideas about cooperativity [148,149]. One approach is to assume that many-body interactions modify the rate constant k in eqn (1.3) during the course of the relaxation according to [148]

$$k(t) = k_0(t/t_0)^{-1/\beta} \tag{1.56}$$

in which t_0 is a constant. When substituted into eqn (1.3), eqn (1.56) gives stretched exponential decay. When many-body interactions cause the relaxation to deviate from eqn (1.4), the temperature dependence usually exhibits a corresponding deviation from Arrhenius behavior (eqn (1.12)). A spectacular increase in the temperature sensitivity of the dynamics is one of the hallmarks of vitrifying liquids and polymers, prototypical of systems with correlated, multi-body interactions. Understanding the connection between the time and temperature dependences of the dynamics is a major theoretical challenge.

References

1. A. Einstein, *Investigations on the Theory of the Brownian Movement*, (ed. R. Fuerth, translated by A.D. Cowper) Methuen Co., London (1926) and Dover Publ., New York (1956).
2. P. Debye, *Polar Molecules*, Dover, New York (1928).
3. *Rubber & Plastics News*, p. 12, Oct. 20, 2008.
4. J. Meissner, *Rheol. Acta* **10**, 230 (1971).
5. M.H. Wagner and H.M. Laun, *Rheol, Acta* **17**, 138 (1978).
6. C.A. Angell, K.L. Ngai, G.B. McKenna and P.F. McMillan *J. Appl. Phys.* **88**, 3113 (2000).
7. C.M. Roland, S. Hensel-Bielowka, M. Paluch and R. Casalini, *Rep. Prog. Phys.* **68**, 1405 (2005).
8. A.J. Kennedy, *J. Mech. Phys. Solids* **1**, 172 (1953).
9. D. McLean, *Rep. Prog. Phys.* **29**, 1 (1966).
10. T.J. Quinn, C.C. Speake, S.J. Richman, R.S. Davis, and A. Picard, *Phys. Rev. Lett.* **87**, 111101 (2001).
11. P.G. Santangelo and C.M. Roland, *Macromolecules* **31**, 3715 (1998).
12. R. de L. Kronig, *J. Opt. Soc. Am.* **12**, 547 (1926); H.A. Kramers, *Atti del Congresso Internazionale dei Fisici* **2**, 545 (1927).
13. N.W. Tschoegl, *The Phenomenological Theory of Linear Viscoelastic Behavior*, Springer-Verlag, Berlin (1989).
14. M. Wubbenhorst and J. Van Turnhout, *J. Non-Cryst. Sol.* **305**, 40 (2002).
15. Y.I. Shimanskii, *J. Eng. Phys. Thermophys.* **13**, 301 (1967).
16. J.D. Ferry, *Viscoelastic Properties of Polymers*, 3rd edition, Wiley, New York (1980).
17. R. Richert, *J. Non-Cryst. Solids* **172-174**, 209 (1994).
18. R. Kohlrausch Ann. *Phys. Chem. (Poggendorff).* **91**, 179 (1854).
19. G. Williams and D.C. Watts, *Trans. Faraday Soc.* **66**, 80 (1970).
20. C.P. Lindsey and G.D. Patterson, *J. Chem. Phys.* **73**, 3348 (1980).
21. A.K. Jonscher, *J. Mater. Sci.* **16**, 2037 (1981).
22. P.J. Flory, *Proc. Roy. Soc. London* **A234**, 60 (1956).
23. P.J. Flory, *Adv. Polym. Sci.* **59,** 1 (1984).
24. V.G. Baranov, *Fiber Chem.* **9**, 223 (1978).
25. G.K. Elyashevich, *Adv. Polym. Sci.* **43**, 205 (1982).
26. C.M. Roland and M.F. Sonnenschein, *Poly. Eng. Sci.* **31**, 1434 (1991).
27. R. Huisman and H.M. Heuvel, *J. Appl. Polym. Sci.* **37**, 595 (1989).
28. M.J. Schroeder, K.L. Ngai, and C.M. Roland, *J. Polym. Sci. Phys. Ed.* **45** 342 (2007).
29. D.A. McQuarrie, *Statistical Mechanics*, 2nd edition, Chap. 11, University Science Books, Sausalito, CA (2000).
30. C. Kittel, *Quantum Theory of Solids*, 2nd edition, Chap. 19, Wiley, New York (1987).
31. U. Buchenau, M. Prager, N. Nucker, A.J. Dianoux, N. Ahmad, and W.A. Phillips, *Phys. Rev. B* **34**, 5665 (1986).

32. V.K. Malinovsky and A.P. Sokolov. *Solid State Commun.* **57**, 767 (1986).
33. S. Etienne, L. David, A.J. Dianoux, L. Saviot, and E. Duval, *J. Non-Cryst. Sol.* **307–310**, 109 (2002).
34. K.L. Ngai, A. Sokolov, and W. Steffen, *J. Chem. Phys.* **107**, 5268 (1997).
35. G.G. Naumis and H.M. Flores-Ruiz, *Phys. Rev. B* **78**, 094203 (2008).
36. H. Shintani and H. Tanaka, *Nature Mater.* **7**, 870 (2008).
37. C.M. Roland, M.J. Schroeder, J.J. Fontanella and K.L. Ngai, *Macromolecules* **37**, 2630 (2004).
38. S.A. Lusceac, C. Gainaru, M. Vogel, C. Koplin, P. Medick, and E.A. Rössler, *Macromolecules* **38**, 5625 (2005).
39. A.P. Sokolov, A. Kisliuk, V.N. Novikov, and K.L. Ngai, *Phys. Rev. B* **63**, 172204 (2001).
40. G. Caliskan, A. Kisliuk, A.P. Sokolov, and V.N. Novikov, *J. Chem. Phys.* **114**, 10189 (2001); A. Kisliuk, V.N. Novikov, and A.P. Sokolov, *J. Polym. Sci. Polym. Phys. Ed.* **40**, 201 (2002).
41. A. Rivera, A. Leon, C.P.E. Varsamis, A.D. Chryssikos, K.L. Ngai, C.M. Roland, and L.J. Buckley, *Phys. Rev. Lett.* **88**, 125902 (2002).
42. P. Lunkenheimer, A. Pimenov, and A. Loidl, *Phys. Rev. Lett.* **78**, 2995 (1997).
43. R. Casalini, K.L. Ngai, and C.M. Roland, *J. Chem. Phys.* **112**, 5181 (2000).
44. O. Kanert, R. Kuchler, P.C. Soares, and H. Jain, J. Non-Cryst. Sol. **307**, 1031 (2002).
45. A. Burns, G.D. Chryssikos, E. Tombari, R.H. Cole, and W.M. Risen, *Phys. Chem. Glasses* **30**, 264 (1989).
46. H. Cang, V.N. Novikov, and M.D. Fayer, *J. Chem. Phys.* **118**, 2800 (2003).
47. K.L. Ngai, *J. Phys. Condens. Matt*er **15**, S1107 (2003).
48. C. Gainaru, A. Rivera, S. Putselyk, G. Eska, and E.A. Rössler, *Phys. Rev. B* **72**, 174203 (2005).
49. C. Hansen and R. Richert, *J. Phys. Condens. Matter* **9**, 9661 (1997).
50. C.M. Roland and R. Casalini, *Macromolecules* **40**, 3631 (2007).
51. J. Williams and A. Eisenberg, *Macromolecules* **11**, 700 (1978).
52. Y. Aoki and J.O. Brittain, *J. Appl. Polym. Sci.* **20**, 2879 (1976).
53. S. Corezzi, M. Beiner, H. Huth, K. Schroter, S. Capaccioli, R. Casalini, D. Fioretto, and E. Donth, *J. Chem. Phys.* **117**, 2435 (2002).
54. H.W. Starkweather, *Macromolecules* **14**, 1277 (1981).
55. H.W. Starkweather and J.R. Barkley, *J. Polym. Sci. Polym. Phys. Ed.* **19**, 1211 (1981).
56. O. Yano and Y. Wada, *J. Polym Sci. Polym. Phys. Ed.* **9**, 669 (1971).
57. M. Cook, G. Williams, and T.T. Jones, *Polymer* **16**, 835 (1975).
58. S.S.N. Murthy and M. Shahin, *Eur. Polym. J.* **42**, 715 (2006).
59. B.E. Read and G. Williams, *Polymer* **2**, 239 (1961).
60. K. Adrjanowicz, M. Paluch, and K.L. Ngai, *J. Phys. Condens. Matter* **22**, 125902 (2010).
61. R. Casalini and C.M. Roland, *J. Non-Cryst. Solids*, **357**, 282 (2011).
62. R. Casalini and C.M. Roland, *Phys. Rev. Lett.* **102** 035701 (2009).
63. V.A. Bershstein and V.M. Yegorov, *Polym. Sci. USSR* **27**, 2743 (1985).
64. K.L. Ngai and M. Paluch, *J. Chem. Phys.* **120**, 857 (2004).
65. G.P. Johari and M. Goldstein, *J. Phys. Chem.* **74**, 2034 (1970).
66. K. Schmidt-Rohr, A.S. Kulik, H.W. Beckham, A. Ohlemacher, U. Pawelzik, C. Boeffel, and H.W. Spiess, *Macromolecules* **27**, 4733 (1994).

67. S.C. Kuebler, D.J. Schaefer, C. Boeffel, U. Pawelzik, and H.W. Spiess, *Macromolecules* **30**, 6597 (1997).
68. K.L. Ngai, T.R. Gopalakrishnan, and M. Beiner, *Polymer* **47**, 7222 (2006).
69. G.P. Johari, G. Power and J.K. Vij, *J. Chem. Phys.* **116**, 5908 (2002).
70. K.L. Ngai, *J. Non-Cryst. Solids* **351**, 2635 (2005).
71. E. Helfand, *Science* **226**, 647 (1984).
72. I. Bahar, B. Erman, and L. Monnerie, *Macromolecules* **24**, 3618 (1991).
73. P.J. Flory, *J. Am. Chem. Soc.* **67**, 2048 (1945).
74. P.J. Flory, *Ind. Eng. Chem.* **38**, 417 (1946).
75. P.J. Flory, *Chem. Rev.* **35**, 51 (1944).
76. A. Barbieri, D. Prevosto, M. Lucchesi, and D. Leporini, *J. Phys. Condens. Matter* **16**, 6609 (2004).
77. T.G. Fox and P.J. Flory, *J. Polym. Sci.* **14**, 315 (1954).
78. D.T. Turner, *Polymer* **19**, 789 (1978).
79. F. Danusso, M. Levi, G. Gianotti, and S. Turri, *Polymer* **34**, 3687 (1993).
80. K. Ueberreiter and G.J. Kanig, *Colloid Sci.* **7**, 569 (1952).
81. C.G. Robertson and C.M. Roland, J. *Polym. Sci. Polym. Phys. Ed.* **42**, 2604 (2004).
82. A.T. Di Benedetto and L. Di Landro, *J. Polym. Sci.* **27**, 1405 (1989).
83. J.E.L. Roovers and P.M. Toporowski, *J. Appl. Polym. Sci.* **18**, 1685 (1974).
84. K.L. Wooley, C.J. Hawker, J.M. Pochan, and J.M.M. Frechet, *Macromolecules* **26**, 1514 (1993).
85. A. Kisliuk, Y. Ding, J. Hwang, J.S. Lee, B.K. Annis, M.D. Foster, and A.P. Sokolov, *J. Polym. Sci. Polym. Phys. Ed.* **40**, 2431 (2002).
86. J.C. Mitchell and D.J. Meier, *J. Polym. Sci. A2* **6**, 1689 (1968).
87. Z.R. Glaser and F.R. Eirich, *J. Polym. Sci. C* **31**, 275 (1970).
88. G.J. Lake, C.C. Lawrence, and A.G. Thomas, *Rubber Chem. Technol.* **73**, 801 (2000).
89. G. Adam and J.H. Gibbs, *J. Chem. Phys.* **43**, 139 (1965).
90. J.D. Stevenson, J. Schmalian, and P.G. Wolynes, *Nature Phys.* **2**, 268 (2006).
91. H. Sillescu, *J. Non-Cryst. Solids* **243**, 81 (1999).
92. R. Richert, *J. Phys. Condens. Matter* **14**, R703 (2002).
93. L. Berthier, G. Biroli, J.-P. Bouchaud, L. Cipelletti, D. El Masri, D. L'Hote, F. Ladieu, and M. Pierno, *Science* **310**, 1797 (2005).
94. R. Bohmer, G. Diezemann, G. Hinze, and E. Rössler, *Prog. Nucl. Magn. Reson. Spectr.* **39**, 191 (2001).
95. P.G. Santangelo, K.L. Ngai, and C.M. Roland, *Macromolecules* **26**, 2682 (1993).
96. K.L. Ngai and D.J. Plazek *Rubber Chem. Technol.* **68**, 376 (1995).
97. D.J. Plazek, X.D. Zheng, and K.L. Ngai, *Macromolecules* **25**, 4920 (1992).
98. K.L. Ngai, D.J. Plazek, and C.A. Bero, *Macromolecules* **26**, 1065 (1993).
99. D.J. Plazek, I.-C. Chay, K.L. Ngai, and C.M. Roland, *Macromolecules* **28**, 6432 (1995).
100. H. Yamakawa, *Modern Theory of Polymer Solutions*, Harper & Row, New York (1971).
101. B. Erman and J.E. Mark, *Structure and Properties of Rubberlike Networks*, Oxford University Press, Cambridge (1997).
102. O. Johnson, *Information Theory and the Central Limit Theorem*, Imperial College Press, London (2004).
103. P.J. Flory, *Statistical Mechanics of Chain Molecules*, Hanser Publishers, NY (1989).
104. *Polymer Data Handbook*, (ed. J.E. Mark) Oxford University Press, Cambridge (1999).

105. F. Bueche, *J. Chem. Phys.* **20**, 1959 (1952).
106. C.M. Roland, L.A. Archer, P.H. Mott, and J. Sanchez-Reyes, *J. Rheol.* **48**, 395 (2004).
107. L. Harnau, R. G. Winkler, and P. Reineker, *J. Chem. Phys.* **106**, 2469 (1997).
108. R. Faller and F. Müller-Plathe, *Chem. Phys. Chem.* **2**, 180 (2001).
109. M. Bishop, J.H.R. Clarke, A. Rey, and J.J. Freire, *J. Chem. Phys.* **94**, 4009 (1991).
110. D.N. Theodorou and U.W. Suter, *Macromolecules* **18**, 1206 (1985).
111. G. Wei and B.E. Eichinger, *J. Chem. Phys.* **93**, 1430 (1990).
112. T. Minato and A. Hatano, *Macromolecules* **14**, 1035 (1981).
113. G. Wei, *Physica A,* **222**, 155 (1995).
114. S.J. Sciutto, *J. Phys. A: Math. Gen.* **29**, 5455 (1996).
115. C.M. Roland in *The Science and Technology of Rubber*, 3rd edn (eds. J.E. Mark, B. Erman, and F.R. Eirich) Chapter 3, Elsevier, New York (2005).
116. R.P. Quirk and D.L. Gomochak Pickel, in *The Science and Technology of Rubber*, 3rd edition (eds. J.E. Mark, B. Erman, and F.R. Eirich) Chapter 2, Elsevier, New York (2005).
117. G.C. Berry and T.G Fox, *Adv. Polym. Sci.* **5**, 261 (1968).
118. H. Watanabe, T. Sakamoto, and T. Kotaka, *Macromolecules* **18**, 1008 (1995).
119. J.P. Monfort, G. Mann, and P. Monge, *Macromolecules* **17**, 1551 (1984).
120. D.J. Plazek, C. Seoul, and C.A. Bero, *J. Non-Cryst. Solids* **131–133**, 570 (1991).
121. P.-G. de Gennes, *Scaling Concepts in Polymer Physics*, Cornell University, Ithaca (1979).
122. M. Doi and S.F. Edwards, *The Theory of Polymer Dynamics*, Clarendon Press, Oxford (1986).
123. T.P. Lodge, N.A. Rotstein, and S. Prager, *Adv. Chem. Phys.* **79**, 1 (1990).
124. K. Binder, T.P. Lodge, H. Sillescu, K. Schweizer, and H. Yu, *J. Non-Cryst. Solids* **131–133**, 742 (1991).
125. T.C.B. McLeish, *Adv. Phys.* **51**, 1379 (2002).
126. S.-Q. Wang, J. Polym. *Sci. Polym. Phys. Ed.* **41**, 1589 (2003).
127. D.S. Pearson, L.J. Fetters, and W.W. Graessley, *Macromolecules* **27**, 711 (1994).
128. J.F. Vega, S. Rastogi, G.W.M. Peters, and H.E.H. Meijer, *J. Rheol.* **48**, 663 (2004).
129. R.H. Colby, L.J. Fetters, and W.W. Graessley, *Macromolecules* **20**, 2237 (1987).
130. M. Abdel-Goad, W. Pyckhout-Hintzen, S. Kahle, J. Allgaier, D. Richter, and L.J. Fetters, *Macromolecules* **37**, 8135 (2004).
131. I. Muller and P. Strehlow, *Rubber and Rubber Balloons*, Springer, Berlin (2004).
132. E. Verron and G. Marckmann, *Thin-Walled Struc.* **41**, 731 (2003).
133. Y. Levin and F.L. da Silveira, *Phys. Rev. E* **69**, 051108 (2004).
134. H. Nyquist, *Phys. Rev.* **32**, 110 (1928).
135. J.B. Johnson, *Phys. Rev.* **32**, 97 (1928).
136. R. Zwanzig, *Ann. Rev. Phys. Chem.* **16**, 67 (1965).
137. R.G. Gordon, *Adv. Magn. Res.* **3**, 1 (1968).
138. L. Cugliandolo, J. Kurchan, and L. Peliti, *Phys. Rev. E* **55**, 3898 (1997).
139. L. Bellon, S. Ciliberto and C. Laroche, *Eur. Phys. Lett.* **53**, 511 (2001).
140. N. Greinert, T. Wood, and P. Bartlett, *Phys. Rev. Lett.* **97**, 265702 (2006).
141. M.H. Cohen and G.S. Grest, *Phys. Rev. B* **24**, 4091 (1981).
142. R.G. Palmer, D.L. Stein, E. Abrahams and P.W. Anderson, *Phys. Rev. Lett.* **53**, 958 (1984).

143. J.T. Bendler and M.F. Shlesinger, *Macromolecules* **18**, 591 (1985).
144. J. Klafter and A. Blumen, *Chem. Phys. Lett.* **119**, 377 (1985).
145. C. DeDominicis, H. Orland, and F. Lainee, *J. Phys. Lett.* **46**, L463 (1985).
146. P.-G. de Gennes, *Macromolecules* **35**, 3785 (2002).
147. M.N. Berberan-Santos, E.N. Bodunov, and B. Valeur, *Chem. Phys.* **315**, 171 (2005).
148. K.L. Ngai, R.W. Rendell, A.K. Rajagopal, and S. Teitler, *Ann. NY Acad. Sci.* **484**, 150 (1986).
149. M.M. Kubat, P. Riha, R.W. Rychwalski, and S. Uggla, *Mech. Time Dep. Mater.* **3**, 31 (1999).

2
Cooperative local dynamics – the glass-transition zone

Among the many intriguing phenomena of the physical world, the spectacular change in the behavior of liquids undergoing vitrification is among the most fascinating. Cooling or pressurizing a liquid causes various property changes, the most significant being the enormous increase in the structural relaxation time – a one degree change in temperature can shift τ_α by decades, corresponding to apparent activation energies that are two or more orders of magnitude larger than van der Waals bond energies. Structural relaxation involves diffusion and reorientation of the molecules of a liquid or local segmental motion in polymers. At the pressure-dependent T_g these relaxation processes effectively cease, giving rise to a glassy material, in which the dynamics are confined to vibrations and localized secondary dynamics. Vitrification to form a disordered solid lacking translational symmetry is a general phenomenon, found in organic, inorganic, metallic, and colloidal materials. Synthetic production of glass began as early as the fifth millennium BC in the form of glazed (vitreous) surfaces, with synthetic silicate glasses appearing in the middle of the fourth millennium. A verse in one of the oldest books of the bible (*ca.* 2000 BC) refers to the value of vitreous material in antiquity: *"Gold and glass cannot equal [wisdom]"* (Job 28:17, ASV). Modern glass making is highly developed, used to produce metallic glasses, optical fibers, many plastics and ceramics, etc. There are various routes to the glassy state besides cooling or compression of a liquid: condensation of gas at low temperature (which is the origin of the water in comets), solvent evaporation or sublimation, irradiation of crystalline material to disrupt the unit cell, and chemical methods, such as polymerization, hydrolysis of organosilicon compounds, or condensation of chemically reacting vapors.

Notwithstanding the long history of glass manufacture, the glass transition itself remains an unsolved problem. Theoretical efforts to describe the phenomenon are still at the model-building stage. In part, this reflects the wealth of the associated phenomena, the many decades spanned by the molecular dynamics of vitrifying materials, and the consequent need to account for a variety of properties. The dramatic changes in macroscopic behavior upon glass formation occur without conspicuous changes on the molecular level. The microscopic long-range disorder is retained as a liquid acquires the properties of a solid. This is seen in the near invariance of the static structure factor, $S_f(\tilde{k})$, defined as the variance in the density ρ

$$S_f(\tilde{k}) = N^{-1}(\langle \rho_{\tilde{k}}^2 \rangle - \langle \rho_{\tilde{k}} \rangle^2) \tag{2.1}$$

where $\tilde{k}$ is the wave vector.[1] The static structure factor can be measured by neutron diffraction, where $S_f(k)$ is proportional to the flux scattered at momentum transfer Q. As shown for

[1] $\tilde{k}$ is the vector representation of a wave having a magnitude equal to the wave number, $|\tilde{k}| \equiv 2\pi/\lambda$. In scattering experiments the scattering angle, θ, is often expressed as the momentum transfer, $Q = 4\pi \sin(\theta)/\lambda$, equal to the difference in wave number of the scattered and incident waves.

PBD subjected to changes in either temperature (Figure 2.1) or pressure (Figure 2.2), the scattering at large Q (> 5 Å^{-1}), reflecting the short-range, intramolecular correlations of the polymer segments, is essentially the same above and below T_g. Glass formation only affects the position and intensity of peaks at small values of Q (< 2.5 Å^{-1}), due to changes in molar volume altering intermolecular correlations.

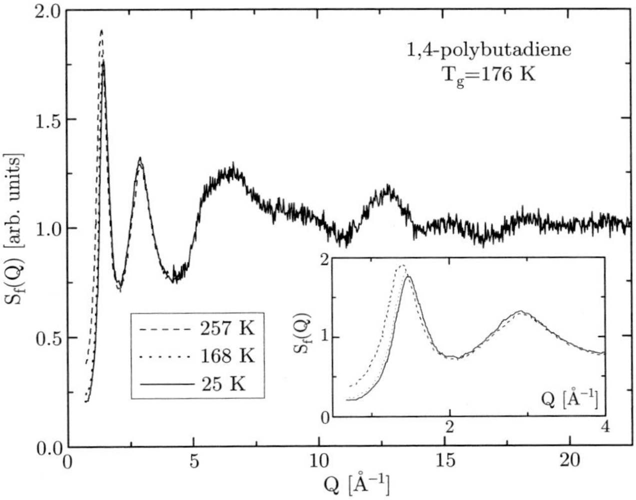

Figure 2.1 Static structure factor of PBD in the liquid state (dashed), just below T_g (dotted), and deeply in the glassy state (solid line). The primary peak at small values of momentum transfer, due to interchain correlations, is shown on an expanded scale in the inset; this is the only region that changes significantly with temperature, reflecting thermal expansion. Data from ref. [1].

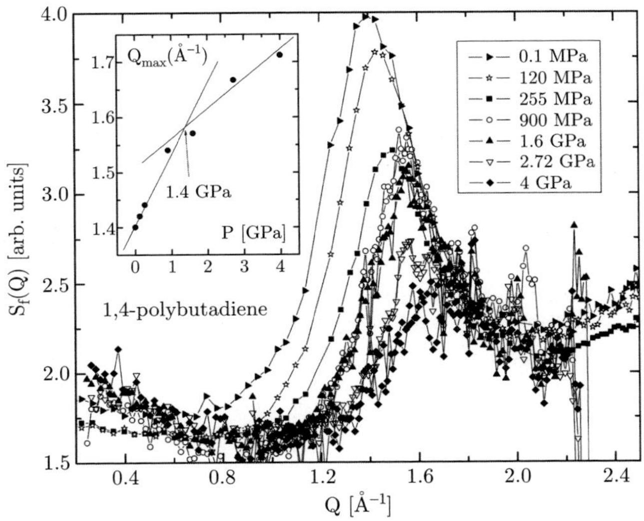

Figure 2.2 Static structure factor of PBD at 295 K and various pressures. The inset shows the increase in momentum transfer for the primary peak with densification. The glass transition at 1.4 GPa is indicated by the change in the compressibility. Data from ref. [2].

Thus, the glass transition is not a transformation from one thermodynamic phase[2] to another, in the sense that the liquid and glassy states cannot be distinguished in terms of an ordering of the lower temperature phase. Nevertheless, models sometimes describe vitrification in terms of an underlying thermodynamic transition, obtained in principle by infinitely slow cooling of a liquid to an ideal glass transition at which the entropy would reach that of the perfect crystalline state[3] [3]. However, the glass transition is not a first-order transition, since there is no discontinuity in the entropy (or other properties) at T_g, nor is there any latent heat associated with glass formation. Vitrification is sometimes referred to as a "second-order transition", since second derivatives of the volume and enthalpy exhibit jumps at T_g. The Prigogine–Defay ratio, Π, combines the step changes at T_g in the isobaric thermal expansivity, $\Delta\alpha_P$, isobaric specific heat, Δc_P, and isothermal compressibility, $\Delta\kappa_T$, [4]

$$\Pi = \frac{\Delta c_P \Delta\kappa_T}{T_g(\Delta\alpha_P)^2} \tag{2.2}$$

These steps in the susceptibilities are not sharp and are evaluated over the glass-transition interval. For a thermodynamic second-order transition, $\Pi = 1$; however, for the glass transition it is found experimentally that $1 < \Pi < 5$ [5,6], indicating a decoupling of entropy (ΔC_P) and volume ($\Delta\kappa_T$ and $\Delta\alpha_P$) fluctuations. If the Prigogine–Defay ratio were equal to unity, properties of the glass could be defined by a single parameter, such as the density, but this is not the case. Indeed, glasses can be obtained via different thermodynamic pathways, leading to equivalence of some properties but not others [7–9].

From an experimental viewpoint the glass transition is a kinetic process, with the immediate cause of vitrification being the divergence of τ_α and η with decreasing temperature at constant pressure, as molecular motions can no longer "keep up" with the imposed thermodynamic changes. A pressure-dependent transition temperature is defined as the condition at which the characteristic time for reorientional motion or diffusion exceeds the experimental time-scale; e.g., $\tau_\alpha(T_g) = 100$ s. Traditionally in the study of lubricants, T_g is defined as the temperature at which the viscosity reaches the value of 10^{12} Pa s. Since the viscosity and other transport coefficients are governed by related molecular motions (see eqns (2.19) and (2.20)), this characteristic viscosity can be related to a relaxation time via the Maxwell relation [10]

$$\eta = G_\infty \tau_\alpha \tag{2.3}$$

The high-frequency shear modulus, G_∞, is on the order of a few GPa, so that T_g defined for $\eta = 10^{12}$ Pas corresponds to τ_α in the range 100 – 1000 s. When the determination of T_g involves scanning of temperature, the transition does not occur at a well-defined temperature; both T_g and the structure of the obtained glass are rate dependent. Nevertheless, the value of τ_α at T_g remains unchanged as P is varied, for any given experimental protocol used to determine $T_g(P)$. For many materials, especially when the pressure is not too large, T_g varies linearly with P. This pressure coefficient of T_g for organic materials is in the range from 150 to 400 K/GPa, although it is significantly smaller for hydrogen-bonded materials [11]. Values for polymers are collected in Table 2.1. Note that the density at the glass transition increases with increasing pressure (Figure 2.3 [12–14]).

[2] A phase in the thermodynamic sense is a physically distinct, mechanically separable region of space in which all physical properties of a material are uniform. A thermodynamic phase transition transforms the material into a state having new properties; however, the marked differences between a liquid and glass are rate dependent, rather than being defined by the state of the material.

[3] The implication is that below the ideal glass temperature the entropy would become negative, in contradiction to the third law of thermodynamics. This enigma is known as the Kauzmann paradox [3].

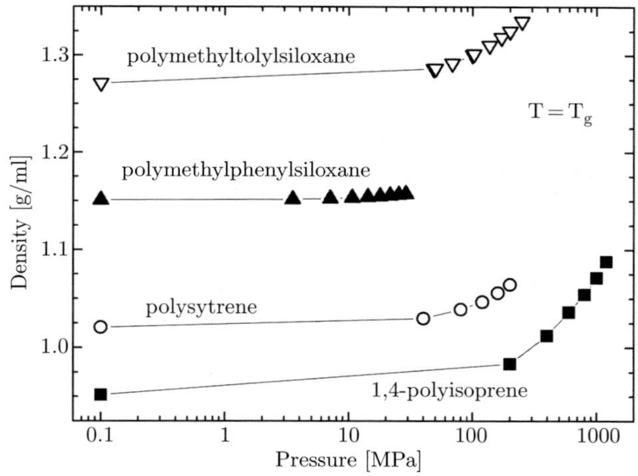

Figure 2.3 Mass density at the glass transition as a function of the hydrostatic pressure for four polymers. Data from refs. [12] (siloxanes), [13] (PS), and [14] (PI).

Table 2.1 Pressure coefficient of T_g in the limit of low pressure.

Polymer	T_g (K)	dT_g/dP (K/GPa)
polymethyltolylsiloxane	261	340
poly(2-vinylpyridine)	337	340
polystyrene	373	332
bisphenol-a polycarbonate	413	320
polymethylphenylsiloxane	248	289
polyvinylacetate	302	250
diglycidylether of bisphenol A	335	244
polyisobutene	198	240
1,2-polybutadiene	253	240
polycyclohexylmethacrylate	380	224
poly(4-vinylphenol)	473	222
polyvinylethylether	241	215
poly(n-butylmethacrylate)	293	204
polypropyleneglycol	210	192
polyvinylchloride	352	189
oligomeric epoxy (4,4'-methylene-bis(N,N-diglycidylaniline))	257	180
1,4-polyisoprene	201	178
polyvinylmethylether	247	177
polymethylmethacrylate	390	170
atatic polypropylene	252	158
polyethylvinylacetate	258	158
polybisphenol A-*co*-epichlorohydrin, glycidyl endcapped	258	156
polyoxybutylene	199	155
poly(phenyl glycidy ether)-co-formaldehyde	258	154
polydimethylsiloxane	148	143
nylons	339	100–160
polychlorotrifluoroethylene	368	100

2.1 Non-exponential and non-Arrhenius behavior

When the frequency of the mechanical, electrical, or other probe corresponds to that of any molecular or segmental motions, there is a peak in the out-of-phase component of the spectral response; the term "loss peak" is used since there is energy dissipation. Loosely speaking, the energy absorption is caused by the molecular motions being out-of-phase and hence resistive to the imposed motion. At higher frequency a mean perturbation is experienced, which is ineffective in displacing molecules, and at lower frequency the perturbing field can be accommodated with minimal resistance. The dispersion (in mechanical, dielectric, light scattering, etc. spectra) due to structural relaxation is illustrated in Figure 2.4 [15,16]. With few exceptions the structural relaxation peak cannot be described by a simple exponential time correlation function; the loss peak is asymmetrically broadened toward higher frequencies. The Kolhrausch stretched exponential function (given for the shear modulus in eqn (1.7)) has only one shape parameter (β_K) and generally gives a good fit to experimental data.[4] The loss

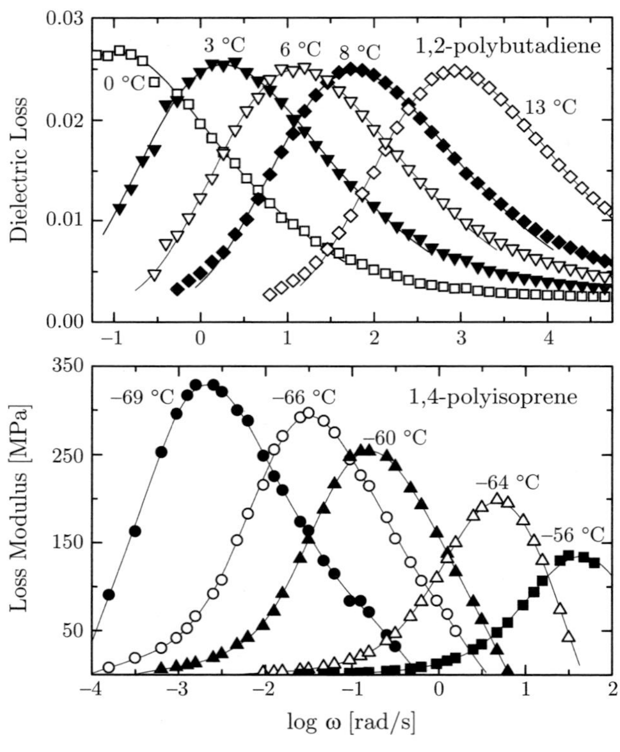

Figure 2.4 Local segmental relaxation dispersion at the indicated temperatures for (top) $\varepsilon''(\omega)$ of PVE and (bottom) $G''(\omega)$ of PI. Solid lines are the fitted Kohlrausch functions. Data from refs. [15,16].

[4] The applicability of the Kohlrausch equation extends beyond structural relaxation of liquids and polymers. It provides an almost universal description of slow relaxations in strongly interacting systems, notwithstanding the different mechanisms underlying the diverse phenomena [17–19]. However, if the relaxation function is a stretched exponential, the corresponding retardation function of linear viscoelasticity cannot also have the KWW form [20].

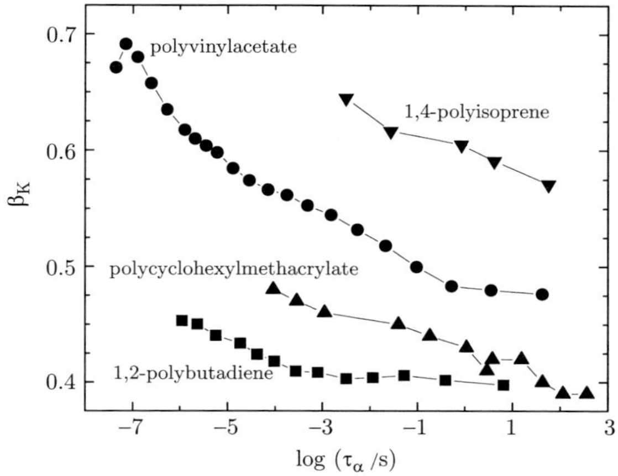

Figure 2.5 Kohlrausch stretch exponent as a function of local segmental relaxation time for four polymers. A smaller β_K corresponds to a broader relaxation peak. Data from ref. [22, 23].

peak broadens as the glass transition is approached (Figure 2.5) [22,23], reflecting the increasing degree of intermolecular cooperativity and consequently stronger many-body effects.

Non-exponentiality of the relaxation function is accompanied by a non-Arrhenius temperature dependence of the relaxation times. This temperature-dependence can be represented in the vicinity of the glass transition by the Vogel–Fulcher–Tammann–Hesse (VFTH) equation [24–26]

$$\log \tau_\alpha(T) = \log \tau_\infty + \frac{B}{T - T_0} \tag{2.4}$$

in which τ_∞, B, and T_0 are constants. The latter is referred to as the Vogel temperature; at T_0 the dynamics would diverge ($\tau_\alpha \to \infty$) if not for intervention of the glass transition.[5][3,27] In Figure 2.6 this non-Arrhenius behavior is illustrated for the local segmental relaxation times of PVME [28]. The VFTH equation reduces to Arrhenius behavior as $T_0 \to 0$ or $T \to \infty$. The non-linearity of $\tau_\alpha(T^{-1})$ can also be expressed as a temperature-dependent (isobaric) activation energy[6]

$$E_a = R\frac{d \ln \tau_\alpha}{dT^{-1}}\bigg|_P = \frac{B \log(e)}{(1 - T_0/T)^2} \tag{2.5}$$

which is illustrated for PVME in the inset to Figure 2.6. Since the data conform to eqn (2.4), they can be linearized by plotting the model-independent Stickel function [29,30]

[5] T_0 is often taken to be coincident with the Kauzmann temperature [27], at which the extrapolated liquid entropy equals the crystal entropy. This causes an ostensible violation of the 3rd law of thermodynamics ("Kauzmann paradox" [3]), because at lower temperatures the entropy would become negative. The resolution of this paradox is much debated, possibilities including error in extrapolation of the heat capacity from T_g to T_0 or intervention of a phase transition in the liquid.

[6] Since the volume changes during an isobaric temperature change, work can be done; thus, strictly speaking E_a measured at constant pressure is an activation enthalpy.

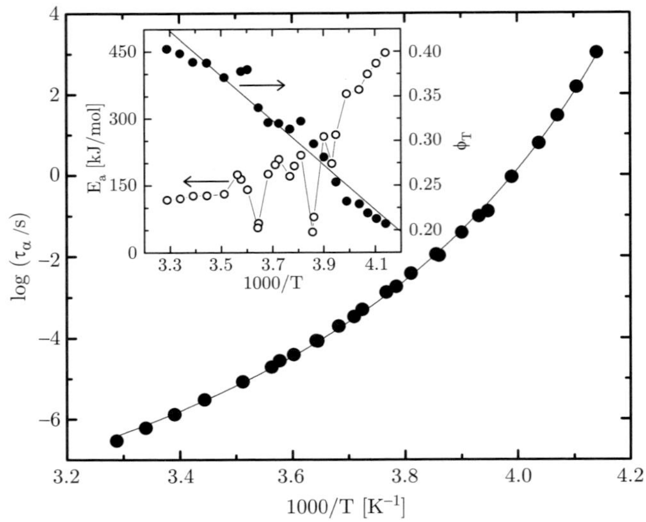

Figure 2.6 Arrhenius plot of local segmental relaxation times for PVME. The solid line is the fit to the VFTH equation (eqn (2.4)) with $\log \tau_\infty = -11.77$ s, $B = 1216$ K, and $T_0 = 205.7$ K. Inset shows the corresponding activation energies (with large scatter) and the Stickel derivative (eqn (2.6)). Data from ref. [28].

$$\phi_T(\tau_\alpha) = \{[d\log(\tau_\alpha)/d(1000/T)]\}^{-1/2} \tag{2.6}$$

This equation is particularly useful in determination of the onset of Arrhenius behavior or other changes in the dynamics (see Chapter 6).

The non-Arrhenius temperature dependence of τ_α precludes direct comparison of T dependences among polymers having different glass-transition temperatures. The "fragility" classification scheme is a means to circumvent this problem. Fragility refers to the steepness of semilogarithmic plots of relaxation time (or viscosity) versus the dimensionless variable T_g/T; it is proportional to the activation energy normalized by the thermal energy at T_g. By defining the glass-transition temperature to be a fixed value of τ_α (or η), fragility plots for different materials have the same origin, with the steepness of the decrease in τ_α as T increases serving as a metric of the temperature sensitivity. The terms fragile and strong imply a connection of the dynamics to the local liquid order – fragile behavior implies the short-range structural environment, and hence τ_α, is rapidly degraded thermally, while strong liquids are resistant to changes in structure as temperature increases through the glass-transition region. However, the utility of the fragility scheme transcends any particular concept of the glass-transition phenomenon. T_g-normalized Arrhenius plots were first proposed by Oldekop [31] to explain the behavior of various inorganic glass-forming materials. Subsequently, Laughlin and Uhlmann [32] presented fragility plots for three organic liquids and seven inorganic oxide glass formers, from which they proposed three classes of behavior; Arrhenius, nearly Arrhenius, and fragile. The idea that the T_g-normalized temperature dependence is a key metric in describing glass-transition properties was further developed by Angell [33], with the fragility parameter (or steepness index)

$$m = \left. \frac{d \log(\tau_\alpha)}{d(T_g/T)} \right|_{T=T_g} \tag{2.7}$$

shown to correlate with many properties, including the non-exponentiality of the relaxation function [34–36], the chemical structure of polymers [37–40], diffusion [41,42], and non-linear behavior in the glassy state [43–45], including properties having timescales much shorter than the structure relaxation time [42,46–53].

Figure 2.7 shows fragility plots for several polymers. The general trend is that an increasing degree of intermolecular cooperativity is associated with more fragile relaxation. Thus, polymers with smooth, compact, symmetrical-chain backbones have τ_α exhibiting a more nearly Arrhenius temperature dependence, reflecting less-constrained dynamics. Polymers having more rigid backbones or sterically hindering pendant groups have steeper T/T_g-temperature dependences, a consequence of segmental relaxation that is more intermolecularly cooperative. The data for polymers parallel those for molecular liquids, whose chemical structure similarly governs the strength of the intermolecular coupling, and thus the temperature dependence of the relaxation times near T_g [33,39].

This interpretation follows from the coupling model (Chapter 4), which characterizes the strength of the intermolecular cooperativity and its effect on various properties. According to the model the breadth of the relaxation function is determined by the degree of intermolecular cooperativity, with the specific prediction that broader relaxation functions (smaller β_K) are associated with a greater temperature-sensitivity of τ_α. This prediction of an inverse correlation between β_K and m is borne out approximately by experimental data (Figure 2.8)[54–64]. Note that the general trend of larger β_K and less-fragile dynamics in more flexible polymers that lack bulky side groups is similar to the expected influence of these same structural features on the glass-transition temperature; however, the correlation between fragility and T_g is poorer than that between m and β_K [65].

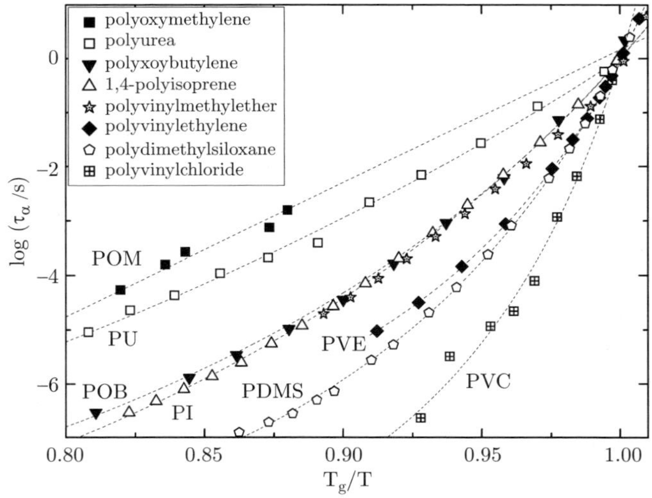

Figure 2.7 T_g-normalized Arrhenius plots of the local segmental relaxation for various polymers, with the glass-transition temperature defined by the condition $\tau_\alpha = 1$ s. The steepness of the curve ("fragility") parallels the degree to which the chemical structure engenders cooperativity with the motion of neighboring segments.

Molecular dynamic (MD) simulations reveal that the fragility of liquids increases with the degree of anharmonicity of the potential; that is, a steeper potential energy surface, having many shallow minima, gives rise to larger m [33,66–68]. Such an interpretation is consistent with a connection to the degree of intermolecular cooperativity, as governed primarily by the repulsive interactions.

As discussed below and in Chapter 6, the preponderance of intramolecular bonds in polymers distinguishes their relaxation dynamics from that of molecular liquids. This suggests an effect of molecular weight on properties such as fragility, similar to that seen for T_g (e.g., Figure 1.13). For polymers having smooth, flexible backbones, such as the PDMS [56], PIB [64], and PMPS [69] (Figure 2.9), there is essentially no effect of M on fragility. However, generally m increases with chain length [56,57,68,69,70,71] (Figure 2.10).

An interesting aspect of the data in Figure 2.8 and Figure 2.10 is the comparison between dielectric and mechanical results. For polymers such as polybutadiene and polyisoprene, fragilities measured by the two methods are essentially equivalent. This is shown in Figure 2.11 for PI. Dielectric peak frequencies are always smaller than the corresponding frequency of the peak in the loss modulus; however, their temperature dependences are the same. In contrast, for a polymer such as polystyrene, in which most of the monomer unit resides pendant to the backbone, m is much larger when measured mechanically than by dielectric spectroscopy (Figure 2.10).

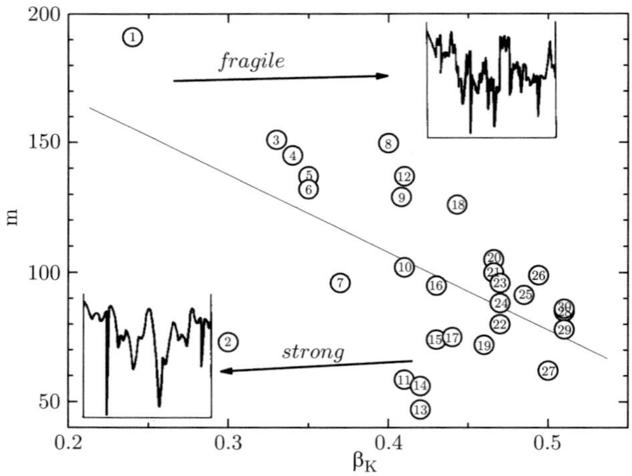

Figure 2.8 Isobaric fragility at ambient pressure *versus* Kohlrausch stretch exponent for: 1 – PVC, 2 –poly (α-methylstyrene), 3 – epoxy resin, 4 – PMMA, 5 – *atactic*-polypropylene, 6 – polycarbonate, 7 – diglycidylether of bisphenol A, 8 – PVE (*dielectric*), 9 – PVE (*mechanical*) 10 – polymethacrylate, 11 – polycyclohexyl methacrylate, 12 – PS (mechanical), 13 – PIB (dielectric), 14 – PIB (mechanical), 15 – PS (dielectric), 16 – PVAc, 17 – PVME, 18—PVE (86% vinyl), 19 – poly(2-chlorostyrene), 20 – PVE (70% vinyl), 21 – PVE (53% vinyl), 22 – PVME, 23 – polypropylene glycol (PPG), 24–1,4-polybutadiene (dielectric), 25 – PDMS, 26 – polybutadiene (mechanical), 27—PI, 28 – polymethyltolylsiloxane, 29 – polyoxybutylene, 30 – polymethylphenylsiloxane. A linear fit approximately describes the data, $m = 227(\pm 36) - 300\beta_K$; however, the correlation coefficient is only –0.7. Data from refs. [28,35,54–64]. Representations of the potential energy surfaces from refs. [33, 68].

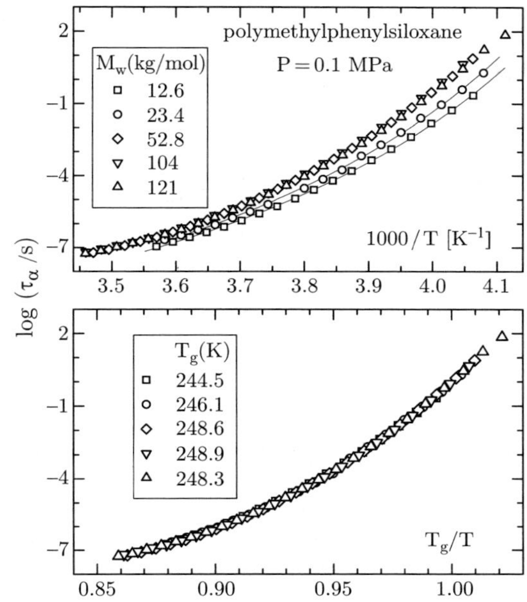

Figure 2.9 Dielectric relaxation times for PMPS having the indicated molecular weights (upper) measured at $P = 0.1$ MPa and (lower) replotted *versus* temperature normalized by T_g, defined as the value of T at which $\tau_\alpha = 1$ s. Data from ref. [69].

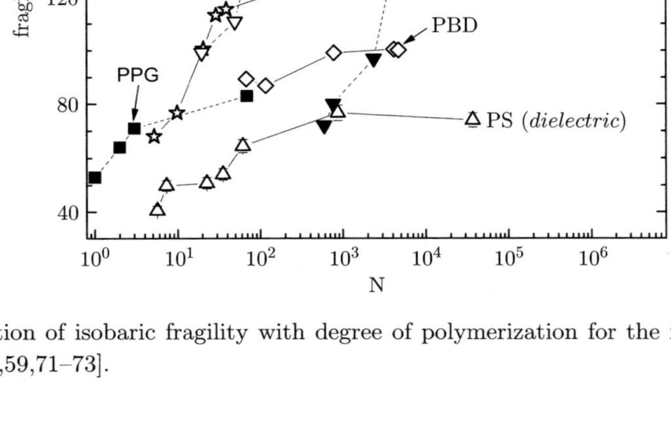

Figure 2.10 Variation of isobaric fragility with degree of polymerization for the indicated polymers. Data from refs. [58,59,71–73].

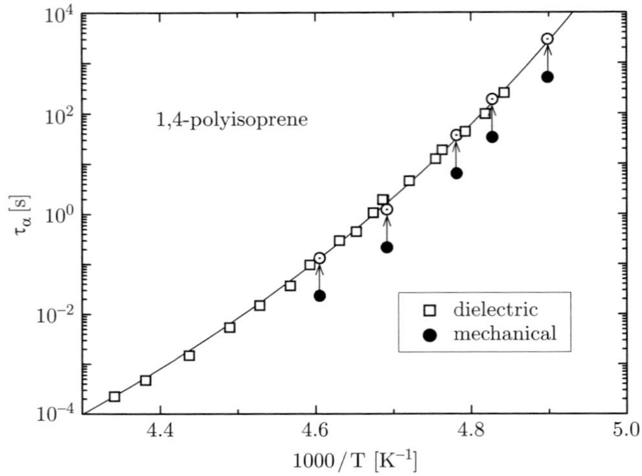

Figure 2.11 Local segmental relaxation times of PI determined as the inverse of the frequency of the maximum in the dielectric loss (open squares) and the loss modulus (filled circles). Multiplying the latter by a factor of 5.7 (open circles) brings them into coincidence with the dielectric τ_α. Data from ref. [15].

2.2 Temperature and density effects on τ_α

Classical models of the glass transition identify the factor governing the divergence of τ_α on approach to T_g as either the unoccupied volume [74,75–77] or the configurational entropy [78,79]. The progressive slowing of molecular motions upon lowering temperature has two direct causes: the decrease of volume, which enhances steric constraint (jamming) of molecular segments, and the reduced thermal energy with consequently greater difficulty for surmounting potential barriers. Free-volume models focus on molecular packing and according to the treatment of Grest and Cohen [75], mobility requires each liquid-like molecule to have at least two near neighbors that are also liquid-like. Relaxation commences above a percolation temperature associated with continuity of the local empty space. Landscape models [80–82], on the other hand, address the progressive trapping of molecules within the potential-energy surface, implying thermal activation as the dominant mechanism. The important distinction between these two approaches is the thermodynamic quantity, volume or temperature, considered to be the *control variable* for the glass transition. However, the dichotomy between the free-volume and energy-landscape models is not so unambiguous. Variations on free-volume models have been proposed that incorporate temperature dependences into the model parameters [83,84] and the potential-energy barriers at the heart of landscape models are affected by the density. For example, the mode-coupling theory of the glass transition [85], which expresses the dynamics in terms of the density correlation function, is based on a static structure factor governed by both temperature and density. Thus, different models of the glass transition are not necessarily incompatible with a varying influence from density and temperature.

The relative degree to which density and thermal energy govern the dependence of τ_α on temperature cannot be determined from isobaric measurements alone; to deconvolute these two effects requires experiments in which either pressure or volume is isothermally varied. As first pointed out by Hoffman *et al.* [86], the isochoric (constant volume) activation energy,

$E_V(T, V) = R\frac{\partial \ln \tau}{\partial T^{-1}}\big|_V$, is always less than the constant pressure activation energy, E_a (eqn (2.5)), with the ratio E_V/E_a serving as a direct measure of the relative contribution of temperature and volume to the variation of τ_α with temperature [87,88]. The determination of this ratio is illustrated in Figure 2.12 for PMTS. Using the equation of state, relaxation times measured as a function of both temperature and pressure are replotted versus volume, from which τ_α at constant volume can be obtained and used to calculate E_V. The latter varies with volume, so the ratio E_V/E_a is conventionally reported as the (interpolated) value at the atmospheric pressure T_g.

In Table 2.2 this ratio at T_g is listed for various polymers. With one exception (polyphenylene oxide, which has an unusually high T_g (= 189°C) for a flexible chain polymer), $E_V/E_a > 0.5$, implying that temperature more strongly influences the segmental dynamics than does density. This contrasts with molecular liquids (Table 2.3), which tend to have smaller activation energy ratios, reflecting a stronger effect of density on $\tau_\alpha(T)$. The reason

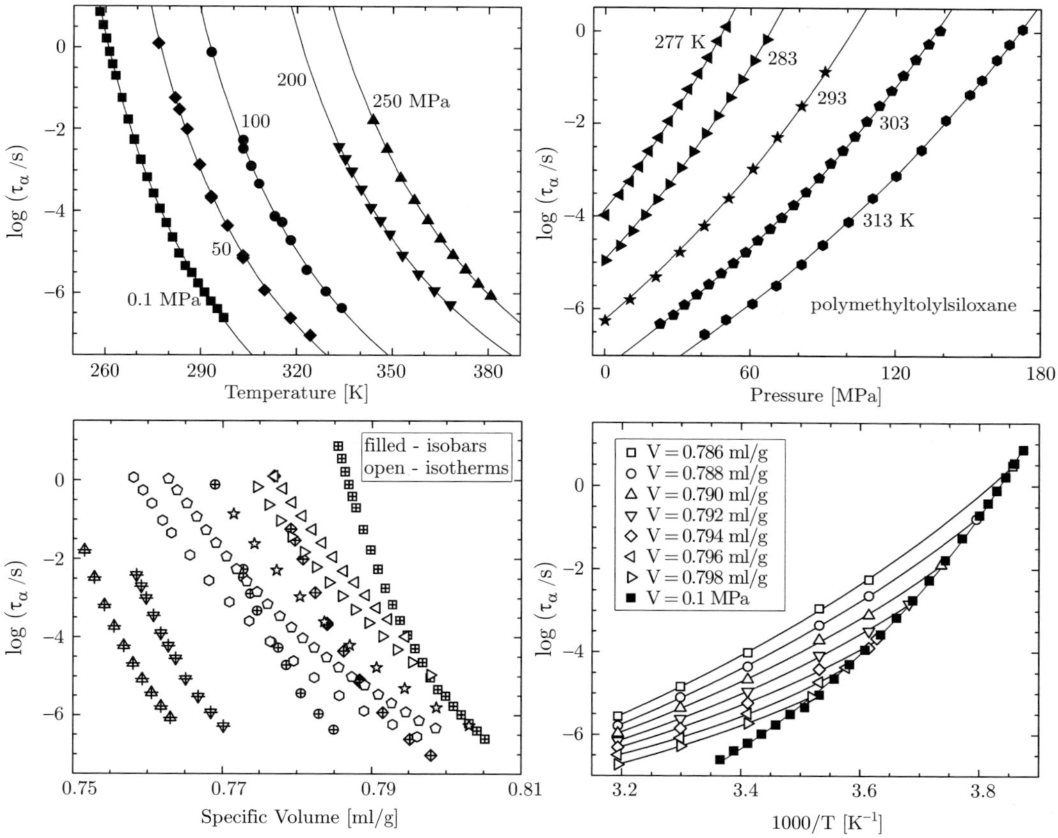

Figure 2.12 Dielectric relaxation times of PMTS (upper left) measured as a function of T at the indicated pressures; (upper right) measured as a function of P at the indicated temperatures; (lower left) replotted as a function of V; (lower right) Arrhenius plots of isochoric τ_α, along with the values measured at ambient pressure, the slopes of which yield the isochoric (E_V = 230 kJ/mol) and isobaric (E_a = 420 kJ/mol) activation energies; thus, $E_V/E_a = 0.55$.

Table 2.2 Activation energy ratio for polymers ($T = T_g$).

Polymer	E_V/E_a
poly(phenylene oxide)	0.25
polymethylphenylsiloxane	0.52
polymethyltolylsiloxane	0.55
polyvinylacetate	0.6
diglycidylether of bisphenol A	0.6
poly(phenol glycidyl ether)-co-formaldehyde	0.63
polystyrene	0.64
poly[(o-cresyl glycidyl ether)-co-formaldehyde]	0.65
polypropylene glycol	0.67
polyvinylmethylether	0.69
1,2-polybutadiene	0.70
polycyclohexylmethacrylate	0.72
poly(2-vinylpyridine)	0.72
polymethylmethacrylate	0.74
polyoxybutylene	0.75
1,4-polyisoprene	0.76
polyvinylphenol	0.77
poly(4-vinylpyridine)	0.77
polyvinylethylether	0.81

Table 2.3 Activation energy ratio for molecular liquids ($T = T_g$)

	E_V/E_a
62% chlorinated biphenyl	0.38
1,1'-bis(p-methoxyphenyl)cyclohexane	0.39
1,1'-di(4-methoxy-5-methylphenyl)cyclohexane (BMMPC)	0.41
cresolphthalein-dimethylether	0.49
54% chlorinated biphenyl	0.50
phenylphthalein-dimethylether	0.53
ortho-terphenyl	0.55
42% chlorinated biphenyl	0.56
propylene carbonate	0.64
α-tocopherol acetate	0.66

for the different behavior is that pressure has a negligible effect on the dimensions of a polymer chain (e.g., its radius of gyration); thus, interactions among directly bonded segments, which are profuse in polymers, are almost unaffected by changes in volume. In molecular liquids, on the other hand, most near-neighbor interactions are intermolecular, and these are sensitive to density. The influence of intramolecular bonds is brought out clearly in the increase in E_V/E_a with degree of polymerization for polymers lacking hydrogen bonds (Table 2.4 [70,73,89]).

Table 2.4 Effect of degree of polymerization on E_V/E_a.

Liquid	N	E_V/E_a
polypropylene glycol	1	0.89
	2	0.73
	3	0.77
	69	0.67
polymethylmethacrylate	3	0.63
	4	0.63
	10	0.63
	1500	0.74
polystyrene	132	0.52
	231	0.54
	332	0.64
	865	0.66
poly 4-vinylpyridine	19	0.62
	152	0.77

Table 2.5 Activation energy ratio for H-bonded liquids ($T = T_g$).

Liquid	E_V/E_a
salol	0.47
polyurea	0.81
sorbitol	0.85
glycerol	0.94
water	~1

The exception in Table 2.4 to an increase in E_V/E_a with molecular weight is polypropylene glycol. In this case there is a competing effect, the reduction in hydrogen bonding with chain length. Temperature exerts the more significant effect on associated materials, such as hydrogen-bonded liquids; these exhibit large values of E_V/E_a (Table 2.5). Thermal energy reduces the degree of H-bonding, which amplifies the direct effect that temperature has on molecular mobility. On the other hand, higher pressure tends to reduce hydrogen bonding [90–92], since less unoccupied space makes directional interactions more difficult. Fewer H-bonds facilitates relaxation, countervailing the direct effect of pressure on τ_α. The consequence is that temperature is the dominant control parameter for structural relaxation in liquids with substantial hydrogen bonds. Only salol in Table 2.5 departs from this pattern; the hydrogen bonding in salol is primarily intramolecular.

All values of E_V/E_a in Tables 2.2–2.5 are for $T = T_g$. At higher temperatures the magnitude of this ratio decreases, as shown in Figure 2.13 for some molecular liquids. Pressure has the opposite effect, increasing E_V/E_a, which indicates that the influence

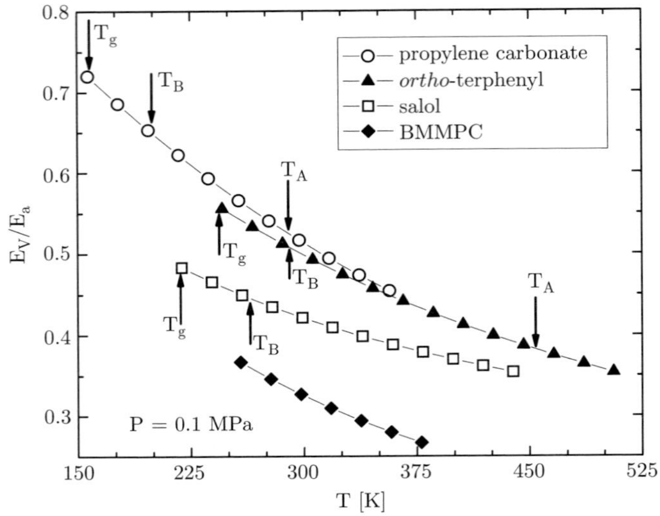

Figure 2.13 Temperature dependence of the activation energy ratio for 4 molecular liquids. The arrows denote the temperature of the glass transition, the dynamic crossover at T_B, and the onset of Arrhenius behavior at T_A.

of volume on the dynamics is enhanced as V increases, whether the change is brought about by higher T or lower P. At sufficiently high temperatures, the behavior becomes Arrhenius; interestingly, as seen in Figure 2.13, the onset of activated dynamics occurs as the value of E_V/E_a approaches zero. Also, there is no change in the nature of the variation of E_V/E_a with temperature through the dynamic crossover, which is generally associated with development of strong intermolecular cooperativity (see Chapter 6).

The fragility measured under constant volume conditions, m_V, is related to the isobaric value as

$$m_V = m\frac{E_V}{E_a} \tag{2.8}$$

Thus, the isochoric fragility is smaller than m and it is also independent of volume [93]. The latter is a consequence of the scaling property of structural relaxation times (Chapter 6). If temperature is the dominant control variable, $m_V \sim m$, whereas m_V becomes negligible when $\tau_\alpha(T)$ depends mainly on V. A plot of m_V versus m for different materials (Figure 2.14 [68]) shows that there is a linear correlation between the isochoric and isobaric fragilities. From these data the limiting values of fragility can be obtained: $m = 37 \pm 3$ for $m_V = 0$ and $m = 231 \pm 72$ for $m_V = m$. These are in accord with the range of experimental m for atmospheric pressure (e.g., Figure 2.8). The isobaric fragility varies only weakly with pressure, experimentally found to decrease about 10% for pressures up through 500 MPa (Figure 2.15) [93].

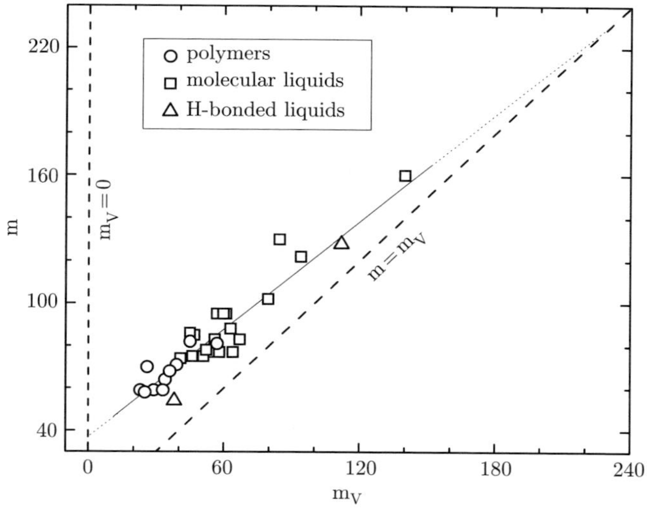

Figure 2.14 Isobaric fragility m at atmospheric pressure *vs.* the fragility at constant volume in order of increasing m_V: polychlorinated biphenyl (62% chlorine), BMMPC, 1,1'-bis(p-methoxyphenyl)cyclohexane (BMPC), polychlorinated biphenyl (54% chlorine, PCB), polychlorinated biphenyl (42% chlorine), cresolphthalein-dimethylether (KDE), salol, glycerol, phenylphthalein-dimethylether, polypropylene oxide, PMPS, o-terphenyl (OTP), polyepichlorhydrin, PMTS, PVME, PVAc, PS, polypropylene glycol, propylene carbonate (PC), diglycidyl ether of bisphenol A, PI, poly[(phenol glycidyl ether)-co-formaldehyde], PVAc, 1,2-polybutadiene, 1,4-polybutadiene, polyethylacrylate, polymethylacrylate (two values), sorbitol, and PVC. The solid line through the data is the least squares fit, $m = 0.84m_V + 37$. The vertical dashed line represents volume-dominated dynamics, and the 45° dashed line corresponds to temperature-dominated dynamics. Data from ref. [68].

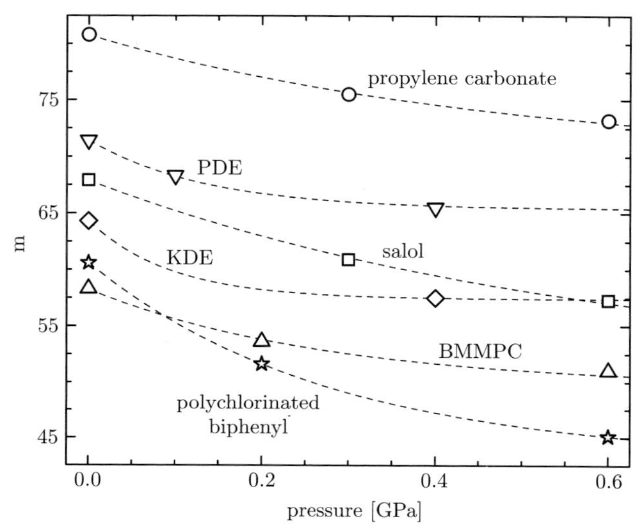

Figure 2.15 Isobaric fragility of several molecular liquids *versus* pressure. Data from ref. [93].

2.3 Dispersion of relaxation times and dynamic heterogeneity

2.3.1 Segmental relaxation dispersion

As discussed above, the breadth of the α-dispersion, reflected inversely in the exponent β_K, varies among glass formers. For most materials the peak broadens with proximity to the glass transition [23,94–97]. (In some molecular liquids the peak breadth is invariant with T [98–100], while in 1,4-polybutadiene the α-peak broadens with increasing $T-T_g$, an artifact of the increasing encroachment of an intense secondary peak [72].) This broadening is shown in Figure 2.16 for polycyclohexylmethacrylate (PCHMA) [101]; as temperature is lowered, the Kohlrausch exponent decreases by 0.09, corresponding to an increase in the peak width at the half-intensity points from less than 2.3 decades to 2.6 decades. Pressure similarly affects β_K, so that in general longer τ_α implies a broader peak. However, if the relaxation time is maintained constant as T increases by simultaneously elevating the pressure, the shape of the structural relaxation dispersion remains constant. This is illustrated in Figure 2.17, wherein the dispersion for PCHMA is shown for combinations of T and P such that the frequency of the α-loss maximum is constant. The peak height depends on the thermodynamic state, but its breadth is determined only by τ_α. This temperature-pressure superpositioning of the normalized α-dispersion at fixed τ_α is general, having been found for molecular liquids and polymers of diverse chemical structure [102,103]. Examples for four polymers are shown in Figure 2.18.

2.3.2 Dynamic correlations

The non-Debye nature of structural relaxation is a consequence of the many-body dynamics (involving more than one molecule or chain segment), which is necessarily intermolecularly correlated. Regions of fast-moving molecules coexist with slow-moving regions, and these

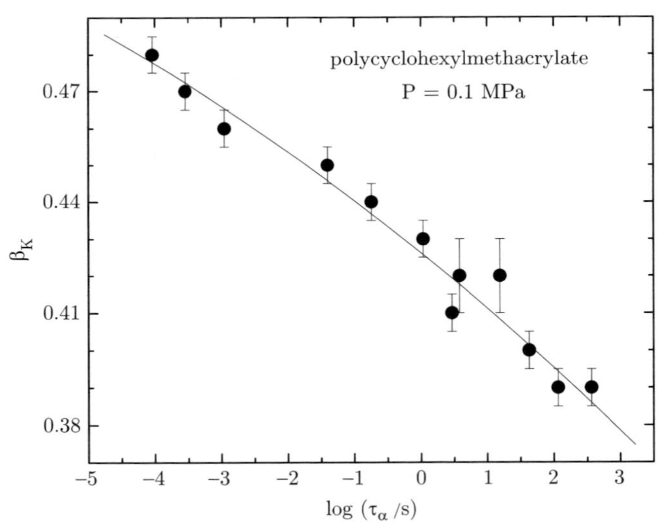

Figure 2.16 Kohlrausch exponent for the segmental dispersion peak in the dielectric loss of PCHMA. The corresponding peak breadths change by more than 0.3 decades over this temperature range from 349.8 to 406.1 K. Data from ref. [101].

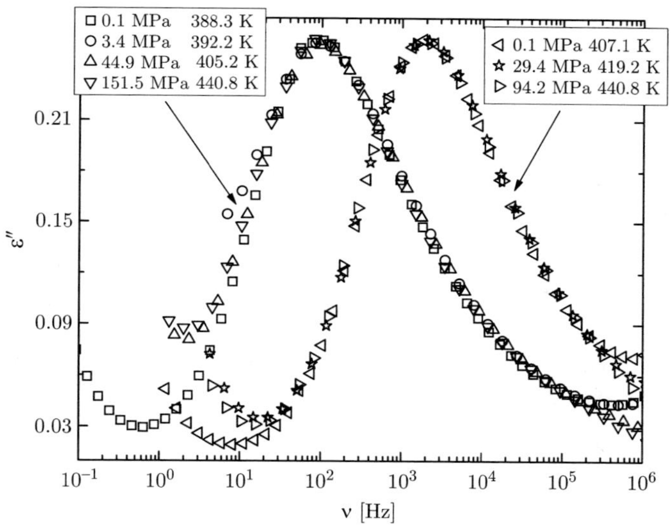

Figure 2.17 Dispersion in the dielectric loss of PCHMA for conditions of T and P for which τ_α is essentially constant. The rise at the lowest frequencies is due to ionic conduction. Data from ref. [101].

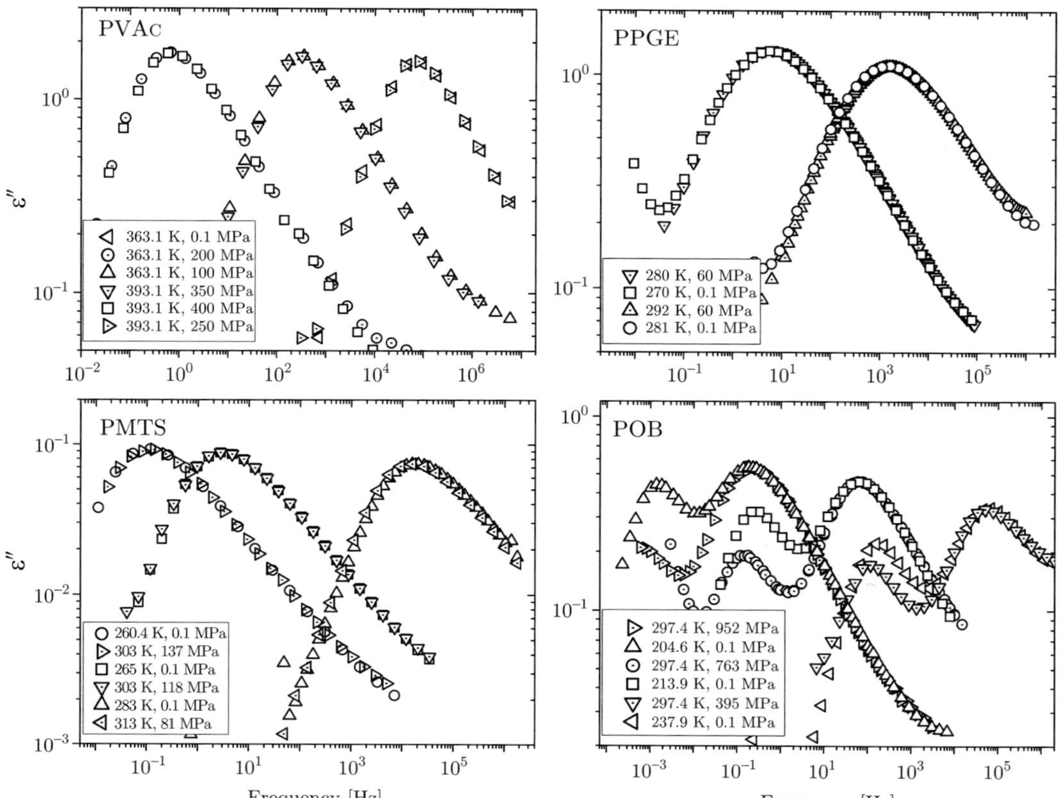

Figure 2.18 Dielectric α-loss peaks at various state points as indicated for which τ_α is constant, and thus the normalized dispersion also has a constant shape. (upper left) polyvinylacetate; (lower left) polymethyltolylsiloxane; (upper right) poly(phenyl glycidyl ether)-*co*-formaldehyde; (lower right) polyoxybutylene. The peak height for spectra measured at higher P are normalized to the maximum of the corresponding loss peak at atmospheric pressure. Data from ref. [103].

spatial variations in mobility persist for times commensurate with τ_α. The heterogeneous dynamics defines a length scale, ξ, within which molecular rearrangements (rotations or translations) cannot occur independently of the local environment. As temperature is reduced or the density increases, ξ becomes larger in tandem with τ_α. Theories of the glass transition such as the classical Adam-Gibbs entropy [78] and Cohen–Grest free volume [75] models, as well as other treatments [104–109], invoke dynamic heterogeneity and the growth of a dynamic correlation length as the cause of the slowed dynamics near T_g. Ultimately, vitrification occurs due to a divergence in the number of molecules that must be accommodated to permit motion. Increased heterogeneity implies more disperse dynamics and hence a broader (more non-Debye) relaxation peak (Figure 2.5 and Figure 2.16). Thus, it follows from the T–P superpositioning of the peak at fixed τ_α that ξ should also be constant at fixed τ_α.

Experimentally quantifying the size of the dynamic heterogeneities is difficult because both spatial and temporal correlations are involved [110]. Information is required about the density simultaneously at two positions and two times, whereas conventional relaxation experiments measuring a linear susceptibility provide only an average two-point time-correlation function. The spatial extent of the fluctuations over a time span t can be characterized using a four-point dynamic susceptibility defined as [109,111]

$$\chi_4(t) = \int \langle \rho(r_1, 0)\rho(r_1 + r_2, 0)\rho(r_1, t)\rho(r_1 + r_2, t)\rangle_{r1} dr_2 \tag{2.9}$$

This quantity describes the probability that if correlation between particles is lost over a time t at position r_1, the same decorrelation transpires within this time interval at r_2. As molecules escape over time from "cages" formed from neighboring molecules, strong dynamic correlations develop and $\chi_4(t)$ grows. At $t \sim \tau_\alpha$ $\chi_4(t)$ exhibits a maximum, $\chi_4^{\max}$, that is proportional to the number of molecules (or repeat units), N_c, dynamically correlated over this time span. Commonly it is assumed that $\xi \sim N_c^{1/3}$, although the dynamic correlation volume has a complex shape that cannot be adequately described by a single length scale.

$\chi_4(t)$ can by computed from MD simulations [112–115], for example as the variance of the self-intermediate scattering function $F_s(\tilde{k}, t)$,[7] defined by

$$F_s(\tilde{k}, t) = N^{-1}\langle \rho_{\tilde{k}}(t)\rho_{-\tilde{k}}(0)\rangle \tag{2.10}$$

The four-point density susceptibility is

$$\chi_4(t) \sim [\langle f_s^2(\tilde{k}, t)\rangle - F_s^2(\tilde{k}, t)] \tag{2.11}$$

where $f_s(\tilde{k}, t)$represents the instantaneous value, $F_s(\tilde{k}, t) = \langle f_s(\tilde{k}, t)\rangle$. Figure 2.19 [115] shows $F_s(\tilde{k}, t)$ from simulation of a Kob–Andersen [116] binary mixture of particles. Regimes of $F_s(\tilde{k}, t)$ representing different physical behavior [117] are evident, from initial diffusive motion, followed by decay of the correlation function as particle interactions intensify. Structural relaxation gives rise to the peak in $\chi_4(t)$ shown in the inset. The growth of the dynamic susceptibility reflects the growth of dynamic correlations.

The connection expected between the magnitude of the dynamic heterogeneities and the dispersion of the relaxation means that the correlation volume should be invariant for state

[7] $F_s(\tilde{k}, t)$ describes the time evolution of a particle's position, deriving its name from a connection to the incoherent part of the inelastic scattering cross-section.

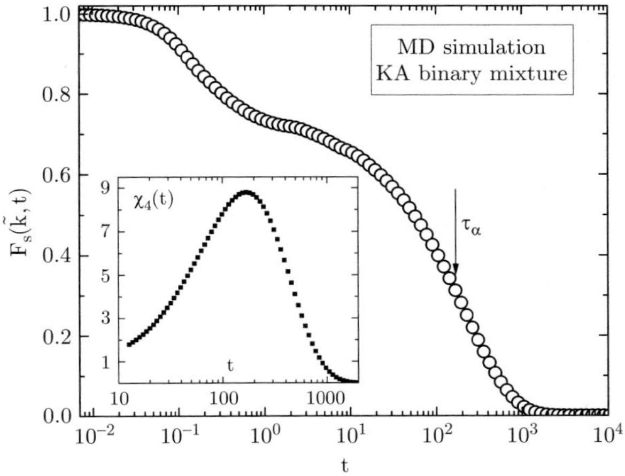

Figure 2.19 Self-intermediate scattering function for a Kob–Andersen binary mixture calculated at $T = 0.4$, $\rho = 1.15$, and $\tilde{k} = 0.7$ (the wave vector matching the position of the first peak in the static structure factor) in Lennard-Jones units. The arrow indicates the $t\,(= \tau_\alpha)$ corresponding to the maximum in $\chi_4(t)$. The inset shows the four-point dynamic susceptibility. Data from ref. [115].

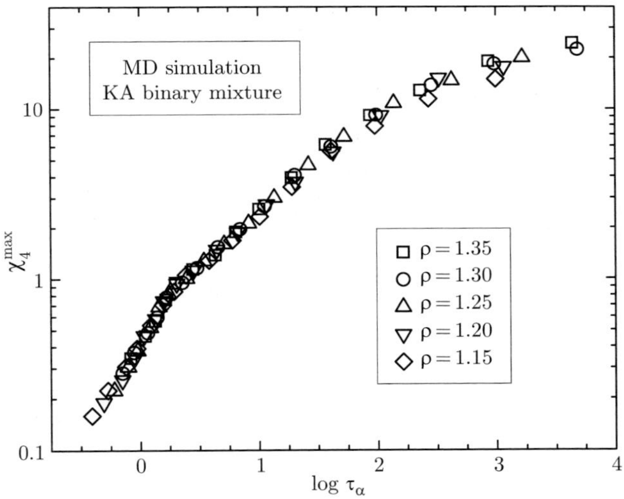

Figure 2.20 Maximum of the four-point dynamic susceptibility as a function of the relaxation time. Each data point represents a different temperature at the density (in Lennard-Jones units) denoted by the symbol type. Data from ref. [115].

points for which τ_α is constant, consistent with the superposition of normalized relaxation peaks (Figure 2.17 and Figure 2.18). MD simulations in which the T and ρ are varied to give state points of constant τ_α bear this out: N_c ($\propto \chi_4^{max}$) is a single function of τ_α (Figure 2.20). Actually, not only does χ_4^{max} depend only on the relaxation time, but the full t dependence of $\chi_4(t)$ from simulations superposes for fixed τ_α.

Table 2.6 Size of dynamic correlations in vicinity of T_g

	T–T_g	β_K	$N_c^{\,1}$	$N_c^{\,2}$	$N_c^{\,3}$	ξ (nm)^1	ξ (nm)^2	ξ (nm)^4
glycerol	10 K	0.55	10	—	97	1.3	2.9	1.4
o-terphenyl	9 K	0.41	76	—	148	2.3	3.0	2.5
polyvinylacetate	9 K	0.45	390	210	—	3.7	3.2	—

[1] NMR [ref. 121]; [2] DSC (eqn (2.12)) [refs. 122, 124]; [3] dielectric relaxation (eqn (2.15)) [ref. 133]; [4] light scattering (eqn (2.13)) [ref. 128]

Experimental determinations of ξ for real materials are limited. Four-dimensional solid-state NMR experiments combine relaxation-time determinations from monitoring the orientation of segments with spin-diffusion measurements to determine a length scale [118,119]. Although limited to temperatures close to T_g, Spiess and coworkers [120,121] used this method to quantify the heterogeneity length scale for several liquids, finding ξ to be in the range 1 to 4 nm, but uncorrelated with the Kohlrausch exponent (Table 2.6).

Donth [122] analyzed dynamic heterogeneity by assuming that the cooperativity volume is large enough to apply thermodynamics and make use of the Landau–Lifshitz fluctuation theory [123]. Fluctuations in temperature occur with a magnitude, δT^2, that depends on the system size, with the "system" regarded as the region of correlated motions. The number of molecules in this volume can be deduced from quantities measured in a calorimetry experiment [122]

$$N_c = k_B T^2 \Delta(c_V^{-1}) / m_0\, \delta T^2 \tag{2.12}$$

Here c_V is the isochoric heat capacity and $\Delta(c_V^{-1})$ is the change in the inverse of c_V through the glass transition. δT^2 is determined as the half-width of the temperature interval of the glass transition. (Note that this breadth of the calorimetric transition is predicted to be inversely related to the size of the dynamic correlation region, whereas a broader loss peak in a relaxation spectrum implies more heterogeneous dynamics.) The method of Donth and coworkers [124] yields estimates of the cooperativity length scale consistent with those from NMR determinations (Table 2.6).

Kremer *et al.* [125] characterized ξ by measuring structural relaxation of samples contained in pores of various size. When the confinement size is sufficiently small, the temperature dependence of τ_α changes from the usual VFTH to Arrhenius behavior. This change implies a loss of intermolecular cooperativity, with the pore size equated to ξ. Dynamic heterogeneity length scales deduced by this method are somewhat smaller than those from multidimensional NMR experiments.

The cessation of relaxation below the glass transition freezes in structural inhomogeneities, manifested as an excess density of states (Figure 1.4) and responsible for the Boson peak in the inelastic scattering of amorphous solids. The frequency of the Boson peak, ν_B, is related to the length scale of the heterogeneities [126,127]

$$\xi \sim \hat{c}/\nu_B \tag{2.13}$$

where $\hat{c}$ is the transverse sound speed. Hong *et al.* [128] have suggested that the structural heterogeneities intrinsic to the glassy state can be identified with the dynamic correlations observed above T_g. Table 2.6 shows that there is at least qualitative agreement between ξ from the Boson frequency in disordered solids and more direct characterizations on liquids.

Berthier and coworkers [129,130] proposed an approximation of $\chi_4(t)$ in terms of the temperature derivative of the experimentally accessible, two-point dynamic correlation function, $C(t)$

$$\chi_T(t) = \frac{\partial C(t)}{\partial T} \leq T^{-1}\left(\frac{\Delta c_P}{k_B}\chi_4(t)\right)^{1/2} \tag{2.14}$$

Thus, a lower bound on the number of correlating molecules is obtained as

$$N_c \geq \frac{k_B}{\Delta c_P} T^2 (\chi_T^{\max})^2 \tag{2.15}$$

where $\chi_T^{\max}$ represents the maximum value of $\chi_T(t)$ for any given state point. Comparisons of simulation results for χ_4 and χ_T indicate a reasonable correspondence between N_c from the two methods (Figure 2.21 [114]), especially near T_g, where to a good approximation the inequality in eqn (2.15) can be replaced by an equal sign [130,131]. The $\chi_T(t)$ approximation has been used to evaluate the N_c for different liquids [132,133] and different thermodynamic conditions [134]. Although some correspondence between N_c and ξ is expected (Table 2.6), the heterogeneous regions are not spherical [135] and so ξ does not specify the volume.

In Figure 2.22 N_c determined at different temperatures and pressures is plotted versus τ_α for four liquids [134]. For each material the relaxation time defines N_c, independently of T and P, the same result seen in the MD simulation results of Figure 2.20. Since dynamic heterogeneity is a defining feature of the supercooled dynamics, and given the established correlation of the relaxation time distribution with various relaxation properties, it might be expected that N_c would likewise correlate with these same properties. However, the situation does not appear to be so simple; for example, the size of the dynamic heterogeneity length scale does not correlate with the fragility [114,128,136], whereas β_K does (Figure 2.5 and Figure 2.16). However, as seen in Figure 2.22, at the dynamic crossover (Chapter 6), the number of dynamically correlated molecules does not vary much among materials.

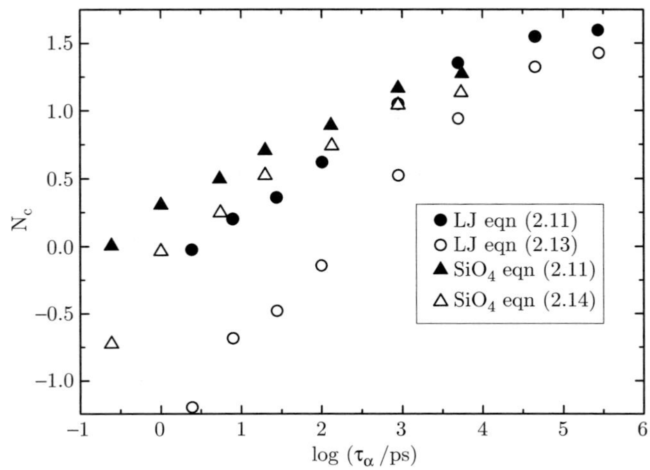

Figure 2.21 Maximum in the dynamic susceptibility (eqn (2.11), filled symbols) and its approximation (eqn (2.14), open symbols) for a binary Lennard-Jones mixture (circles) and for the Si ion dynamics in simulated silica (triangles). The simulations were done for isochoric conditions, so that Δc_V is used for the change in heat capacity at T_g rather than Δc_P. Data from ref. [114].

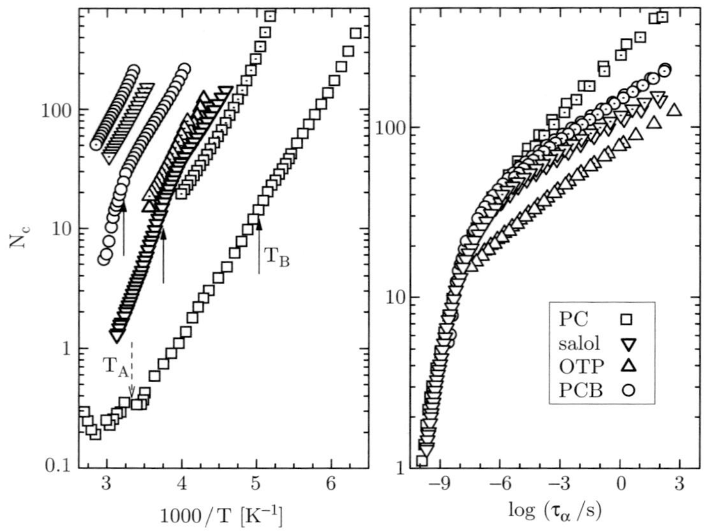

Figure 2.22 Number of molecules in correlation volume plotted: (left) in Arrhenius form and (right) *versus* the α-relaxation time. Pressures (in MPa) are 0.1 (open squares) and 500 (dotted squares) for PC; 0.1 (open circles) and 200 (dotted circles) for PCB; 0.1 (open inverted triangles) and 400 (dotted inverted triangles) for salol; and 0.1 (open triangles) and 28.8 (dotted triangles) for OTP. The up arrows indicate the temperature of the dynamic crossover, T_B, and the down-arrow the onset temperature for Arrhenius behavior, T_A. Data from ref. [134].

2.4 Johari–Goldstein secondary relaxations

There are a number of different secondary dynamics found primarily in dielectric, but also other relaxation spectra at frequencies higher than the α-relaxation peak. The slowest and most important of these is known as the Johari–Goldstein process. The JG relaxation, involving small-angle reorientation of a molecule or the repeat unit in a polymer chain, is a universal feature of the dynamics of glass-forming liquids and polymers. Falling intermediate in time between the caging due to the intermolecular repulsive forces and the rotational and translational diffusion transpiring at longer times, the JG secondary process plays a role in the development of structural relaxation.

Figure 2.23 shows dielectric loss spectra for PI at various temperatures through T_g [137]. On the high frequency flank of the segmental dispersion there is a weak peak identified as the JG process. Typically, it has a weaker dielectric strength than the α-relaxation, because the JG has a smaller amplitude. The relaxation times for both processes in PI [137] and PBD [72] are plotted in Figure 2.24. Unlike the more T-sensitive τ_α, the JG relaxation times have an Arrhenius temperature dependence, with $E_a = 38.2$ and 36.5 kJ/mol for PI and PBD respectively. This magnitude for E_a is within the range of the activation energies for non-JG processes in polymers (Table 1.1).

Near T_g the JG and α-relaxations become closer in frequency, eventually merging above the glass-transition temperature [138,139]. Estimates of this merging temperature have often been made by extrapolating measured JG relaxation times to temperatures above T_g, with the assumption that the Arrhenius temperature dependence of $\tau_{JG}(T)$ in the glassy state persists

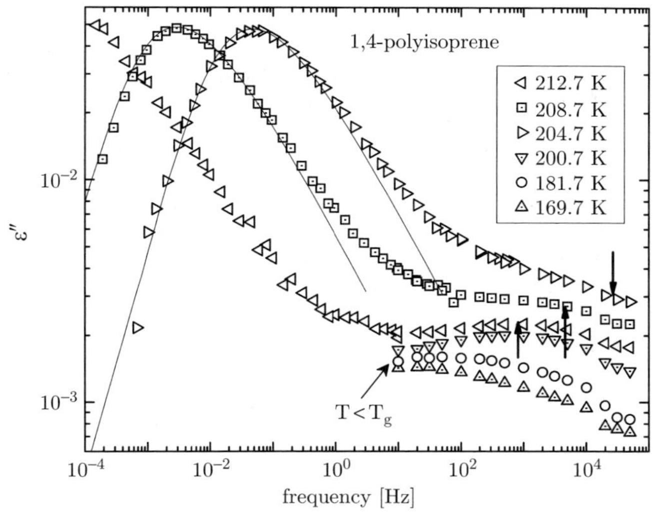

Figure 2.23 Dielectric loss spectra of PI ($M_w = 500$ kg/mol; $T_g = 207.4$K) at the indicated temperatures. The lower three curves are sufficiently below T_g that only the JG peak falls within the experimental window. The vertical arrows mark the primitive relaxation frequencies of the coupling model calculated from β_K for the α-relaxation. Data from ref. [137].

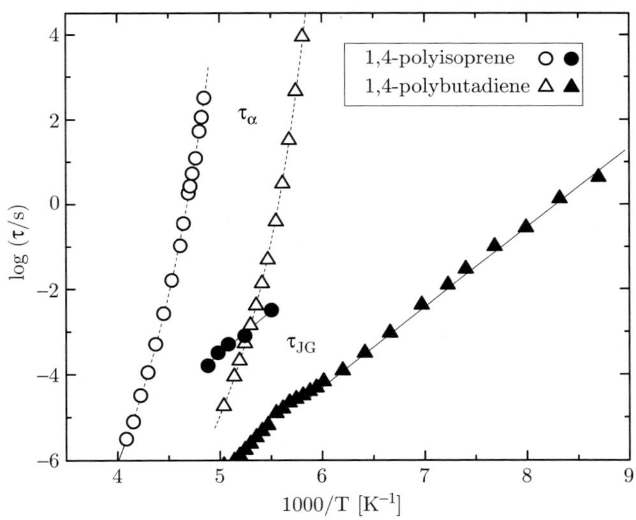

Figure 2.24 Local segmental (open symbols) and JG secondary (filled symbols) relaxation times for PI ($M_w = 500$ kg/mol; $T_g = 207.4$K) and PBD ($M_w = 45.9$ kg/mol; $T_g = 174.8$K). Data from refs. [72,137].

in the equilibrium liquid state. The validity of this assumption is difficult to assess because of the need to deconvolute the overlapping peaks, as illustrated for PBD in Figure 2.25 [72]. There are two common approaches: One is to assume the α- and JG processes contribute in additive fashion to the frequency-dependent loss [140,141], with the symmetric JG peak (symmetric versus log frequency) fitted to the Cole–Cole function [142]

$$\varepsilon''(\omega) = \frac{\Delta\varepsilon(\omega\tau_{cc})^{\alpha_{CC}} \cos\left(\pi(1-\alpha_{CC})\right)}{1 - 2(\omega\tau_{CC})^{\alpha_{CC}} \sin\left(\pi(1-\alpha_{CC})/2\right) + (\omega\tau_{CC})^{2\alpha_{CC}}} \tag{2.16}$$

where $\Delta\varepsilon$, τ_{CC}, and α_{CC} are fitting parameters. Equation (2.16) in combination with a Kohlrausch function for the segmental peak successfully fit the spectra in Figure 2.25. However, if the two processes are statistically independent, they will be additive in the time domain, which implies that the frequency response is a convolution, rather than simple sum, of their respective contributions [143,144]. In this approach the relaxation of the dipoles is considered to involve both the α- and β-processes simultaneously, leading to a dielectric relaxation function having the form [145],

$$\varepsilon(t) = f_\alpha \varepsilon_\alpha(t) + (1 - f_\alpha)\varepsilon_\alpha(t)\varepsilon_{JG}(t) \tag{2.17}$$

in which ε_α and ε_{JG} represent the respective α- and JG relaxation functions, and f_α is the relative strength of the former. According to this interpretation, when the α-peak is not well separated in frequency, the observed secondary relaxation would not be the pure JG process. If the timescales of these relaxations are very different, the assumption of statistical independence is warranted. However, NMR experiments show definitively that there is correlation between the α- and JG relaxations [146], so that the various deconvolution schemes are only approximations.

The difference in τ_{JG} determined using either assumption – frequency- or time-domain additivity – becomes significant only near the merging region and only when the JG relaxation is relatively weak. For example, for PBD (Figure 2.25), which has an intense secondary peak, the difference in τ_{JG} for the two methods of deconvolution is negligible. In

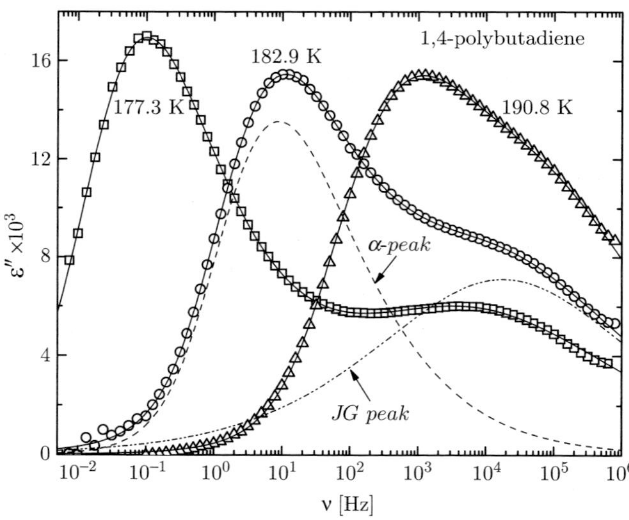

Figure 2.25 Dielectric loss spectra for PBD ($M_w = 284$ kg/mol; $T_g = 173.7$K) at the indicated temperatures. The solid lines are the fits obtained by addition of a KWW and Cole–Cole function; these are shown for $T = 182.9$ K as the respective dashed and dotted-dashed lines. Data from ref. [72].

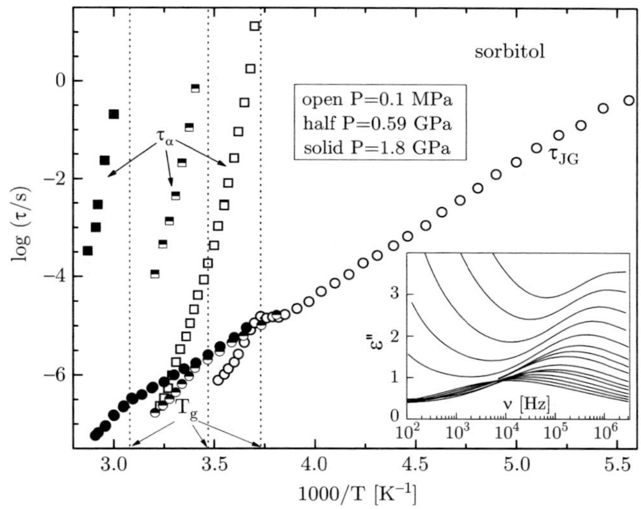

Figure 2.26 Isobaric α-relaxation times (squares) at 0.1 MPa (hollow symbols), 0.59 MPa (half-filled symbols), and 1.8 GPa (filled symbols), along with the corresponding JG relaxation times (circles) for sorbitol. The activation energy of τ_{JG} is independent of pressure, although it becomes smaller above the glass transition. The inset shows the JG peak in the dielectric loss at $P = 1.8$ GPa for temperatures from 273 to 343 K, in 5 increments (bottom to top). The α-peak falls at lower frequencies.

Figure 2.24 the activation energy for PBD decreases near T_g. This behavior is general and does not appear to be an artifact of separating out the contribution from the α-process. This can be shown from measurements under hydrostatic pressure. τ_α becomes larger but the effect of P on τ_{JG} is weak. The result is an increasing separation in frequency of the two peaks with pressure. As seen in Figure 2.26 for the polyalcohol sorbitol, the decreasing degree of peak overlap with pressure does not change the result that τ_{JG} exhibits a weaker T dependence near T_g.

The expectation of a JG process in all liquids and polymers comes from the idea that all molecular motions are unconstrained at short times. Intermolecular cooperativity requires time for the unbalanced forces and torques from the environment to build up sufficiently to affect the dynamics. Nevertheless, in some materials no JG peak is evident in the relaxation spectrum; instead, only an excess intensity is observed on the high frequency side of the structural relaxation peak. This "excess wing" is an unresolved JG peak, appearing as a power-law deviation from the (steeper) power law characterizing the α-process beyond the peak maximum.

An excess wing is found in materials with weak intermolecular constraints, since this means that τ_α is not much larger than τ_{JG} [71,147,148]. Various experiments that increase the degree of intermolecular coupling preferentially shift the α-relaxation toward lower frequencies, and thereby transform an excess wing into a distinct JG peak by reducing their overlap. For example, physical aging, which shifts the α-peak to lower frequency, causes the excess wing in polypropylene glycol [149], propylene carbonate [150],

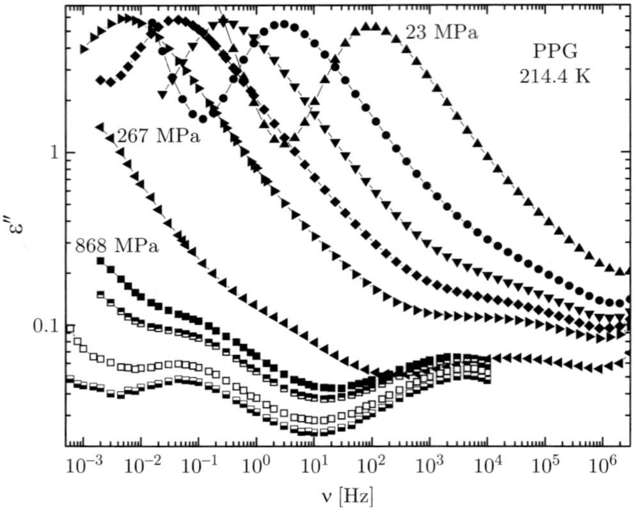

Figure 2.27 Dielectric loss for the trimer of propylene glycol at pressures equal to (top to bottom) 23, 51, 92, 130, 160, 267, and 868 MPa. The lowest three curves are at $P = 868$ MPa after aging for 2.2, 36.9, and 139.3 h. The excess wing transforms to a distinct peak with increasing aging. There is an additional peak around 10^4 Hz that is relatively insensitive to density. Data from ref. [152].

and glycerol [150] to become a nascent secondary peak. Pressure has the same effect [149,151], as shown in Figure 2.27 [152] for propylene glycol trimer. The JG relaxation, present as an excess wing, is obscured at ambient pressure by a higher-frequency secondary relaxation, but becomes evident in the spectrum at higher pressure. The emergence of the JG peak can also be effected by suppressing the intensity of the α-relaxation through geometrical confinement, as demonstrated for propylene carbonate and propylene glycol contained within the galleries of vermiculite clay[8] [153].

The coupling model of relaxation offers a quantitative prediction for the non-cooperative relaxation time prevailing at short times from τ_α and β_K for the α-relaxation [154,155]. By identifying the relaxation time of this non-cooperative motion with the JG process, the following equation is obtained for JG relaxation times [147]

$$\tau_{JG} = t_c^{1-\beta_K}\ \tau_\alpha^{\beta_K} \tag{2.18}$$

in which t_c (~ 1 ps) is the time at which distortions from the intermolecular potential begin to perturb the dynamics. In Figure 2.23 the predicted location of the JG process is seen to fall near the maximum of the experimental peak. τ_{JG} for various polymers is plotted versus β_K in Figure 2.28 [147], showing that in general eqn (2.18) provides an estimate, although the agreement is not quantitative.

[8] In vermiculite clay the silica or alumina crystals are arranged in octahedral sheets separated by layers of water molecules, providing nm-size space that can be filled by organic liquids.

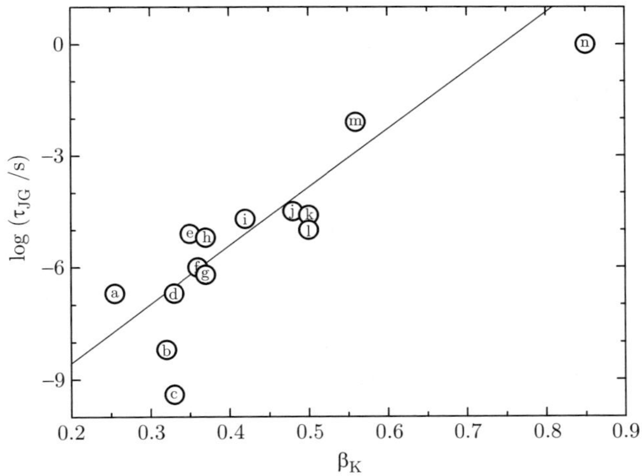

Figure 2.28 JG relaxation times at T_g (defined as $\tau_\alpha = 10^4$ s) as a function of the Kohlrausch stretch exponent: (a) polyvinylchloride (PVC); (b) poly(aryl ether ether ketone) (PEEK); (c) polysulfone; (d) liquid-crystal elastomer with polyethylene backbone; (e) polycarbonate; (f) poly(2-chlorocyclohexyl acrylate); (g) PCHMA; (h) PMMA; (i) PBD; (j) PVAc; (k) PI; (l) PPG; (m) PDMS; (n) poly(laurylmethacrylate). The line through the data is eqn (2.18). Data from ref. [147].

2.5 Decoupling phenomena

The motions underlying structural relaxation and the glass transition comprise both diffusion and reorientation. For a macroscopic object, the translational (D) and rotational (D_{rot}) diffusion constants are described respectively by the Stokes–Einstein (SE) relation (*cf.* eqn (1.34))[9]

$$D = \frac{kT}{6\pi\eta\bar{r}} \tag{2.19}$$

and the Debye–Stokes–Einstein (DSE) equation

$$D_{rot} = \frac{kT}{8\pi\eta\bar{r}^3} \tag{2.20}$$

where $\bar{r}$ is the spherical radius. These equations can be applied to molecular motions in liquids. At high temperatures the prediction that diffusion constants will have the same temperature dependence as T/η is borne out by experiments. However, almost invariably (tricresyl phosphate being an exception [156]), as the temperature is lowered towards T_g, there are deviations from simple proportionality between the diffusion constant and T/η. Translational diffusion exhibits a weaker dependence on temperature than does viscosity [157–161], as

[9] This form of the Stokes–Einstein equation assumes "stick" boundary conditions, whereby the fluid at the surface has the same velocity as the object. If "slip" conditions are assumed, so that there are no tangential forces on the object, the diffusion constant is 3/2 times larger.

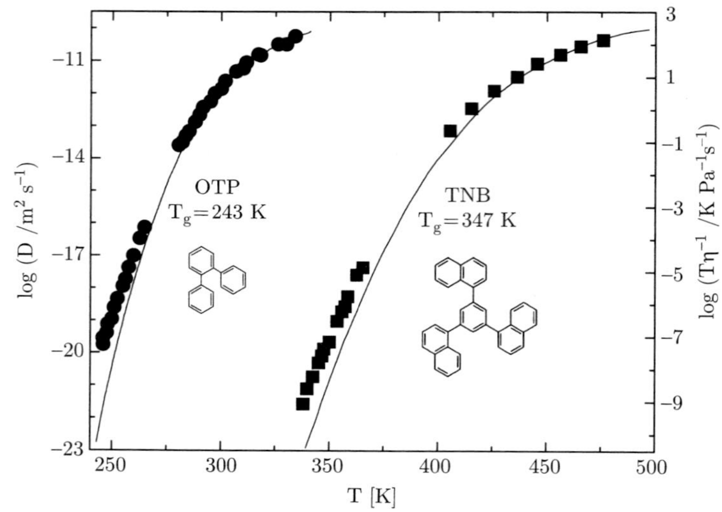

Figure 2.29 Self-diffusion coefficients of *o*-terphenyl (circles) and *tris*-naphthylbenzene (squares – values shifted upward by 0.16) compared to the ratio of temperature to the viscosity (solid lines). The right ordinate was adjusted to give superposition with the diffusion coefficients at high T, illustrating the enhancement of translational diffusion in the supercooled regime. Chemical structures are shown. Data from refs. [162,163].

illustrated in Figure 2.29 for *ortho*-terphenyl (OTP) [162] and *tris*-naphthylbenzene (TNB) [163]. Within a few degrees of T_g, self-diffusion is enhanced by 2 or 3 orders of magnitude from the behavior expected from the viscosity. Similar results are found for probe molecules having a size on the order of the host molecules [159,164,165].

Making use of eqn (2.3) gives alternate forms for the SE

$$\frac{D\tau_a}{T} = constant \tag{2.21}$$

and DSE relations

$$\frac{D_{rot}\tau_a}{T} = constant \tag{2.22}$$

which can be used to compare diffusion constants to reorientational relaxation measured by dielectric spectroscopy or dynamic light scattering. The Nernst–Einstein equation relates the DC conductivity[10] of ions, σ_{DC}, to their mobility

$$\sigma_{DC} \sim \sum_i z_i c_i \hat{\mu}_i \tag{2.23}$$

where the sum is over all ion types, having charge (valency) z_i, concentration c_i, and mobility $\hat{\mu}_i$. The translational diffusion constant is proportional to the ion mobility

[10] At low concentration ions move independently, giving frequency-independent DC conduction. At higher concentrations the ion motion becomes mutually correlated, leads to an AC conductivity that decreases with frequency.

$$D = kT\hat{\mu} \tag{2.24}$$

From eqns (2.21) and (2.23)

$$\sigma_{DC}\ \tau_a = constant \tag{2.25}$$

Equation (2.25) assumes that the ion population is constant; that is, there is no change in the degree of electrolytic dissociation with temperature [166,167].

Experiments show that while eqn (2.25) holds at higher T, σ_{DC} and τ_α also decouple as T_g is approached; the ion conduction has a weaker T dependence than the reorientational motion (Figure 2.30 [60]). This behavior can be described using an empirical power law [168–170]

$$\sigma_{DC}\tau_a^{k_{DSE}} = constant \tag{2.26}$$

where the exponent $k_{DSE} \leq 1$ is a measure of the degree of decoupling. Similar results are obtained when the dynamics are altered by changing the pressure instead of temperature (Figure 2.31 [171] and Figure 2.32 [89]). Not only is there a correspondence between the decoupling induced by temperature and that resulting from pressure changes, but in Figure 2.32 k_{DSE} is also seen to be independent of molecular weight for a series of propylene glycols [89]. Since dielectric relaxation times can be measured over a broader range of conditions, eqn (2.26) can be used to deduce conductivities, and by extension viscosities, when these quantities cannot be obtained directly [172,173].

Different mechanisms have been proposed to explain the decoupling of translation and reorientation. There is no quantitative, predictive model for the phenomenon, beyond it being a consequence of the strong intermolecular cooperativity and dynamic heterogeneity developing near T_g [164,174–176].

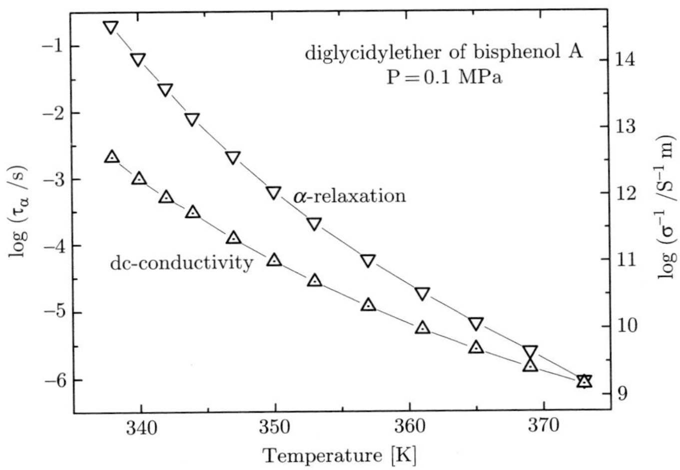

Figure 2.30 Dielectric relaxation times (down triangles) and inverse DC conductivity (triangles) for an oligomeric epoxy having about 5 repeat units per chain. The fractional DSE exponent (eqn (2.26)) $k_{DSE} = 0.645$. Data from ref. [60].

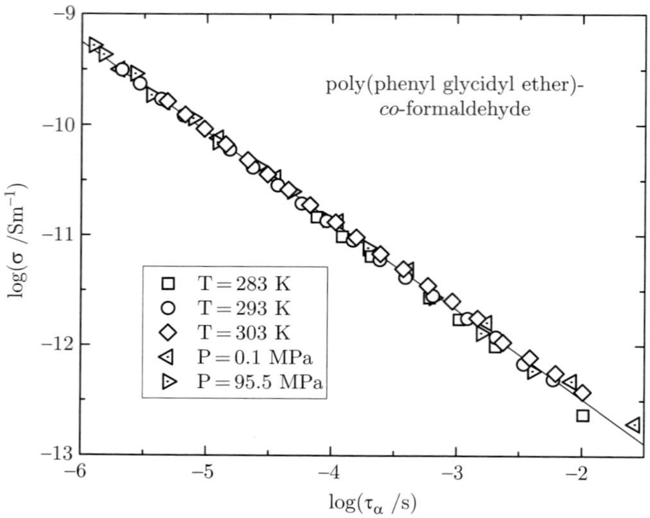

Figure 2.31 Conductivity *versus* the dielectric relaxation time for an oligomeric epoxide ($M_w = 345$ g/mol) measured as a function of temperature at various P and as a function of pressure at various T. The isothermal and isobaric results are indistinguishable; the straight line represents eqn (2.26) with $k_{DSE} = 0.81$. Data from ref. [171].

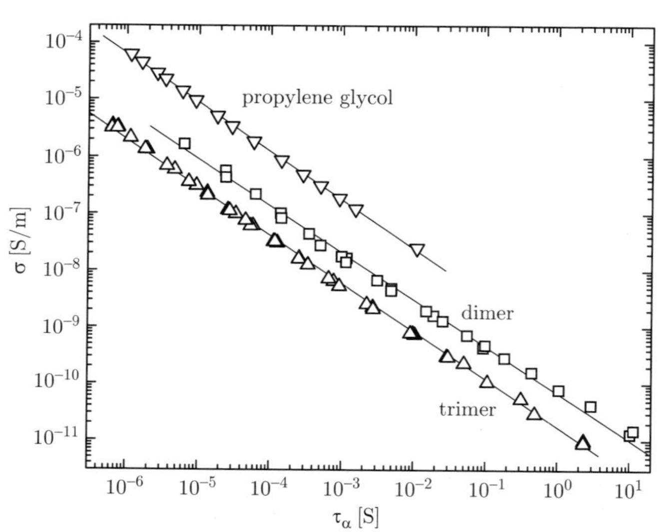

Figure 2.32 Double logarithmic plot of DC conductivity *versus* τ_α for polypropylene glycol with $N = 1, 2,$ and 3. The data for all three molecular weights are parallel and for various combinations of T and P are described using a single value of the fractional DSE exponent, $k_{DSE} = 0.84$. Data from ref. [89].

2.6 Applications

If the deformation of an elastomer occurs at a sufficiently high strain rate, the material response can extend to frequencies beyond the rubbery plateau regime of the viscoelastic spectrum. This means that rearrangements of polymer segments over a significant length scale are precluded and the material exhibits "leathery" behavior, as it begins to transition to a glassy state. This glass-transition zone is associated with large mechanical hysteresis, which is exploited in various applications reliant on energy dissipation. The common feature is that rapid straining induces a reversible transition of the elastomer to the glassy state.

2.6.1 Wet skid resistance of tires

Surface conditions exert a dominant role on the friction of rubber [177,178], with friction coefficients varying by four orders of magnitude depending on the interface [179]. Adhesive forces (primarily van der Waals bonding) contribute to the friction of rubber on dry and very smooth surfaces. On wet or rough surfaces, bulk energy dissipation becomes dominant as the rubber is deformed by surface asperities, which can range in size from *ca.* 0.01 to 1 mm [180]. This bulk mechanical hysteresis governs the friction of sliding tires [181–183], and it is a material, rather than a surface, property. (An exception might be sliding on ice, for which the structure of the surface ice is significant [184–186].) During sliding on a wet surface, anti-lock braking systems increase the contribution from surface adhesion, which otherwise is negligible, by ensuring that the tire continues to roll and tractively interact with the road surface. Persson *et al.* [183] has shown that wet sliding reduces the contact area of the tire, with pooled water isolating the tread from the high-frequency deformations caused by road asperities.

Since wet skid resistance is determined by the amount of energy dissipated in the rubber, the response of optimal tire tread compounds should fall within the viscoelastic glass transition zone, where hysteresis is maximum. The frequency associated with wet skidding is in the range from 10^3 to 10^6 Hz [187,188]; thus, rubbers having high glass-transition temperatures are expected to exhibit superior wet skid resistance. However, since T_g as conventionally measured corresponds to low frequencies ($\tau_\alpha^{-1} \leq 0.1$ s^{-1}) and extrapolation of time-temperature shift factors in the transition zone is ambiguous (Chapter 6), deviations from a correlation between T_g and wet skid resistance are found unless a high frequency is used to define the transition [189,190]. For example, Takino *et al.* [189] found good correlation of wet skid behavior with the temperature of the peak of the mechanical loss tangent measured at 10 Hz (Figure 2.33), although this is lower than the rate of the local deformations associated with wet sliding of rubber.

2.6.2 Sound transmission and damping

Rubber finds various uses in acoustics, including decouplers, which prevent sound passage; anechoic coatings, which attenuate sound reflections; and acoustic windows, which transmit sound waves. Although commercial materials are usually proprietary, compilations of acoustic properties for many elastomers have been published [191]. One important property is the attenuation of sound as it passes through an elastomer. The diminution of the sound pressure is quantified by an attenuation coefficient, $\hat{\alpha}$, a material property defined

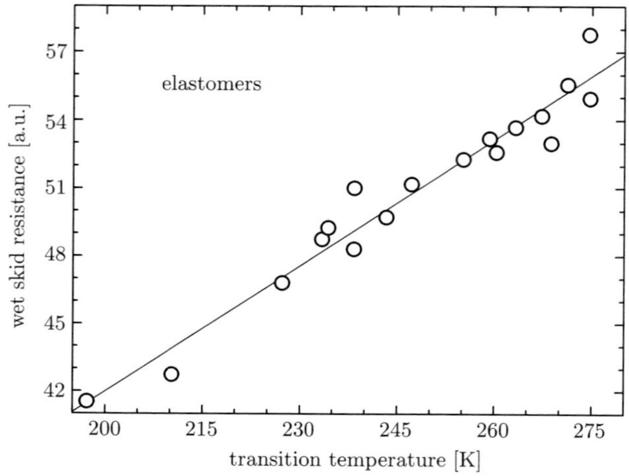

Figure 2.33 Laboratory wet skid resistance *versus* the temperature of the maximum in the mechanical loss tangent measured at 10 Hz for rubbers having calorimetric glass transitions in the range 167 to 253 K. Data from ref. [189].

as $\hat{\alpha} \equiv \frac{-dp/p_0}{dx}$ where p_0 is the initial pressure and x the propagation distance. For small values of the loss tangent [74]

$$\hat{\alpha} = \frac{\pi tan\delta}{\lambda} \quad (\hat{\alpha} << 1) \qquad (2.27)$$

For longitudinal waves (oscillating in the direction of the sound propagation), the loss tangent in eqn (2.27) refers to the bulk modulus, for which tan δ is quite small ($<10^{-3}$) in rubber.

When sound damping is the objective, longitudinal waves can be converted into shear waves [192], for which loss tangents are larger. The greatest attenuation will be achieved if tan δ is maximized by ensuring that the elastomer undergoes its rubber-to-glass transition at the sonic frequency. For this reason the selection of damping materials for sound attenuation is governed primarily by their T_g. The objective is to have the peak of the loss tangent centered within the sound frequencies at the operating temperature. Variation of the latter means that a broad glass transition is desirable, which has led to the use of blends and polyurethanes or polyureas for many applications [191,192]. These same consideration apply to vibration-isolation systems: rubber having a large and broad loss tangent most effectively suppresses transmitted forces [193,194].

2.6.3 Impact protection

A recent development in military technology is the use of rubber coatings for protection from ballistic and explosive assault [195,196]. In the case of a bomb blast, coatings can minimize damage to the underlying structure and collateral damage from fragmentation of the structure. For ballistic assault from either a bomb fragment or bullet, elastomeric coatings can increase the penetration resistance of the substrate. A variety of mechanisms contribute to the performance of elastomer coatings [197–199]; however, an important factor is the energy dissipation associated with an induced glass transition [200,201]. The strain rate for bullets and bomb fragments striking a hard surface, estimated as the ratio of the projectile speed to the coating thickness, is on the order of 10^5 Hz. If the rubbery coating has a high enough T_g, the impact induces a transition from the rubbery to a glassy state, with consequent attenuation in the impact energy.

This behavior is demonstrated by a comparison of results for PBD (calorimetric $T_g = 182$ K) and a polyurea (PU) [200]. The soft segment domains of PU, which provide the elastomeric properties, are amine-terminated polybutadiene, having a $T_g = 213$ K. (The hard domains in the PU have a glass-transition temperature exceeding 400 K and thus are not affected by the rate of the impact strain.) The segmental relaxation times for the two materials are plotted in Figure 2.34, along with fits to eqn (2.4). From these data the frequency of the maximum in the dielectric loss at 296 K is determined to be 500 MHz for PBD and 0.9 MHz for PU. This defines the mean frequency for transition to the glassy state under ambient conditions, with ν_{max} for PBD much larger than for the PU.

In Figure 2.35 are high-speed photographs of these coatings during projectile impact. The PBD deforms in typical rubbery fashion – a high level of strain with delocalized deformation. The PU behavior is quite different; it shatters with minimal stretching. Consistent with a transition to the glassy state of the PU during impact and consequently greater energy absorption, the penetration velocity of the underlying steel substrate increased > 50% with the PU coating, versus only 40% for steel coated with the PBD.

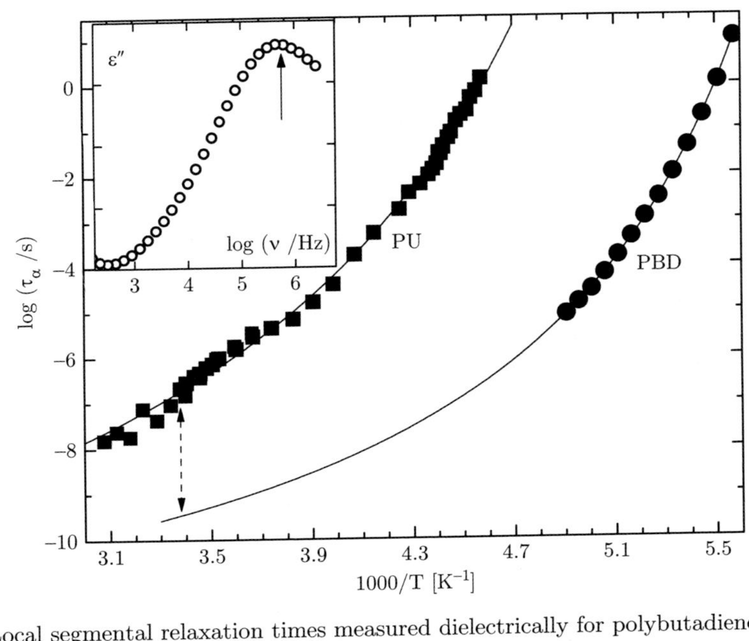

Figure 2.34 Local segmental relaxation times measured dielectrically for polybutadiene and polyurea. Solid lines are fits to eqn (2.4). The inset shows the dielectric loss of the PU at 296 K (corresponding to the dashed-arrow in the main figure), with the peak frequency corresponding to τ_α indicated by the arrow. Data from ref. [200].

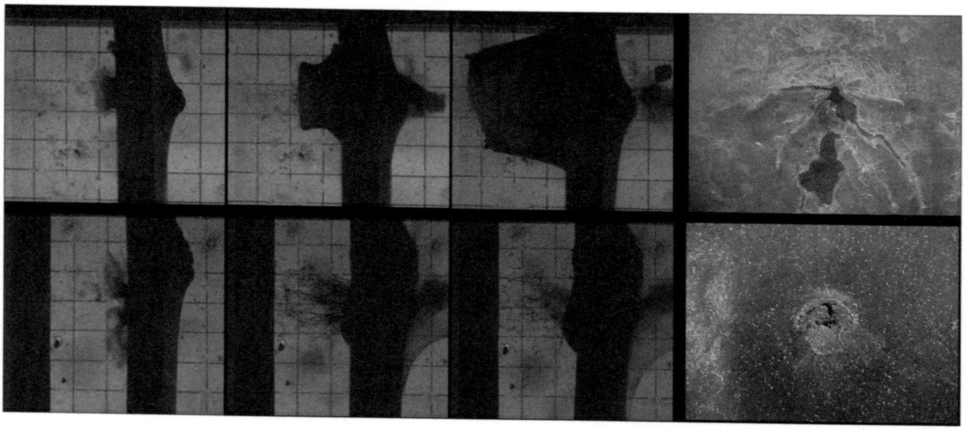

Figure 2.35 High-speed photographs of the impact response of elastomeric coatings on front (impact) side of steel substrate; projectile arrives from the left (speed ~ 900 m/s). The grid lines are spaced 25 mm apart. The pre-existing protrusions evident in the substrate are from previous impacts. (*top*) PBD exhibits rubbery behavior, with substantial stretching and tearing, leading (far right) to extensive cracks in the rubber. (*bottom*) PU undergoes an impact-induced glass transition, with consequent brittle failure via shear-dominated separation of a cylinder of material. The damage to the PU coating (far right) is limited to an area having a cross-section (~ 13 mm) matching that the projectile. From refs. [200 and 201].

References

1. B. Frick, C. Alba-Simionesco, K.H. Andersen, and L. Willner, *Phys. Rev. E* **67**, 051801 (2003).
2. A. Cailliaux, C. Alba-Simionesco, B. Frick, L. Willner, and I. Goncharenko, *Phys. Rev. E* **67**, 010802 (2003).
3. W. Kauzmann, *Chem. Rev.* **43**, 219 (1948).
4. I. Prigogine and R. Defay, *Chemical Thermodynamics* (Longman, London, 1954).
5. C.T. Moynihan, P.B. Macedo, C.J. Montrose, P.K. Gupta, M.A. Debolt, J.F. Dill, B.E. Dom, P.W. Drake, A.J. Easteal, P.B. Witerman, R.P. Moeller, H. Sasabe, and J.A. Wilder., *Ann. N.Y. Acad. Sci.* **279**, 15 (1976).
6. E. Donth, *The Glass Transition* (Springer-Verlag, New York, 2001), Table 2.11.
7. R. Casalini and C.M. Roland, *J. Chem. Phys.* **131**, 114501 (2009).
8. G.W. Scherer, *Relaxations in Glass and Composites* (Academic, New York, 1986).
9. A. J. Kovacs, *Fortsch. Hochpolym.-Forsch.* **3**, 394 (1963).
10. J.C. Maxwell, *Philos. Trans. R. Soc. Lond. Ser. A*, **157**, 49 (1867).
11. C M Roland, S Hensel-Bielowka, M Paluch, and R Casalini, *Rep. Prog. Phys.* **68**, 1405 (2005).
12. M. Paluch, R. Casalini, A. Patkowski, T. Pakula, and C.M. Roland, *Phys. Rev. E* **68**, 031802 (2003).
13. C.M. Roland and R. Casalini, *J. Chem. Phys.* **119**, 1838 (2003).
14. L. Hong, B. Begen, A. Kisliuk, V.N. Novikov, and A.P. Sokolov, *Phys. Rev. B* **81**, 104207 (2010).
15. P.G. Santangelo and C.M. Roland, *Macromolecules* **31**, 3715 (1998).

16. R. Casalini and C.M. Roland, *J. Polym. Sci. Polym. Phys. Ed.* **48**, 582 (2010).
17. H. Scher, M.F. Shlesinger and J.T. Bendler, *Phys. Today,* **41**, 26 (1991).
18. K.L. Ngai, *Comm. Solid State Phys.* **9**, 127 (1979); **9**, 141 (1980).
19. W.F. Knoff, I.L. Hopkins, and A.V. Tobolsky, *Macromolecules* **4**, 750 (1971).
20. G.C. Berry and D.J. Plazek, *Rheol. Acta* **36**, 320 (1997).
21. M. Cardona, R. V. Chamberlain, and W. Marx, *Ann. Phys. (Leipzig)* **16**, 842 (2007).
22. K. Trachenko, C.M. Roland, and R. Casalini, *J. Phys. Chem. B* **112**, 5111 (2008).
23. R. Casalini, K.L. Ngai, and C.M. Roland, *Phys. Rev. B* **68**, 014201 (2003).
24. H. Vogel, *Phys. Z.* **222**, 645 (1921).
25. G.S. Fulcher, *J. Am. Ceram. Soc.* **8**, 339 (1923).
26. V.G. Tammann and W.Z. Hesse, *Anorg. Allg. Chem.* **156**, 245 (1926).
27. C.A. Angel, *Pure Appl. Chem.* **63**, 1387 (1991).
28. R. Casalini and C.M. Roland, *J. Chem. Phys.* **119**, 4052 (2003).
29. F. Stickel, R. Richert and E.W. Fisher, *J. Chem. Phys.* **102**, 6251 (1995).
30. C. Hansen, F. Stickel, R. Richert, and E. W. Fischer, *J. Chem. Phys.* **108**, 6408 (1998).
31. V.W. Oldekop, *Glastechnisch Berichte* **30**, 8 (1957).
32. W.T. Laughlin and D.R. Uhlmann, *J. Phys. Chem.* **76**, 2317 (1972).
33. C.A. Angell, *J. Non-Cryst. Solids* **131–133**, 13 (1991).
34. D.J. Plazek and K.L. Ngai, *Macromolecules* **24**, 1222 (1991).
35. R. Bohmer, K.L. Ngai, C.A. Angell, and D.J. Plazek, *J. Chem. Phys.* **99**, 4201 (1993).
36. A. Heuer, *J. Phys. Condens. Matter* **20**, 373101 (2008).
37. C.M. Roland, *Macromolecules* **27**, 4242 (1994).
38. C.M. Roland and K.L. Ngai, *Macromolecules* **24**, 5315 (1991); **25**, 1844 (1992).
39. K.L. Ngai and C.M. Roland, *Macromolecules* **26**, 6824 (1993).
40. C.M. Roland, *Macromolecules* **25**, 7031 (1992).
41. C.A. Angell, P.H. Poole, and J. Shao, *Nuovo Cimento* **16**, 883 (1994).
42. C.M. Roland and K.L. Ngai, *J. Chem. Phys.* **104**, 2967 (1996).
43. I.M. Hodge, *Macromolecules* **20**, 2897 (1987).
44. I.M. Hodge, *J. Non-Cryst. Solids* **203**, 164 (1996).
45. C.M. Roland and K.L. Ngai, *J. Non-Cryst. Solids* **212**, 74 (1997).
46. C.A. Angell, *Polymer* **38**, 6 (1997).
47. T. Scopigno, G. Ruocco, F. Sette, and G. Monaco, *Science* **302**, 849 (2003).
48. U. Buchenau and R. Zorn, *Europhys. Lett.* **18**, 523 (1992).
49. K.L. Ngai, *Philos. Mag.* **84**, 1341 (2004).
50. V.N. Novikov and A.P. Sokolov, *Nature* **431**, 961 (2004).
51. J.C. Dyre, *Nature Mater.* **3**, 749 (2004).
52. J.C. Dyre, N.B. Olsen, and T. Christensen, *Phys. Rev. B* **53**, 2171 (1996).
53. U. Buchenau and A. Wischnewski, *Phys. Rev. B* **70**, 092201 (2004).
54. C.M. Roland, K.J. McGrath, and R. Casalini, *Macromolecules* **39**, 3581 (2006).
55. M. Paluch, S. Pawlus, and C.M. Roland *Macromolecules* **35**, 7338 (2002).

56. C.M. Roland and K.L. Nagi, *Macromolecules* **29**, 5747 (1996).
57. C.M. Roland and R. Casalini, *Macromolecules* **40**, 3631 (2007).
58. R. Casalini and C.M. Roland, *Macromolecules* **38**, 1779 (2005).
59. P.G. Santangelo and C.M. Roland, *Macromolecules* **31**, 4581 (1998).
60. M. Paluch, C.M. Roland, J. Gapinski and A. Patkowski, *J. Chem. Phys.* **118**, 3177 (2003).
61. R. Zorn, G.B. McKenna, L. Willner, and D. Richter, *Macromolecules* **28**, 8552 (1995).
62. C.G. Robertson and C.M. Rademacher *Macromolecules* **37**, 10009 (2004).
63. C.G. Robertson and C.M. Roland, *J. Polym. Sci. Pt. B-Polym. Phys.* **42**, 2604 (2004).
64. K. Kunal, M. Paluch, C.M. Roland, J.E. Puskas, Y. Chen, and A.P. Sokolov, *J. Polym. Sci. Pt. B-Polym. Phys.* **46**, 1390 (2008).
65. Q. Qin and G.B. McKenna, *J. Non-Cryst. Solids* **352**, 2977 (2006).
66. T. Scopigno, G. Ruocco, F. Sette, and G. Monaco, Science **302**, 849 (2003).
67. P. Bordat, F. Affouard, M. Descamps, and K. L. Ngai, *Phys. Rev. Lett.* **93**, 105502 (2004).
68. R. Casalini and C.M. Roland, *Phys. Rev. E* **72**, 031503 (2005)
69. S. Pawlus, S.J. Rzoska, J. Ziolo, M. Paluch, and C.M. Roland, *Rubber Chem. Technol.* **76**, 1106 (2003).
70. R. Casalini, C.M. Roland, and S. Capaccioli, *J. Chem. Phys.* **126**, 184903 (2007).
71. C. Leon, K.L. Ngai, and C.M. Roland, *J. Chem. Phys.* **110**, 11585 (1999).
72. R.B. Bogoslovov, T.E. Hogan, and C.M. Roland, *Macromolecules* **43**, 2904 (2010).
73. P.G. Santangelo, unpublished.
74. J.D. Ferry, 1980, *Viscoelastic Properties of Polymers*, 3rd edition (New York: Wiley).
75. G.S. Grest and M.H. Cohen, *Adv. Chem. Phys.* **48**, 455 (1981).
76. J.T. Bendler, J. Fontanella and M. Schlesinger, *Phys. Rev. Lett.* **87**, 195 503 (2001).
77. G. Dlubek, J. Wawryszczuk, J. Pionteck, T. Gowrek, H. Kaspar, and K.H. Lochhass, *Macromolecules* **38**, 429 (2005).
78. G. Adam and J.H. Gibbs, *J. Chem. Phys.* **43**, 139 (1965).
79. D. Prevosto, M. Lucchesi, S. Capaccioli, R. Casalini, and P.A. Rolla, *Phys. Rev. B* **67**, 174202 (2003).
80. M. Goldstein, *J. Chem. Phys.*, **51**, 3728 (1969).
81. C. A. Angell, *Science*, **267**, 1924 (1995).
82. S. Sastry, *Nature*, **409**, 164 (2001).
83. P.B. Macedo and T.A. Litovitz, *J. Chem. Phys.* **42**, 245 (1965).
84. T. Pakula, *J. Molec. Liq.* **86**, 109 (2000).
85. W. Gotze and L. Sjogren, *Rep. Prog. Phys.* **55**, 241 (1992).
86. J.D. Hoffman, G. Williams, and E. Passaglia, *J. Polym. Sci.,* **C14**, 173 (1966).
87. G. Williams, *Trans. Faraday Soc.*, **60**, 1548 (1964); **61**, 1564 (1965).
88. M. Naoki, H. Endou, and K. Matsumoto, *J. Phys. Chem.*, **91**, 4169 (1987).
89. R. Casalini and C.M. Roland, *J. Chem. Phys.* **119**, 11951 (2003).
90. M. Naoki and S. Katahira, *J. Phys. Chem.* 95, 431 (1991).
91. P.H. Poole, F. Sciortino, T. Grande, H.E. Stanley, and C.A. Angell, *Phys. Rev. Lett.*, **73**, 1632 (1994).
92. R.E. Cook, H.E. King, and D.G. Peiffer, *Phys. Rev. Lett.*, **69**, 3072 (1992).
93. R. Casalini and C.M. Roland, *Phys. Rev. B* **71**, 014210 (2005).

94. F. Stickel, E.W. Fischer, and R. Richert, *J. Chem. Phys.* **102**, 6251 (1995); **104**, 2043 (1996).
95. K.L. Ngai, and C.M. Roland, *Polymer* **43**, 567, 2002.
96. J. Colmenero, A. Alegria, P.G. Santangelo, K. L. Ngai, and C. M. Roland, *Macromolecules* **27**, 407 (1994).
97. D. Cangialosi, A. Alegría, and J. Colmenero, *J. Chem. Phys.* **130**, 124902 (2009).
98. R. Richert, *J. Chem. Phys.* **123**, 154502 (2005).
99. R. Richert, K. Duvvuri, and L.-T. Duong, *J. Chem. Phys.* **118**, 1828 (2003).
100. J.R. Rajian, W. Huang, R. Richert, and E.L. Quitevis, *J. Chem. Phys.* **124**, 014510 (2006).
101. C.M. Roland and R. Casalini, *J. Non-Cryst. Solids* **353**, 3996 (2007).
102. C.M. Roland, R. Casalini, and M. Paluch, *Chem. Phys. Lett.* **367**, 259 (2003).
103. K. L. Ngai, R. Casalini, S. Capaccioli, M. Paluch, and C. M. Roland, *J. Phys. Chem. B* **109**, 17356 (2005).
104. H. Sillescu, *J. Non-Cryst. Solids* **243**, 81 (1999).
105. P.G. Debenedetti and F.H. Stillinger, *Nature* **410**, 259 (2001).
106. V. Lubchenko and P.G. Wolynes, *Ann. Rev. Phys. Chem.* **58**, 235 (2007).
107. I. Avramov, *J. Non-Cryst. Solids* **351**, 3163 (2005).
108. S. Franz and G. Parisi, *J. Phys. Condens. Matter* **12**, 6335 (2000).
109. D. Chandler, J.P. Garrahan, R.L. Jack, L. Maibaum, and A.C. Pan, *Phys. Rev. E* **74**, 051501 (2006).
110. R. Bohmer, R.V. Chamberlin, G. Diezemann B. Geil, A. Heuer, G. Hinze, S.C. Kuebler, R. Richert, B. Schiener, H. Sillescu, H.W. Spiess, U. Tracht, and M. Wilhelm, *J. Non-Cryst. Solids* **235–237**, 1 (1998).
111. S. Franz, C. Donati, G. Parisi, and S.C. Glotzer, *Philos. Mag. B* **79**, 1827 (1999).
112. K. Vollmayr-Lee, W. Kob, K. Binder, and A. Zippelius, *J. Chem. Phys.* **116**, 5158 (2002).
113. N. Lacevic, F.W. Starr, T. B. Schrøder, and S. C. Glotzer, *J. Chem. Phys.* **119**, 7372 (2003).
114. L. Berthier, G. Biroli, J.-P. Bouchaud, W. Kob, K. Miyazaki, and D.R. Reichman, *J. Chem. Phys.* **126**, 184503 (2007).
115. D. Coslovich and C.M. Roland, *J. Chem. Phys.* **131**, 151103 (2009).
116. W. Kob and H.C. Andersen, *Phys. Rev. E* **51**, 4626 (1995).
117. C. Toninelli, M. Wyart, L. Berthier, G. Biroli, and J.-P. Bouchaud, *Phys. Rev. E* **71**, 041505 (2005).
118. K. Schmidt-Rohr and H.W. Spiess, *Phys. Rev. Lett.* **66**, 3020 (1991).
119. M.T. Cicerone, F. R. Blackburn, and M. D. Ediger, *Macromolecules* **28**, 8224 (1995).
120. U. Tracht, M. Wilhelm, A. Heuer, H. Feng, K. Schmidt-Rohr, and H.W. Spiess, *Phys. Rev. Lett.* **81**, 2727 (1998).
121. S.A. Reinsberg, A. Heuer, B. Doliwa, H. Zimmermann, and H.W. Spiess, *J. Non-Cryst. Solids* **307–310**, 208 (2002).
122. E. Donth, *J. Non-Cryst. Solids* **53**, 325 (1982).
123. L.D. Landau and E.M. Lifshitz, *Statistical Physics* (Pergamon, NewYork, 1977), Vol. 5.
124. E. Hempel, G. Hempel, A. Hensel, C. Schick, and E. Donth, *J. Phys. Chem. B* **104**, 2460 (2000).
125. F. Kremer, A. Huwe, M. Arndt, P. Behrens, and W. Schwieger, *J. Phys. Condens. Matter* **11**, A175 (1999).
126. S.R. Elliott, *Europhys. Lett.* **19**, 201 (1992).

127. W. Schirmacher, B. Schmid, C. Tomaras, G. Villiani, G. Baldii, G. Ruocco, and T. Scopigno, *Phys. Status Solidi C*, **6**, 862 (2008).
128. L. Hong, P.D. Gujrati, V.N. Novikov, and A.P. Sokolov, *J. Chem. Phys.* **131**, 194511 (2009).
129. L. Berthier, G. Biroli, J.-P. Bouchaud, L. Cipelletti, D. El Masri, D. L'Hôte, F. Ladieu, and M. Pierno, *Science* **310**, 1797 (2005).
130. C. Dalle-Ferrier, C. Thibierge, C. Alba-Simionesco, L. Berthier, G. Biroli, J.P. Bouchaud, F. Ladieu, D. L'Hote, and G. Tarjus, *Phys. Rev. E* **76**, 041510 (2007).
131. L. Berthier, G. Biroli, J.P. Bouchaud, W. Kob, K. Miyazaki, and D. Reichman, *J. Chem. Phys.* **126**, 184504 (2007).
132. C. Dalle-Ferrier, S. Eibl, C. Pappas, and C. Alba-Simionesco, *J. Phys.: Condens. Matter* **20**, 494240 (2008).
133. S. Capaccioli, G. Ruocco, and F. Zamponi, *J. Phys. Chem. B* **112**, 10652 (2008).
134. D. Fragiadakis, R. Casalini, and C.M. Roland, *J. Phys. Chem. B* **113**, 13134 (2009).
135. J.D. Stevenson, J. Schmalian, and P.G. Wolynes, *Nature Phys.* **2**, 268 (2006).
136. X. H. Qiu and M. D. Ediger, *J. Phys. Chem.* **107**, 459 (2003).
137. C.M. Roland, M.J. Schroeder, J.J. Fontanella, and K.L. Ngai, *Macromolecules* **37**, 2630 (2004).
138. M. Schulz and E. Donth, *J. Non-Cryst. Solids* **168**, 186 (1994).
139. J. Rault, *J. Non-Cryst. Solids* **271**, 177 (2000).
140. E. Donth, K. Schroter, and S. Kahle, *Phys. Rev. E* **60**, 1099 (1999).
141. F. Garwe, A. Schonhals, H. Lockwenz, M. Beiner, K. Schroter, and E. Donth, *Macromolecules* **29**, 247 (1996).
142. K.S. Cole and R.H. Cole, *J. Chem. Phys.* **9**, 341 (1941).
143. A. Arbe, J. Colmenero, D. Gomez, D. Richter, and B. Farago, *Phys. Rev. E* **60**, 1103 (1999).
144. R. Bergman, F. Alvarez, A. Alegria, and J. Colmenero, *J. Chem. Phys.* **109**, 7546 (1998).
145. G. Williams, *Adv. Polym. Sci.* **33**, 60 (1979).
146. R. Bohmer, G. Diezemann, B. Geil, G. Hinze, A. Nowaczyk, and M. Winterlich, *Phys. Rev. Lett.* **97**, 135701 (2006).
147. K.L. Ngai, *J. Chem. Phys.* **109**, 6982 (1998).
148. M. Paluch, K.L. Ngai, and S. Hensel-Bielowka, *J. Chem. Phys.* **114**, 10872 (2001).
149. R. Casalini and C.M. Roland, *Phys. Rev. B* **69**, 094202 (2004).
150. U. Schneider, R. Brand, P. Lunkenheimer, and A. Loidl, *Phys. Rev. Lett.* **84**, 5560 (2000).
151. R. Casalini and C.M. Roland, *Phys. Rev. Lett.* **91**, 015702 (2003).
152. C.M. Roland and R. Casalini in *Slow Dynamics in Complex Systems*, AIP Conf. Proc., vol. 708 (M. Tokuyama and I. Oppenheim, eds.), Amer. Inst. Phys., Melville, NY (2004).
153. R. Bergman, J. Mattsson, C. Svanbert, G.A. Schwartz, and J. Swenson, *Europhys. Lett.* **64**, 675 (2003).
154. K.L. Ngai, A.K. Rajagopal, and S. Teitler, *J. Chem. Phys.* **88**, 5086 (1988).
155. K.L. Ngai and R.W. Rendell, *ACS Symp. Ser.* **676**, 45 (1997).
156. R. Deegan, R.L. Leheny, N. Menon, S.R. Nagel, and D.C. Venerus, *J. Phys. Chem. B* **103** 4066 (1999).
157. T. Keyes and I. Oppenheim, *Phys. Rev. A* **8**, 937 (1973).
158. D. Ehlich and H. Silescu, *Macromolecules* **23**, 1600 (1990).

159. J.H. Magill and D.J. Plazek, *J. Chem. Phys.* **46**, 3757 (1967).
160. G. Heuberger and H. Sillescu, *J. Phys. Chem.* **100**, 15255 (1996).
161. M.G. Mazza, N. Giovambattista, H.E. Stanley, and F.W. Starr, *Phys. Rev. E* **76**, 031203 (2007).
162. M.K. Mapes, S.F. Swallen, and M. D. Ediger, *J. Phys. Chem. B* **110**, 507 (2006).
163. S.F. Swallen, K. Traynor, R.J. McMahon, M. D. Ediger, and T.E. Mates, *J. Phys. Chem. B* **113**, 4600 (2009).
164. M. T. Cicerone and M. D. Ediger, *J. Chem. Phys.* **104**, 7210 (1996).
165. J.R. Rajian and E.L. Quitevis, *J. Chem. Phys.* **126**, 2243506 (2007).
166. G.P. Johari and O. Andersson, *J. Chem. Phys.* **125**, 124501 (2006).
167. G. Power, J.K. Vij, and G.P. Johari, *J. Phys. Chem. B* **111**, 11201 (2007).
168. H. Sasabe and S. Saito, *Polym. J.* **3**, 6254 (1972).
169. T. Koike, *Adv. Polym. Sci.* **148**, 139 (1999).
170. S. Corezzi, E. Campani, P. A. Rolla, S. Capaccioli, and D. Fioretto, *J. Chem. Phys.* **111**, 9343 (1999).
171. M. Paluch, T. Psurek, and C.M. Roland, *J. Phys. Condens. Matter* **14,** 9489 (2002).
172. A. Suzuki, M. Masuko, and T. Nikkuni, *Tribol. Int.* **33**, 107 (2000).
173. M. Masuko, A. Suzuki, S. Hanai, and H. Okabe, *Jpn. J. Tribol.* **42**, 286 (1997).
174. E.W. Fischer, E. Donth, and W. Steffen, *Phys. Rev. Lett.* **68** 2344 (1992).
175. I. Chang and H. Sillescu, *J. Phys. Chem. B.* **101**, 8794 (1997).
176. K.L. Ngai *J. Phys. Chem. B* **103**, 10684 (1999).
177. E.M. Bevilacqua and E.M. Percaprio, *Rubber Chem. Technol.* **41**, 843 (1968).
178. B.E. Sabey and G.N. Lupton, *Rubber Chem. Technol.* **37**, 878 (1964).
179. A.D. Roberts, *Tribol. Int.* **4**, 115 (1977).
180. B.N.J. Persson, *Surf. Sci.* **401,** 445 (1998).
181. G. Heinrich, *Rubber Chem. Technol.* **70**, 1 (1997).
182. M. Kluppel and G. Heinrich, *Rubber Chem. Technol.* **73**, 578 (2000).
183. B.N.J. Persson, U. Tartaglino, O. Albohr, and E. Tosatti, *Phys. Rev. B* **71**, 035428 (2005).
184. S. Futamura, *Rubber Chem. Technol.* **69**, 648 (1996).
185. A. Ahagon, T. Kobayashi, T. Kobayashi, and M. Misawa, *Rubber Chem. Technol.* **61**, 14 (1988).
186. A.D. Roberts, *J. Adhes.* **13**, 77 (1981).
187. R. Bond, G.F. Morton, and L.H. Krol, *Polymer* **25**, 132 (1984).
188. R.R. Rahalkar, *Rubber Chem. Technol.* **62**, 246 (1989).
189. H. Takino, R. Nakayama, Y. Yamada, S. Kohjiya, and T. Matsuo, *Rubber Chem. Technol.* **70**, 584 (1997).
190. G. Heinrich and H.B. Dumler, *Rubber Chem. Technol.* **71**, 53 (1998).
191. R.N. Capps, *Elastomeric Materials for Acoustical Applications*, Naval Research Laboratory, Washington DC (1989).
192. *Sound and Vibration Damping with Polymers*, R.D. Corsaro and L.H. Sperling, eds., *ACS Symposium Series* **424**, American Chemical Society, Washington DC (1990).
193. R.N. Capps, *Rubber Chem. Technol.* **59**, 103 (1986).
194. J.C. Snowdon, *Vibration and Shock in Damped Mechanical Systems*, Wiley, New York (1968).

195. P. Toensmeier, *Defense Tech. Internat.* Sept./Oct., p. 18 (2005).
196. J.S. Davidson, J.R. Porter, R.J. Dinan, M.I. Hammons, and J.D. Connell, *J. Perform. Construct. Facil.* **18**, 100 (2004).
197. Z. Xue and J.W. Hutchinson, *Int. J. Solids Struct.* **45**, 3769 (2008).
198. Z. Xue and J.W. Hutchinson, *Mech. Mater.* **39**, 473 (2007).
199. L.J. Malvar, J.E. Crawford, and K.B. Morrill, *J. Comp. Construct.* **11**, 601 (2007).
200. R. Bogoslovov, C.M. Roland, and R.M. Gamache, *Appl. Phys. Lett.* 90, 221910 (2007).
201. C.M. Roland, D. Fragiadakis, and R.M. Gamache, *Comp. Struct.* **92**, 1059 (2010).

3

Chain dynamics

The motion of polymers is distinguished from that of low molecular weight materials by the slow dynamics; the relaxation properties at timescales longer than the time for backbone torsional transitions are unique to chain molecules. The extended spatial scale, exceeding many repeat units, makes the polymeric behavior largely independent of chemical structure. This universality allows modeling of structural and dynamic properties to ignore many details ("coarse graining"), in order to extract the essential physics underlying general properties of long, flexible macromolecules.

3.1 Unentangled polymers

For chains too short or too dilute, or for times too brief, for the effects of the mutual uncrossability of segments to exert an effect, polymer chains can be depicted as a series of beads connected by randomly oriented springs [1–5]. Each bead–spring unit represents a couple of repeat units, with the beads subject to a drag from the environment described by Stokes law.[1] If these subchains are long enough to exhibit Gaussian behavior, intrachain interactions occur via the Hookean springs. Since the subchains are statistically independent,[2] the cumulative chain motions are expressed as a superposition of independent modes (Fourier decomposition). Of course, there are limitations to neglecting atomistic details in a mechanical analog of real polymers, whose structure can be quite complex. This is seen in Figure 3.1, comparing the single chain structure factor for large momentum transfer of a bead–spring model to that of simulated PBD [6]. For both cases $S(Q)$ varies on the scale of the typical bond (or connector) length, but the more realistic chain also exhibits a faster oscillation, the nature of which depends on the chemical structure. The bead-spring analysis breaks down for modes with a characteristic distance smaller than the subchain size; this commences the non-Gaussian regime discussed in Chapter 1 (see Figure 1.14).

The most popular bead–spring model for melts is the Rouse model [1] (Section 1.2.5), which neglects two effects – excluded volume and hydrodynamic (flow field) interactions – that are important for polymers in dilute solution. Excluded volume refers to the prohibition of simultaneous occupancy of space due to hard-core repulsion; it is the physical origin of entanglements. In solutions the excluded volume can cause an expansion of the chain as segments avoid intrachain

[1] The British mathematician George Stokes discovered in 1851 that the force required to drag a sphere through a liquid is proportional to the product of the sphere's velocity, its radius, and viscosity of the liquid. This relation neglects inertial forces and so is applicable to slow flows; any effects due to a wall or edges will augment the Stokes frictional force.

[2] Statistical independence of the subchains means the joint probability distribution for their conformations is the product of the individual probability distributions. This implies the total energy is the sum of the individual energies, which is only true if there is no enthalpic interaction between subchains.

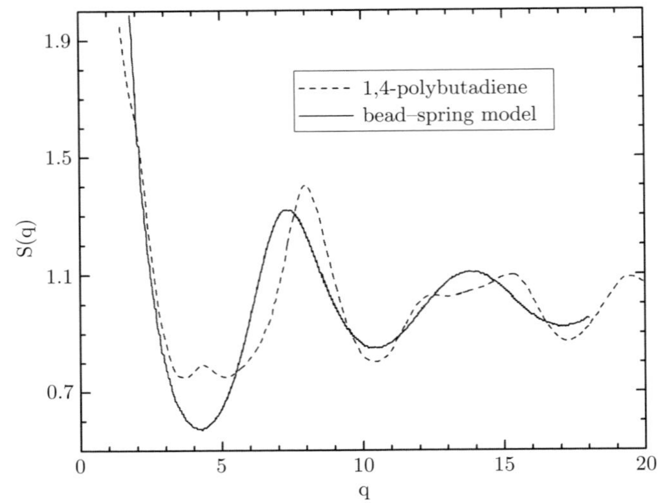

Figure 3.1 Comparison of simulated single-chain structure factor for PBD and a bead–spring chain; the latter lacks the short-ranged structure of the more realistic molecular representation. The abscissa is scaled by the average bond length. Data from ref. [6].

interferences, with the degree of expansion dependent on the nature of the solvent. For Gaussian chains the chain size reflects the random walk result, $<r^2> \sim M$ (eqn (1.22)), whereas for "self-avoiding" random walks (in which paths never intersect) the molecular weight dependency is stronger, $<r^2> \sim M^{6/5}$ [7].

When there is a significant concentration of chains, the intrachain interferences avoided by expansion of the chain are replaced by contacts with neighboring chains; that is, the repulsive forces between monomers of a chain are balanced by competing repulsions from monomers on other chains. The result is screening of excluded-volume effects in concentrated solutions and melts. This is directly demonstrated by neutron-scattering experiments [8–10] that show isotopically labeled chains to have their unperturbed Gaussian dimensions.[3] The screening of the excluded volume is a consequence of the highly interspersed nature of the polymer coils; the monomer density for a chain is substantially less than the total monomer density for the melt (Section 3.2). The omission of excluded volume in the Rouse model, a limitation when applied to dilute solutions, makes it a more appropriate description of polymer melts, at least for chain lengths too short for entanglement interactions.

In the Rouse model the motion of a given bead (or subchain) is unaffected by the motion of other beads; that is, neighboring segments or solvent molecules are unperturbed, so there are no hydrodynamic interactions. This "free-draining" assumption distinguishes the Rouse model from the model of Zimm [2], which considers chains to be impermeable, with friction arising from drag by other beads[4]. The Rouse model assumes the surroundings exert only

[3] The same situation occurs in polymer solutions under Theta conditions, due to a canceling out of repulsive and attractive interactions. This is analogous to gases at their Boyle temperature.

[4] An alternative approach used for polymers in solution, for example for tethered polymers [11], is to model the chain as a series of independent, impenetrable blobs, each having a radius proportional to the square of the number of segments within the blob.

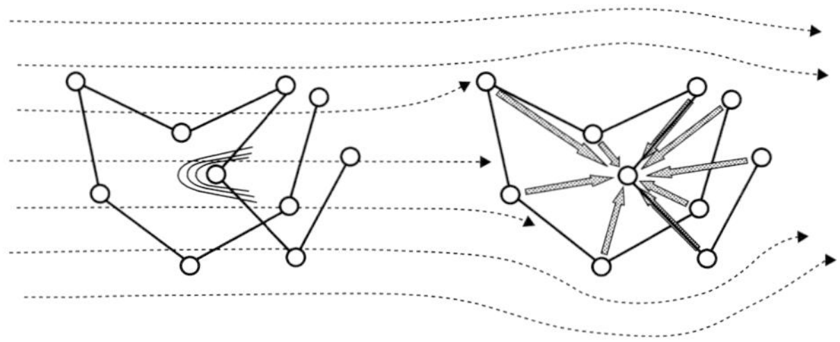

Figure 3.2 (a) Rouse model – free draining with negligible hydrodynamic interaction; (b) Zimm model – impenetrable coil with diffusive momentum transport through fluid.

a uniform frictional force, of magnitude proportional to the number of beads. These two conditions are depicted in Figure 3.2. Hydrodynamic interactions are important in dilute solution, although real chains may be partially draining. At high polymer concentrations, these interactions can be neglected altogether, because the perturbation of the solvent is limited to very small distances due to the presence of many neighboring segments (beads). Thus, similar to its neglect of excluded volume, the absence of hydrodynamic interactions in the Rouse description, a defect in the study of dilute solutions, makes the model more applicable to unentangled polymer melts.

The linearly coupled modes of motion of the Rouse submolecules give rise to a spectrum of relaxation times, which determines the viscoelastic behavior. The relaxation time is equal to the ratio of the friction force to the spring constant, with τ_p for the different modes related according to eqn (1.32), and the storage and loss dynamic moduli given by eqns (1.37) and (1.39). In Figure 3.3 the Rouse model is fitted to dynamic mechanical spectra for unentangled polystyrene [12]. The experimental data are described approximately, but it is a general failing of the model that it cannot quantitatively reproduce the shape of the mechanical spectra.

The longest viscoelastic relaxation time of the Rouse model is related to the relaxation time of any mode by

$$\tau_R = 2p^2\tau_p \tag{3.1}$$

The viscosity is the integral of the relaxation modulus (eqn (1.47))

$$\eta_0 = \int_0^\infty G(t)\,dt = \frac{\pi^2 \rho R T}{6M}\tau_R \tag{3.2}$$

($\rho R/M = ck_B$ in eqns (1.35)–(1.39)) and η_0 is predicted to be directly proportional to M (since $\tau_R \sim M^2$; eqn (1.33)). There is no strain rate dependence of η_0, and any shear-thinning at high rates is unaccounted for by the model. Since the Rouse relaxation time is proportional to ζ/T, the viscosity will depend on the friction coefficient. This complicates assessment of eqn (3.2), since chains short enough to be unentangled have a concentration of chain ends sufficient to affect T_g (Chapter 2) and thus ζ. Measurements must be corrected to approximate "iso-friction" conditions. In Figure 3.4 experimental viscosities for unentangled PBD normalized by the molecular weight are plotted versus the molecular weight [13]. After correction for the change in T_g with M, the dependence of η on M is confirmed, at least over a narrow range of molecular weight. However, in Figure 3.5 [14–16] relaxation times for unentangled PI measured over a broader span of M exhibit an M dependence that is a bit stronger than

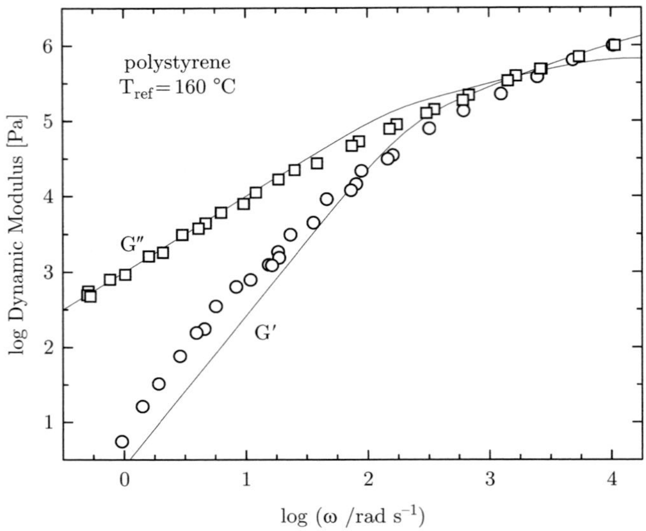

Figure 3.3 Master curve of the dynamic storage (circles) and loss (squares) moduli for PS ($M_w = 14\,800$ g/mol; $M_w/M_n = 1.15$), along with the fits to eqns (1.36) and (1.37). Data from ref. [12].

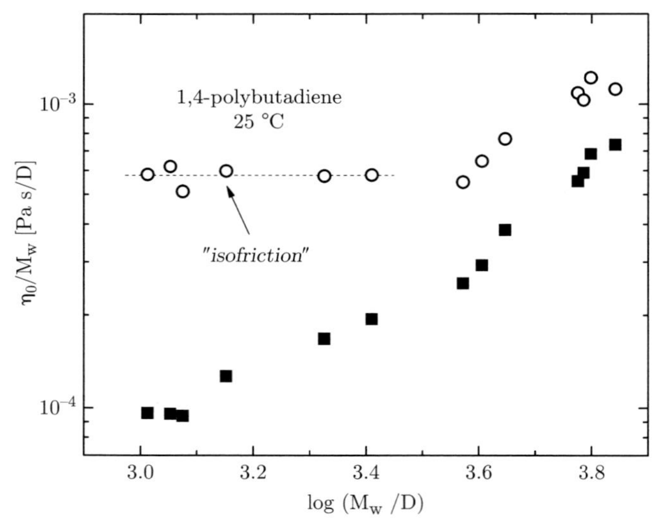

Figure 3.4 Low shear rate viscosity normalized by molecular weight for PBD as measured (filled symbols) and after a correction for the effect of M on T_g (open symbols). The latter conform over a half decade range to the M dependence predicted by the Rouse model, before entanglement effects become apparent for $M > 3000$ g/mol. Data from ref. [13].

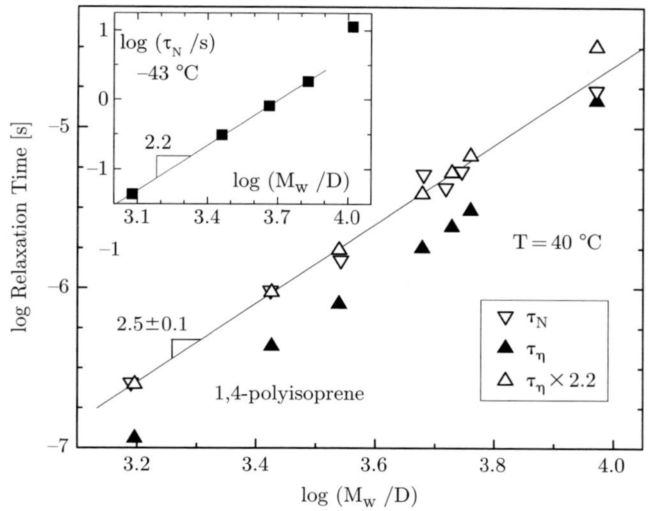

Figure 3.5 Dielectric (open inverted triangles) and mechanical (filled triangles) global relaxation times for unentangled PI as a function of molecular weight. After multiplying by a factor of 2.2 (open triangles), the latter superpose on the former, with the combined data exhibiting a stronger dependence on molecular weight than the M^2 of the Rouse model. Data from refs. [14,15]. In the inset are dielectric relaxation times for another set of PI samples (squares), measured at lower T. Their M dependence is also stronger than quadratic. Data from ref. [16].

quadratic. The uncertainty in analyses such as these is that the usual method to correct the viscosities to iso-friction conditions assumes that the dynamics is governed by free volume and that the thermal expansion coefficient reflects only the variation of free volume with T; however, neither assumption is supported by experiments. Moreover, since the data span less than half a decade, this is not an especially severe test of the model. The range of validity of the model is limited at high M by the onset of entanglements, and at low M due to non-Gaussian statistics.

The equilibrium recoverable compliance, a measure of the elastic deformation during steady-state flow,

$$J_s^0 = \int_0^\infty tG(t)\,dt \Big/ \left(\int_0^\infty G(t)\,dt\right)^2 \tag{3.3}$$

for Rouse chains is

$$J_s^0 = \frac{2}{5}\frac{M}{\rho RT} \tag{3.4}$$

The predicted linear dependence on chain length in the absence of entanglements is consistent with experiments, as seen in Figure 3.6 for PI [17] and PS [18–20]. The actual magnitude of the recoverable compliance is also in accord with experiments, as is illustrated in data for PI [21,22], PS [19,23], and PVE [24], each J_s^0 normalized by the Rouse value (Figure 3.7). Eqn (3.4) assumes monodisperse chains, but J_s^0 is extremely sensitive to the high-M tail of the distribution.

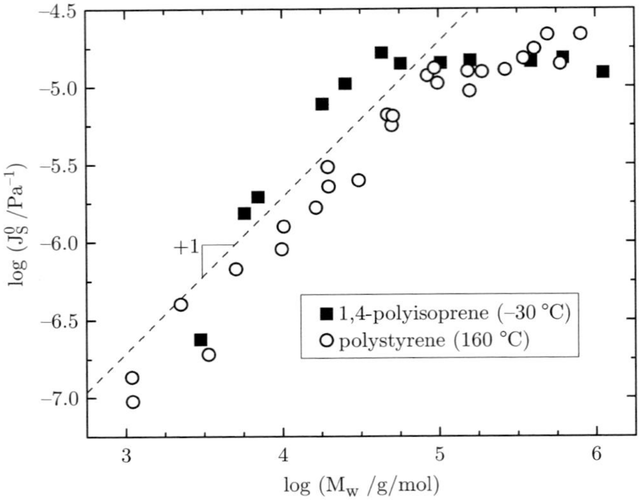

Figure 3.6 Steady-state recoverable compliance of PI (squares) and PS (circles), showing conformance to the Rouse prediction (eqn (3.4), indicated by the line) prior to the onset of entanglement effects. Data from ref. [17] for PI and refs. [18–20] for PS.

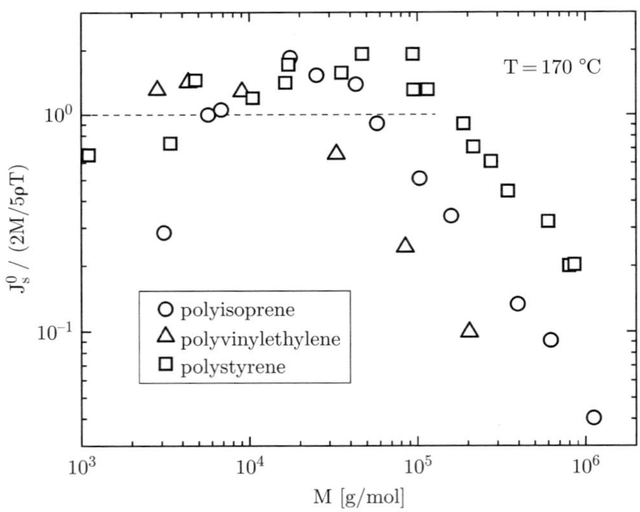

Figure 3.7 Steady-state recoverable compliance of PI [21,22], PS [19,23], and PVE [24] divided by the value of J_s^0 from eqn (3.4). The horizontal line at unity is the Rouse prediction; the strong deviations at higher M are due to entanglements.

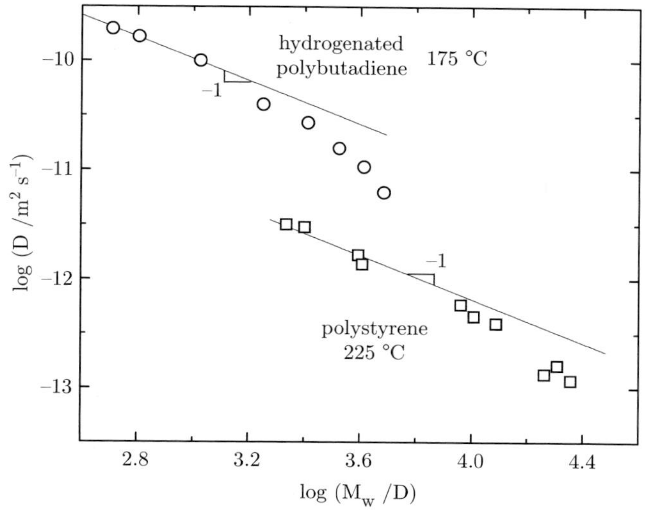

Figure 3.8 Diffusion coefficient for hydrogenated PBD (circles) and PS (squares). For sufficiently low molecular weights, there is a narrow region that can be described by the inverse dependence on *M* predicted by the Rouse model, after which entanglements cause a stronger variation. Data from refs. [25–29].

The $p = 0$ mode measures the position of the chain's center of mass, with the translational relaxation time defined as (*cf.* eqn (1.46))

$$\tau_0 = \frac{\langle r^2 \rangle}{3\pi^2 D} \tag{3.5}$$

The Rouse diffusion constant varies inversely with molecular weight (eqn (1.34)), consistent with experimental results; e.g., for hydrogenated PBD [25,26] and PS [27–29] in Figure 3.8. However, the range of data showing $D \sim M^{-1}$ is very limited due to the onset of entanglement effects at higher *M*.

For chains having a dipole moment parallel to the backbone, such as 1,4-polyisoprene and polypropylene glycol, there is a low-frequency peak in the dielectric spectrum due to fluctuations of the end-to-end chain vector. The relaxation time for this fluctuation, referred to as the normal mode relaxation time,[5] is twice the longest Rouse viscoelastic relaxation time, $\tau_N = 2\tau_R$, although experimentally τ_N is somewhat larger than this Rouse prediction; see Figure 3.5. (Experimental relaxation times are commonly defined from the inverse of the peak frequency, which, unless the peak is very broad, is essentially equal to the relaxation time of the slowest mode.) The dielectric constant due to the normal mode spectrum[6] is [30]

[5] In dielectric spectroscopy "normal mode" relaxation time refers to the longest relaxation time, $\tau_N \approx \sum_p p^{-4}\tau_p^2 / \sum_p p^{-2}\tau_p$.

[6] Strictly speaking, "normal mode" refers to analysis of chain dynamics in terms of decoupled, single-frequency mathematical solutions to the equations of motion for bead–spring models; however, in practice the term is used to refer to dielectric relaxation arising from global chain motions, without reference to any normal mode coordinates.

$$\varepsilon'(\omega) = \varepsilon_0 - \Delta\varepsilon\omega^2 \sum_p g_p \frac{\tau_p^2}{1+\omega^2\tau_p^2} \tag{3.6}$$

and the dielectric loss

$$\varepsilon''(\omega) = \Delta\varepsilon\omega \sum_p g_p \frac{\tau_p}{1+\omega^2\tau_p^2} \tag{3.7}$$

In these equations $\Delta\varepsilon$ is the dielectric strength for the normal mode and ϵ_0 is the static dielectric constant. The weighting factors are given by

$$\begin{aligned} g_p &= 8\pi^{-2}p^{-2}\ (p \text{ odd}) \\ g_p &= 0\ (p \text{ even}) \end{aligned} \tag{3.8}$$

That is, in the usual case only the odd eigenmodes are dielectrically active. For chains synthesized with inverted repeat units at midchain to cause a symmetrical reversal of the dipole moment, the even Rouse modes contribute, while the odd modes are inactive [30].

The shape of the dielectric normal mode is invariant with M and, similar to results for the dynamic modulus (Figure 3.3), experimental spectra deviate from the relaxation function of the Rouse model (Figure 3.9). As this figure shows, the discrepancy is not due to polydispersity; experimental spectra always exhibit more intensity on the high-frequency side [16,30]. This failing is shared by tube models for entangled polymers, as discussed below. Another problem concerns the temperature dependence of the Rouse dynamics, which is determined by the quantity ζ/T. Since ζ is the friction factor governing the local segmental dynamics, the Rouse prediction is that the chain modes and the local segmental modes have the same

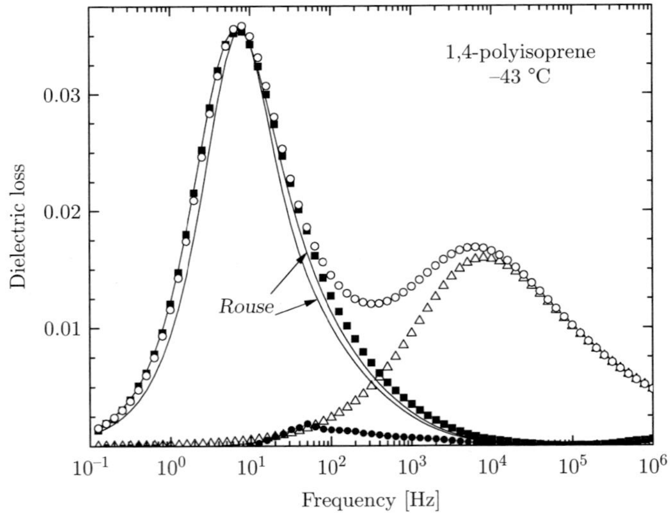

Figure 3.9 Dielectric spectrum of unentangled PI ($M_w = 2.9$ kg/mol; $M_n = 2.7$ kg/mol) (open circles), showing the normal mode (filled squares) resolved by subtraction of the segmental peak (open triangles). The dashed line is the fitted Rouse peak assuming monodisperse M. The solid line is the weighted Rouse contributions summed over the molecular weight distribution. After this correction for polydispersity, a discrepancy remains between the experimental and calculated spectra around 0.01–10 kHz (filled circles), corresponding to the position of the $p = 3$ mode. Data from ref. [16].

T dependence. However, as discussed in Chapter 6, this is not the case near T_g, where another problem arises: the strong decrease of the compliance, in contradiction to eqn. (3.4).

In summary, the Rouse model does remarkably well in describing certain properties of unentangled polymer melts: the molecular weight dependences of J_s^0 and D, and the magnitude of the recoverable compliance. The relaxation time varies more strongly with M than predicted, although the need to correct for the effect of a changing T_g at low M complicates the assessment. The most significant deviations are twofold: the shape of the relaxation spectrum, the Rouse model consistently underestimating the intensity at higher frequencies, and the breakdown of time–temperature superpositioning in the glass-transition zone. Since the reptation model builds on the Rouse model, it inherits these limitations.

3.2 Entangled polymers

Polymer chains cannot intersect one another, which limits transverse motions to small (~0.5 nm) fluctuations. For high polymers these topological interactions are almost all intermolecular because the volume of a given chain is pervaded by other chains. To show this, consider a PBD with $M = 10^5$ g/mol, or $n = 7400$ backbone bonds per chain ($m_0 = 54$ g/mol). The volume of a chain, V_{chain}, that is, the space it pervades, can be estimated from its radius of gyration (*cf.* eqn (1.24))

$$V_{chain} \sim \langle R_g^2 \rangle^{3/2} = \left(\frac{1}{6} C_\infty n l^2 \right)^{3/2} \tag{3.9}$$

Using $C_\infty = 5.1$ [31] and 0.15 nm for the mean polybutadiene bond length, $\langle R_g^2 \rangle \sim 140\ \text{nm}^2$ and $V_{chain} \sim 1700\ \text{nm}^3$. The concentration of repeat units for one chain is $M/(m_0 V_{chain})$, or about 1 PBD monomer per nm^3. Assuming a mass density of 1 g/ml, the concentration of all repeat units is $cM/m_0 = N_A \rho / m_0 \sim 10\text{nm}^{-3}$, which is tenfold higher than that for a given chain. This means that neighboring chains strongly interpenetrate, leading to strong entanglement effects for long chains.

The entanglements are topological, due only to the connectedness and uncrossability of chains. Entanglements have no effect on static properties or the segmental dynamics; however, they strongly affect the low-frequency dynamics. "Optical tweezer" methods, in which a laser beam is used to exert a small (~ pN) force on chromophore-labeled chains, show directly the confining effect of entanglements on the lateral motion of DNA chains [32,33]. Long-range reconfigurations involving substantial longitudinal displacements must proceed by diffusion along the chain contour; i.e., reptation [34,35], a mechanism akin to the motion of snakes. Entanglements can be conceived as a "tube" of lateral constraints, whose diameter is determined by the concentration of entanglements.

At short times, prior to relaxation via longitudinal diffusion, the entanglements act similarly to the crosslinks of a rubbery network. The quasi-elastic response is nearly constant over a range of frequencies known as the rubbery plateau zone. The storage modulus in this plateau zone is called the plateau modulus, G_N, and has a magnitude proportional to the entanglement concentration

$$G_N = \rho RT / M_e \tag{3.10}$$

where M_e is the molecular weight between entanglements. Equation (3.10) is for a neat polymer; in the presence of diluent, G_N is reduced by a factor equal to roughly the square of the polymer volume fraction. Of course, G_N goes to zero when the concentration becomes too low for significant chain overlap. Since the plateau modulus is not strictly independent of frequency, evaluation of G_N must rely on indirect methods. From the Kramers–Kronig relation [36], G_N can be obtained from integration over the terminal dispersion

$$G_N = \frac{2}{\pi} \int_{-\infty}^{+\infty} G''(\omega) d \ln \omega \tag{3.11}$$

In practice, this integration has to be terminated on the high frequency side to avoid contribution from the glass-transition zone. Empirically the plateau modulus has been related to the peak value of the loss modulus in the terminal zone [37,38]

$$G_N = 3.56 G'' \max \tag{3.12}$$

G_N can also be estimated from the storage modulus at the frequency, ω_{min}, of the minimum in the loss modulus [39]

$$G_N = G'(\omega_{\min}) \tag{3.13}$$

These three methods are illustrated in Figure 3.10 for PBD [40]; the obtained plateau modulus values are listed in Table 3.1, along with the average of literature values of G_N for this polymer [41].

There have been many attempts to develop a quantitative relationship between the structure of a polymer and the magnitude of its plateau modulus. Various definitions of an entanglement coupling have been proposed, including a fixed number of binary contacts per entanglement [42] or per entanglement volume [43], and a fixed number of strands per entanglement volume [44–47]. Scaling arguments have been employed, with the underlying idea that the more volume

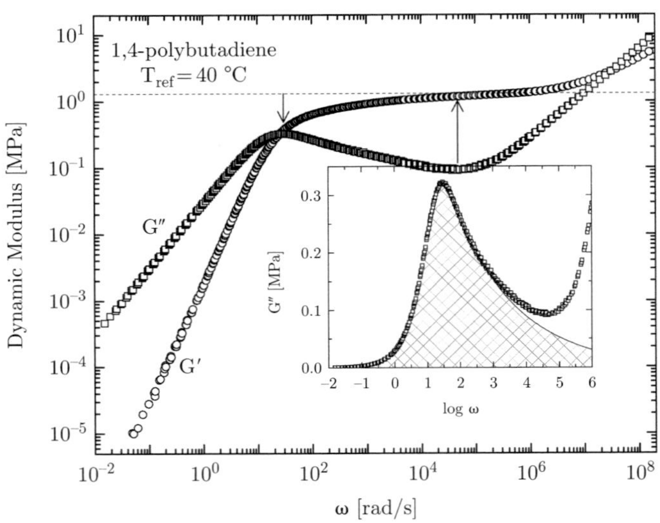

Figure 3.10 Master curve of the dynamic storage (circles) and loss (squares) moduli for PBD (M_w = 99 060 g/mol; M_w/M_n = 1.01). The horizontal dashed line denotes the value of the plateau modulus determined from integration of the loss modulus terminal peak (using a linear ordinate, as shown in inset); a power law is fit to the high frequency side to avoid the overlapping segmental modes. The peak in G' (downward arrow) is a factor of 4.06 less than G_N (cf. eqn (3.12)). The upward arrow at the minimum in G'' indicates that $G' \sim G_N$ at this frequency. Data from ref. [40].

Table 3.1 Plateau modulus of 1,4-polybutadiene.

	G_N (MPa)		
	eqn (3.11)	eqn (3.12)	eqn (3.13)
from Figure 3.10	1.31	1.15	1.29
literature average [41]	1.13 ±0.03	1.18 ±0.07	1.15 ±0.05

a chain permeates, the more numerous will be its contacts with other chains, each encounter potentially serving as an entanglement [48–52]. This is a geometric argument, so the important aspects of the chain structure are physical: the occupation volume of a chain ($= MN_A/\rho$) and its stiffness, which, along with the repeat unit length, determines its spatial extent (V_{chain} in eqn (3.9)). Entanglement interactions arise when the occupied volume becomes a sufficiently large fraction of the pervaded volume; the qualitative expectation is that stiffer polymers with slimmer repeat units lacking bulky side groups will be more entangled. Fetters *et al.* [52] derived a specific prediction for the plateau modulus in terms of the number of chains within V_{chain}

$$G_N = Ak_BT\left(\frac{M}{\rho N_A \langle R_g^2\rangle}\right)^{-3} \tag{3.14}$$

where A is a constant that depends on the value necessary for entanglements of the ratio of V_{chain} to occupied volume. The bracketed quantity in eqn (3.14) is known as the packing length, $l_p = \frac{M}{\rho N_A \langle R_g^2\rangle}$. Defined as the ratio of the occupied volume to the square of the chain radius, l_p is a measure of a chain's cross-sectional size. Figure 3.11 [53] shows the plateau modulus normalized by RT (this ratio is the interaction density; *cf.* eqn (1.35)) plotted versus the packing length. The observed correlation is consistent with $G_N \sim l_{p^{-3}}$. From the value obtained for the coefficient in eqn (3.14), it is estimated that about 1 in 6 "binary contacts" are

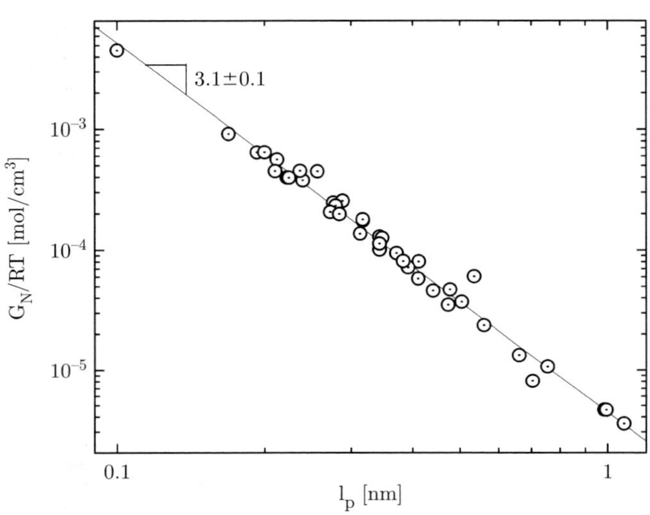

Figure 3.11 Entanglement interaction density *versus* the packing length for various polymer melts; the line is the least squares fit. Data from [47].

Table 3.2 Characteristic molecular weights (kg/mol) [55]].

polymer	M_e	M_c	M_c/M_e
polyethylene	1.15	3.48	3.03
polyethylene oxide	2.00	5.87	2.94
1,4-polybutadiene	2.00	6.38	3.19
ethylene-propylene rubber	3.10	8.10	2.61
styrene-butadiene rubber	2.96	8.21	2.77
1,2-polybutadiene	3.85	8.20	2.13
1,4-polyisoprene	6.40	10.0	1.56
polyvinylacetate	9.10	24.5	2.69
poly α-methylstyrene	13.3	28.0	2.11
polyisobutylene	10.5	17.0	1.62
polymethylmethacrylate	13.6	29.5	2.17
polystyrene	18.1	31.2	1.72
polydimethysiloxane	12.0	24.5	2.04

effective in entangling a chain [54] (or that 6 such contacts are necessary to produce an entanglement).

Related to the molecular weight between entanglements is the characteristic value, M_c, at which the viscosity begins to exhibit a non-linear M dependence. Values of M_e and M_c are tabulated in Table 3.2 [55]; for these thirteen polymers, $M_c = 2.4 \pm 0.6\ M_e$. The steady-state recoverable compliance also exhibits a change in M dependence (Figure 3.7), but at a characteristic molecular weight that is several times larger than M_e.

3.2.1 Reptation dynamics

The Doi and Edwards [56] constitutive theory and related "tube models" for entangled polymers describe the dynamics in two stages. Shortly after a deformation, the chains equilibrate their contour length by Rouse motion, accompanied by small fluctuations around a primitive path defined by the chain contour. At longer times relaxation involves diffusion parallel to the contour by reptation, with transverse motions highly restricted by neighboring segments. In tube theories, the contour length relaxation time is the longest Rouse relaxation time, and the time for a chain to disengage from its tube of constraints, τ_d, is the average molecular orientation relaxation time. τ_d is the longest relaxation time.

There have been many experimental investigations showing that the tube-model approach yields descriptions of the rheology of high molecular weight polymers that are at least semiquantitative [54,57–59]. As discussed in Section 1.2.6, the prediction for the molecular weight dependence of the terminal relaxation ($\sim M^3$) is weaker than the experimental result, for which the exponent is in the range 3.3 to 3.7 (Figure 18 in Chapter 1). Likewise, the self-diffusion constant varies more strongly with M than predicted [60,61], although for trace diffusion D has the correct M^2 dependence [61]. However, even this agreement presents a problem: Since the time for a chain to diffuse via reptation over a distance equal to its contour length should increase as M (eqn (1.46)), if $D \sim M^{-\upsilon}$, then η_0 and τ_d should vary as $M^{1+\upsilon}$; that is, the respective experimental M dependences of the viscosity and diffusion constant are mutually inconsistent. Deviations such as these from the basic tube model, which considers only a single chain constrained by fixed entanglements, have been ascribed to additional relaxation mechanisms, such as constraint release [62,63], due to motion of the chains comprising the tube of entanglements, and contour-length fluctuations [64,65], which refers to changes in the chain contour due to fluctuations in the primitive path. These processes are illustrated in Figure 3.12.

Various refinements of the tube model lead to the expectation that these additional contributions to relaxation dissipate with increasing chain length [13]. This implies that experimental results will be in conformance with the predicted M dependences at sufficiently high M. For the viscosity this M^3 limiting behavior is unattained even at $M/M_e > 10^3$ (Figure 1.18). On the other hand, diffusion coefficients appear to attain the expected M^2 behavior at relatively low M, especially for trace diffusion (Figure 3.13). In the framework of tube theories, the different limiting behavior of η_0 and D requires constraint release and contour length fluctuations to affect chain relaxation and diffusion differently [61,66–68].

Confinement of a Rouse chain to a tube of constraints makes the even modes inactive, so that the reptation prediction for the relaxation modulus is

$$G(t) = G_N \frac{8}{\pi^2} \sum_{p,odd}^{\infty} p^{-2} \exp\left(-\frac{p^2 t}{\tau_d}\right) \tag{3.15}$$

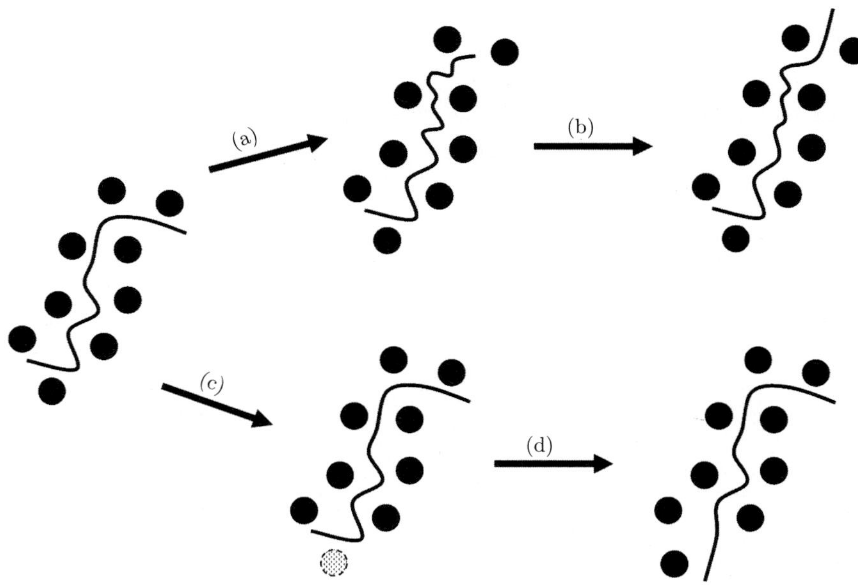

Figure 3.12 Schematic representation of a chain free to move but not through entanglements (denoted as circles). *top: contour length fluctuations* (a) Chain end retracts along its contour path via longitudinal Brownian motion. For a bead–spring model chain, this corresponds to changes in the distance between segments due to compression of the interconnecting springs. (b) The chain end can now explore orientations outside the tube of constraints. *bottom: constraint release mechanism* (c) Movement of a neighboring, entangling chain partially dissolves the tube, allowing (d) the chain to move beyond its original boundaries. A new entanglement forms at a different position.

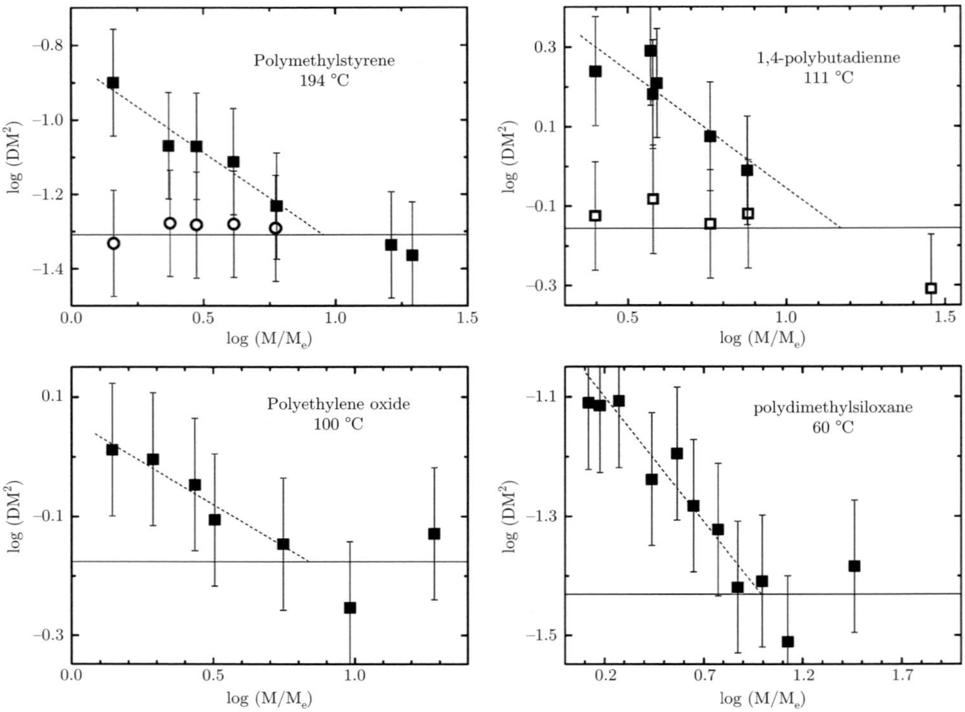

Figure 3.13 Diffusion constants multiplied by the square of M plotted *versus* the number of entanglements per chain. Filled symbols are for self-diffusion and open symbols trace diffusion. The horizontal line denotes the asymptotic quadratic M dependence and the dashed line the variation for M less than about $10M_e$. Data from ref. [61].

and for the loss modulus

$$G''(\omega) = G_N \frac{8}{\pi^2} \sum_{p,odd}^{\infty} p^{-2} \frac{\omega\tau_p}{1+\omega^2\tau_p^2} \tag{3.16}$$

The dielectric response of the tube model is still described by eqns (3.6) through (3.8), which means the Rouse and reptation models predict no change in the dielectric spectrum due to entanglements, in agreement with experimental data (Figure 3.14) [30]. The tube model also predicts that the peak shape for the terminal loss modulus and the normal mode dielectric loss will be the same, apart from their different intensity (the mechanical response is weighted by G_N and the dielectric by $\Delta\varepsilon$). As seen in Figure 3.15 [69,70], however, mechanical spectra are invariably broader than the corresponding dielectric normal mode dispersion.

Similar to the failing of the Rouse model for unentangled polymers, the tube model predicts a narrower spectrum than found experimentally (Figure 3.14 and Figure 3.15). This has been attributed to constraint release, with motion of the tube reducing τ_d and thereby broadening the terminal dispersion. This idea can be tested by dissolving chains at low concentration in a higher molecular weight matrix. The longer relaxation time of the latter means that constraint release is slowed, occurring on a longer timescale than τ_d of the probe chains. The suppression of the relaxation due to constraint release should yield a dispersion for the probe that more

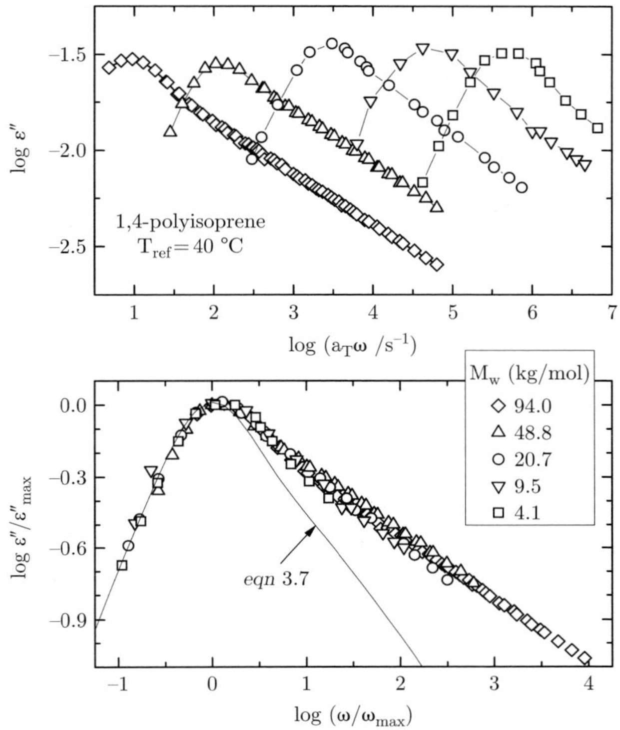

Figure 3.14 (*top*) Master curves of the dielectric loss for monodisperse PI. The data for the lowest molecular weight sample was adjusted to account for its lower T_g. (*bottom*) Superposition of the spectra to show the invariance of the peak shape with molecular weight. At high frequencies the spectra deviate from the Rouse and tube model prediction. Data from ref. [30].

closely approaches the form of eqn (3.15). However, experimental results [71–73] consistently show that the relaxation peak actually becomes broader; that is, further from the tube model prediction (Figure 3.16). This calls into question the relevance of constraint release for explaining the discrepancy between theory and experiment. The broadening of the spectra also indicates coupling of the probe and host chains, as discussed further in Section 5.2.3.

The deviations of the calculated loss curves in Figure 3.14, Figure 3.15 and Figure 3.16 fall mainly toward higher frequencies, similar to the failing of the Rouse model (Figure 3.3 and Figure 3.9). This indicates a problem in both models' description of the fast dynamics. In these models the motion at short times is diffusive, but this is at odds with data from both experiments and simulations [74–77]. The Rouse and tube models neglect cooperative effects and the consequent heterogeneous dynamics. This deficiency is apparent in Figure 3.17, showing the mean square displacement of the center of mass of polyethylene chains [78]. At shorter times the behavior becomes subdiffusive, with the discrepancy with the models increasing with molecular weight (the polyethylene samples in Figure 3.17 have molecular weights spanning $M_e = 830$ g/mol [79]).

The friction factor for entangled chains is the same as ζ for Rouse dynamics; hence, the shortcoming of the Rouse model with regard to the T dependence carries over to tube models.

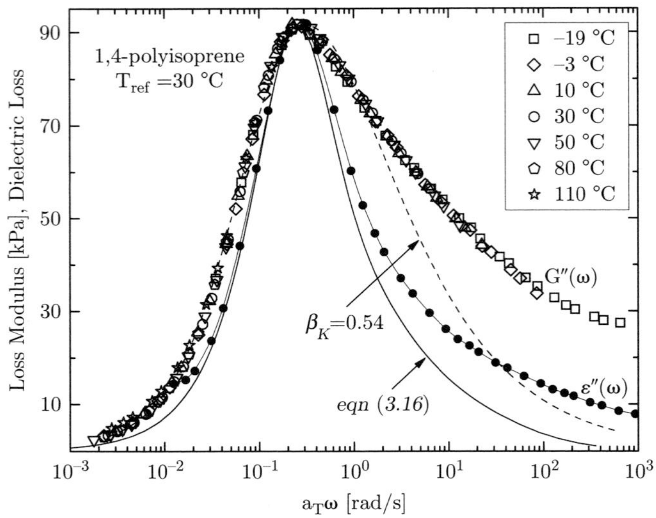

Figure 3.15 Master curve of the loss modulus for PI ($M_w = 357$ kg/mol; $M_w/Mn = 1.08$), along with the fits to eqn (3.16) (solid line) and to the Kohlrausch function (dashed line). Also shown is the corresponding dielectric loss (filled circles), scaled to superpose the peak maxima. Data from refs. [69,70].

This disconnect between local and global temperature dependences is discussed in detail in Chapter 6. In summary, the reptation model offers a description of the complex dynamics of long-chain molecules that is plausible and physically appealing. Its predictions of the M dependences of viscoelastic properties are close but not quantitative. The model also underpredicts the intensity at higher frequencies in the relaxation spectrum and cannot address the breakdown of t–T superpositioning of polymers near T_g. Whether these shortcomings can be rectified within the scope of mean-field theories[7] of single-chain dynamics or require a different approach remains to be seen.

3.2.2 Stretched exponential relaxation

In the early 1970s Tobolsky and coworkers [80,81] reported that stress relaxation in the terminal zone of polymers could be adequately described by the Kohlrausch function (eqn (1.7)), with β_K in the range 0.5 – 0.65. Figure 3.18 shows some typical results for monodisperse PS. Over a limited range the data exhibit the stretched exponential form, although the fit is imperfect, especially toward shorter times. Since the viscosity is the integral of the relaxation modulus (eqn (1.47)), an analytical expression can be obtained for the viscosity

$$\eta_0 = \int_0^\infty \exp(-t/\tau_\eta)^{\beta_K}\, dt = \left[\frac{\Gamma(\beta_K^{-1})}{\beta_K}\right] G_N \tau_\eta \tag{3.17}$$

[7] In a mean (or self-consistent) field treatment the myriad interactions of many, highly interspersed chains are approximated as independent interactions of each individual chain with an external field. This field is the average of the interactions among all monomers, so that the multibody problem becomes a tractable, single-body problem.

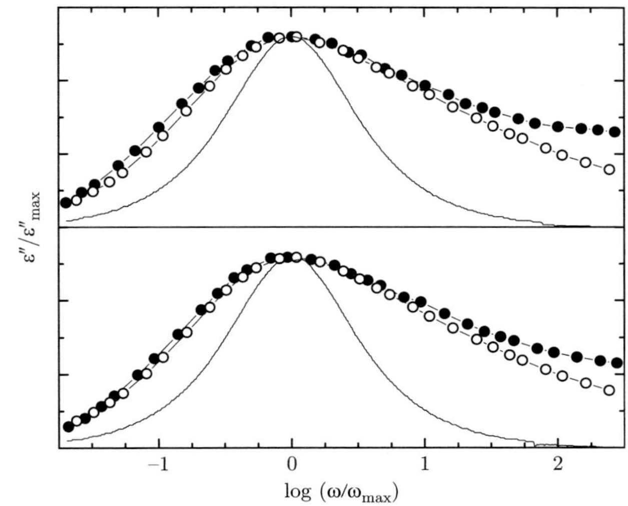

Figure 3.16 The normal mode dispersion measured for PI neat (open circles) and in 15% blends with polybutadiene (filled circles): (top) PI19 in PVE and (bottom) PI19 in PBD. The data have been shifted to superimpose the peaks. When diluted with higher molecular weight polymer, the terminal relaxation of the PI is always broader. All curves are substantially broader than the tube model prediction, eqn (3.7) (solid line). Data from ref. [73].

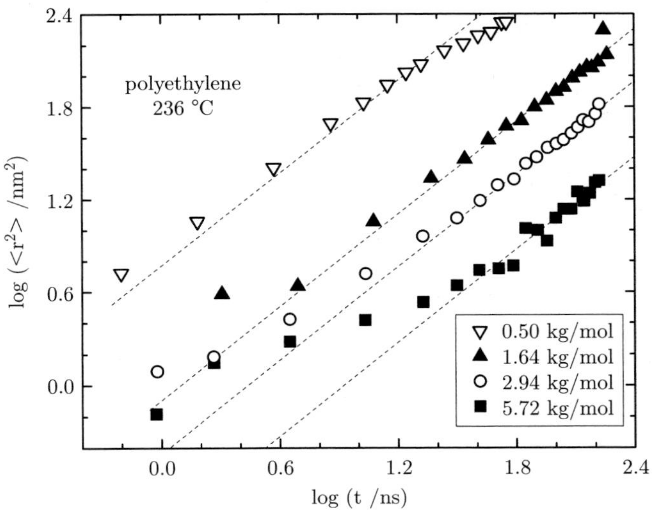

Figure 3.17 Mean square diffusion of the center of mass of polyethylene chains having the indicated M_w; these correspond to a range from unentangled to ∼7 entanglements per chain. The dashed line represents the Brownian diffusion predicted by the Rouse model and tube models at short times. Data from ref. [78].

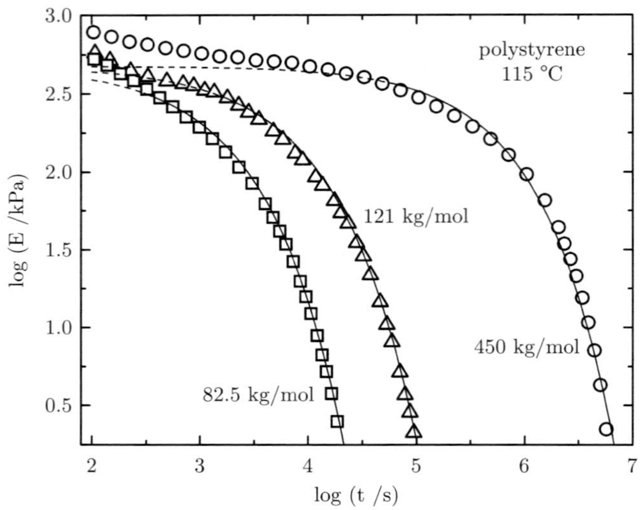

Figure 3.18 Tensile relaxation modulus in the terminal zone for monodisperse PS having the indicated M_w. The solid lines represent the fits of the Kohlrausch function to the data ($\beta_K = 0.62$, 0.61, and 0.65 with increasing molecular weight). The solid lines denote the range of data used for fitting, while the dashed lines show the deviation at shorter times. Data from ref. [81].

where Γ is the gamma function. A plot of $\frac{\Gamma(\beta_K^{-1})}{\beta_K}$ versus $\eta_0 \tau_\eta^{-1}$ should have a slope equal to the plateau modulus. However, as seen in (Figure 3.19) for the PS data in ref. [81], the obtained slope, = 0.4 MPa, is a factor of two larger than the value of $G_N = 0.2$ MPa for polystyrene. Using eqn. (3.3) the equilibrium recoverable compliance for stretched exponential decay is given by

$$J_s^0 = \eta^{-2} \int_0^\infty t \exp(-t/\tau_\eta)^{\beta_K} dt = \frac{\Gamma(2\beta_K^{-1})\beta_K}{[\Gamma(\beta_K^{-1})]^2} G_N \tau_\eta \tag{3.18}$$

Using this expression Knoff *et al.* [81] obtained $J_s^0 G_N = 1.42$ for PS. This product is a measure of the breadth of the terminal zone, and the result from eqn (3.18) is much less than the directly measured experimental values, $J_s^0 G_N = 2.5 \pm 0.5$ [58]. These failings of the Tobolsky analysis [80,81] underscore the approximate nature of the fits in Figure 3.18.

This approach is empirical, relying on no assumptions aside from the use of the Kohlrausch function to fit experimental $G(t)$. The method was extended by Ngai and coworkers [82,83] using the coupling model. (Application of the coupling model to the Johari–Goldstein relaxation and to the dynamics of networks is described, respectively, in Section 2.4 and Section 4.4.) According to the model, after some temperature-insensitive time, t_c, the Rouse modes become mutually inhibited by entanglements to give rise to the Kohlrausch form for the relaxation function. The magnitude of the interaction strength is reflected (inversely) in the magnitude of β_K. Rewriting eqn (2.18) for the terminal relaxation time

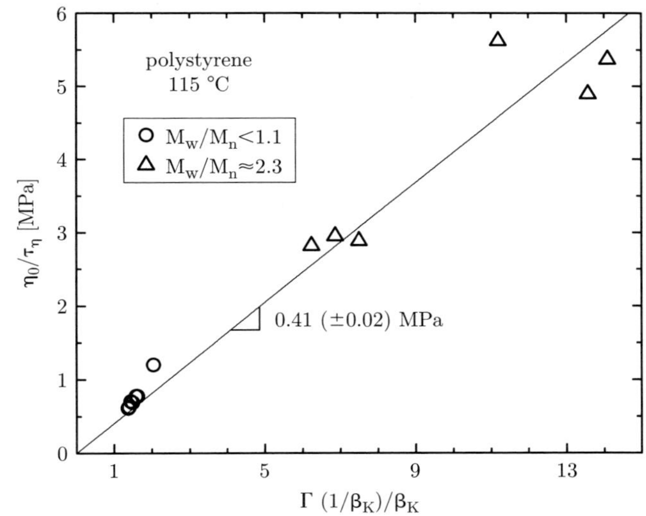

Figure 3.19 Ratio of terminal viscosity to terminal relaxation time for PS plotted *versus* the bracketed quantity on the rhs of eqn (3.17). The lower values are for monodisperse samples (circles) and the larger ones for polydisperse PS (triangles). The least squares linear fit (solid line) has a slope that is significantly larger than the experimental value of G_N for PS. Data from ref. [81].

$$\tau_\eta = (\beta_K t_c^{\beta_K - 1} \tau_R)^{1/\beta_K} \tag{3.19}$$

Equation (3.19) shows how the decrease in β_K due to entanglements amplifies non-linearly the Rouse relaxation times. Since the effect of entanglement constraints on linear polymers is largely independent of their chemical structure, β_K for the terminal dynamics should be the same for all polymers. Thus, the quadratic M dependence of the Rouse relaxation time becomes

$$\tau_\eta \sim \tau_R^{1/\beta_K} \sim (M^2)^{1/\beta_K} \tag{3.20}$$

Consistency with the experiment result $\tau_\eta \sim M^{3.4}$ requires $\beta_K \sim 0.6$; this value is within the range found for monodisperse PS [81].

More generally, best-fit β_K are smaller than 0.6. The problem with the coupling-model description of the terminal dynamics is that relaxation functions depart from the Kohlrausch form. The agreement is only fair in Figure 3.18, although the logarithmic ordinate scale obscures the discrepancies. The deviations become clearer for well-entangled, monodisperse polymers, in which the terminal peak is well separated from the segmental dispersion [84]. Examples are given in Figure 3.15 for PI and Figure 3.20 for PBD.

Another prediction of the coupling model is that the temperature dependence of the Rouse dynamics will be amplified by the power of $1/\beta_K$ (analogous to eqn (3.20) for the M dependence). Since segmental relaxation peaks are broader than the terminal relaxation peak for monodisperse polymers, β_K for segmental relaxation is smaller and therefore the coupling

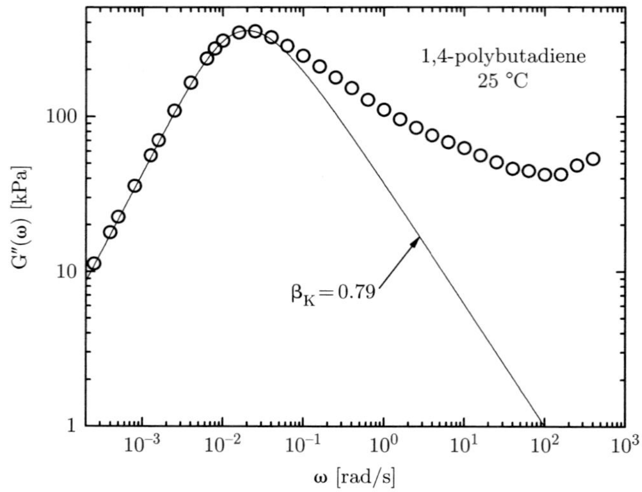

Figure 3.20 Shear relaxation modulus in the terminal zone for monodisperse PBD ($Mw = 925$ kg/mol; $M_w/M_n < 1.1$). The solid line is the best fit to the Kohlrausch function, yielding $\beta_K = 0.79$. There is a substantial departure of the data at higher frequencies, even though the segmental dynamics are well resolved. Data from ref. [84].

model predicts that segmental relaxation times will exhibit stronger T dependences. This result is in qualitative accord with experimental data [85,86] (see Section 6.2).

3.2.3 Branched Polymers

The discussions above pertain to linear macromolecules. Long branches are commonly incorporated in commercial rubbers to reduce "cold flow"; that is, room-temperature creep during storage [87]. In addition to linear side-chains, there are various kinds of branching, giving rise to different shapes and sizes of the molecule (Figure 3.21). In star polymers all branches emanate from a common site. The mobility of star polymers can be lower or higher than the corresponding linear chain, depending on the length of the arms (Figure 3.22 [69,88]. The viscosity and diffusion coefficient of stars depend more strongly on the molecular weight of the branch, M_{Br}, than on the number of branches [103,107]. Dendrimers[8] have regular, tree-like branches issuing from a common core [89,90]. Inspired by biomolecules, their uniform structure, functional surface, and internal cavities suggest potential as a carrier material for drug administration [91–93]. The crowding at the molecular periphery of dendrimers precludes entanglements, leading to Rouse-like behavior, even at molecular weights as high as 10^5 g/mol [94]. Poly(macromonomers) have a similar structure, with a high density of branches attached to a central chain backbone [95]. Arborescent[9] chains are copolymers in which one of the monomer types also initiates polymerization, leading to a random, hyperbranched structure [96,97]. Ring polymers [98,99] and microgels [100–102] lack intermolecular

[8] *Dendron* is the Greek word for tree.
[9] *Arbor* is the Latin word for tree.

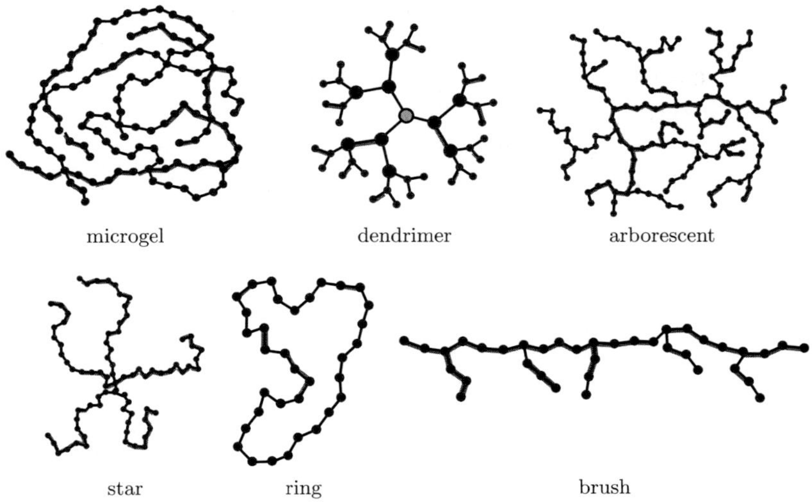

Figure 3.21 Branched architectures: *microgel* formed by simultaneous polymerization and crosslinking of a dispersed phase of monomers; third-generation *dendrimer* formed by cascade synthesis to yield successive tiers; *arborescent* branched structure, similar to dendrimers but lacking their regularity; *star polymer*, in which each arm emanates from a common locus and terminates as a dangling end; *ring* or cyclic polymer; and *brush* or *comb*, in which an assembly of chains are tethered at one end to a linear backbone polymer (as shown) or to a substrate. These pendant chains become extended as the packing density increases, up to as much as one branch per repeat unit.

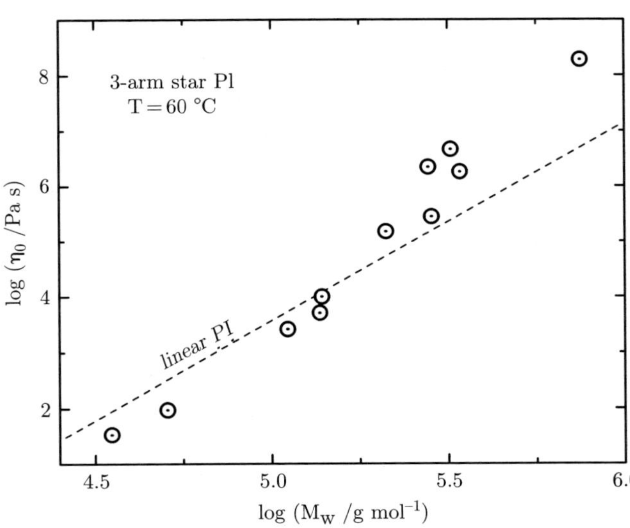

Figure 3.22 Zero-shear rate viscosities of 3-arm star-branched 1,4-polyisoprene (circles). The line represents the behavior of linear PI: $\eta_0 = 5.26 \times 10^{-15} M_w^{3.57}$. Data from refs. [69,88].

entanglements of the type dominating the behavior of long linear chains, yet show rheological properties commonly associated with reptation.

Although for given M, branching reduces the molecular size, which would lower the viscosity, for long branches (defined as having molecular weights at least a few times M_e) this size effect is overwhelmed by the hindering of the long-range, cooperative motions, which causes a strong increase in η and D, as well as their molecular-weight dependences [103–108]. Only a few, randomly situated long branches can have a dramatic effect, for example increasing η_0 by more than an order of magnitude (Figure 3.22).

An interesting effect of branching concerns the temperature dependence of the viscosity. Early studies of polyethylene found that long chain branching significantly increased the temperature-sensitivity of η [105,109]. This is also seen in PI (Figure 3.23 [88]); however, long chain branching does not enhance the T dependence of all polymers. For example, the temperature dependence is the same for linear and branched polyisobutylene [110,111], polydimethysiloxane [112], and hydrogenated 1,2-polybutadiene [113]. Graessley [114] proposed a model to account for an effect of branching on the T dependence, based on the idea that reptation of a branched polymer requires retraction of entangled arms. Such retraction would cause a transient increase in the population of *gauche* ("bent") rotamers, relative to segments having the *trans* conformation. Thus, the prediction is that the energy difference between *gauche* and *trans* rotamers will determine the effect of branching on the temperature coefficient of the terminal relaxation time and viscosity of branched chains. The magnitude of the temperature coefficient of chain expansion

$$\kappa_r = \frac{d \ln\langle r^2\rangle}{dT} \tag{3.21}$$

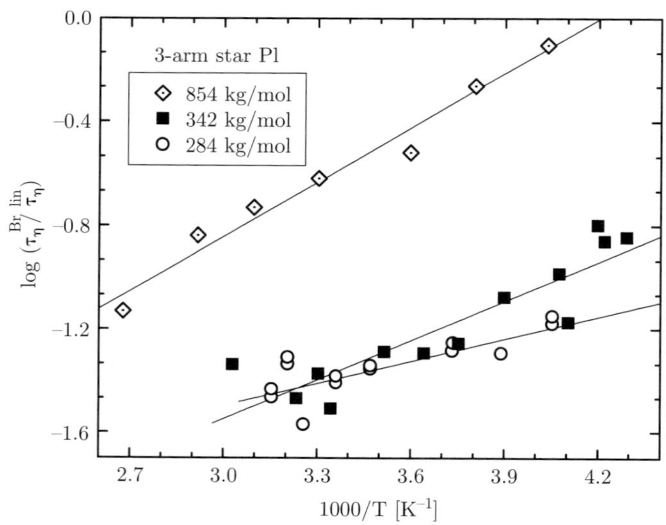

Figure 3.23 Arrhenius plots of the terminal relaxation times of 3-arm star 1,4-polyisoprenes having the indicated molecular weight normalized by τ_η for linear PI having the same M_w. The slopes yield the excess activation energy due to long branches, $\Lambda = 400$ J per mole of branch entanglements. Data from ref. [88].

should be a measure of the enhancement of the T sensitivity of branched polymers. A more negative κ_r means that the *gauche* rotamers are higher in energy, and therefore entangled branches will have a greater effect on $d\eta_0/dT$. The latter can be quantified from comparing the change in apparent activation energy of branched and linear chains. Since longer branches would confer greater enhancement, the relevant metric is the change in activation energy per branch entanglement [113,114]

$$\Lambda = \frac{E_a^{Br} - E_a^{lin}}{M_{Br}M_e^{-1}} \tag{3.22}$$

where E_a^{lin} and E_a^{Br} are the respective activation energies for terminal relaxation of linear and branched chains. Comparisons of this quantity with κ_r are given in Table 3.4. While the model has success in accounting for experimental results for polymers with large negative κ_r, materials with small or positive chain expansion coefficients behave contrary to the predictions.

3.2.4 Entanglement formation during flow

At shear rates higher than associated with Newtonian behavior, the steady-state shear viscosity, η_{ss}, of entangled polymers becomes rate dependent. This is illustrated for several PBD solutions in Figure 3.24 [115]. The minimum shear rate required for a viscosity decrease is on the order of the inverse of the terminal relaxation time. Concentrated solutions and polymer melts show similar behavior, but measurements are difficult for the latter because their greater elasticity makes high strain rate data more susceptible to artifacts, for example due to edge fracture. The shear thinning is ascribed to a loss of entanglements, resulting from flow-induced orientation of the chains [116–120]. Prior to attainment of a steady-state reflecting the extent of entanglement under flow, the stress and normal force both pass through a maximum versus time. Since the segment distribution of flexible chains is not spherically symmetric (Section 1.2.5), unentangled polymers can also exhibit stress overshoots and viscosities that decrease with shear rate [121–123]. The effects are due to orientation and are much weaker than for entangled polymers (and usually can be observed only near T_g, since at higher temperatures the strain rate necessary to induce shear thinning in the absence of entanglements is too high).

Since disentanglement is a physical effect, the viscosity reduction is reversible, provided the flow stress does not cause any chain scission. After cessation of flow, Brownian motion brings the entanglements back to their equilibrium concentration. However, if the time after the initial shearing is not long enough, the magnitude of the stress overshoots will be diminished when the flow is resumed, reflecting incomplete re-entanglement [115,124–130]. This is

Table 3.4 Effect of temperature on coil size and rheology of star branched polymers [88,111,112].

Polymer	M_{Br}/M_e	κ_r ($\times 10^3$)	Λ
hydrogenated PBD	12–39	–1.05	0.25
PVE	12	–1	0.17
hydrogenated PI	23–33	–1.3	0.165
PBD	12–13	0.39	0.11
PI	15–44	0.3	0.07
hydrogenated PVE	5–6	0.5	0
PDMS	—	0.59	0
PIB	4–9	1.1	0

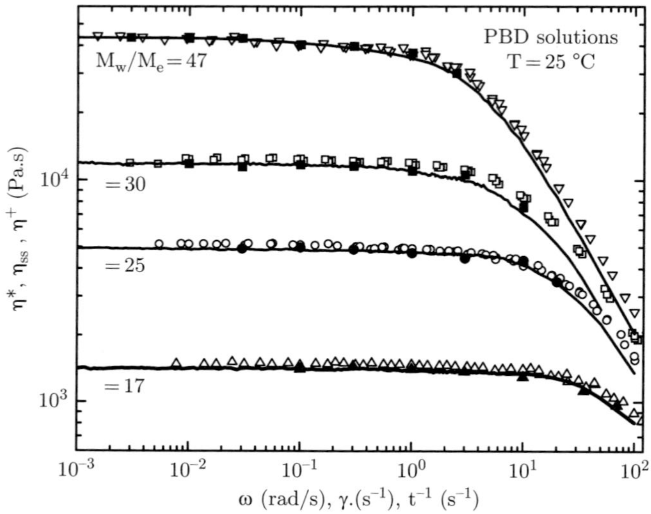

Figure 3.24 Steady-state shear viscosities (solid symbols), dynamic shear viscosities (open symbols), and the transient viscosities (line), the latter determined from the lowest shear rates using the Gleissle relation (eqn (5.46)), plotted *vs.* the shear rate, frequency, or reciprocal of the shear time. The polymer is PBD ($61 \leq M_w$ (kg/mol ≤ 167) at 0.52 volume fraction in an aromatic oil, with the number of entanglements per chain as indicated. Data from ref. [115].

illustrated in Figure 3.25 for one of the PBD solutions from Figure 3.24; the maxima in both the viscosity and first normal stress are lower upon resumption of flow, unless the intervening rest period is sufficiently long. The steady-state values remain the same, since they represent the equilibrium properties at the particular shear rate and temperature.

Figure 3.26 shows the increase of the overshoots measured during subsequent shearing after quiescent times of varying duration. Such recovery data can be fitted using exponential equations proposed by Stratton and Butcher [125] for the viscosity overshoot

$$\eta_{\max}(t) = \eta_{ss} + A_\eta\Big(1 - \exp(-t/\tau_{rec,\eta})\Big) \tag{3.23}$$

and for the overshoot of the first normal stress coefficient

$$\Psi_{\max}(t) = \Psi_1^0 + A_\Psi\Big(1 - \exp(-t/\tau_{rec,\Psi})\Big) \tag{3.24}$$

In these equations η_{ss} and Ψ_{ss} are the steady-state values, $\tau_{rec,\eta}$ and $\tau_{rec,\Psi}$ are time constants for the recovery, and the prefactors A_η and A_Ψ are constants. The dynamic modulus measured for small oscillatory strains during zero-flow recovery also increases over time, reflecting the re-entanglement; however, the timescale for this process, $\tau_{recG'}$, is significantly longer than the time constants for recovery of $\eta_{\max}$ and $\Psi_{\max}$. That is,

$$\tau_{rec,G'} > \tau_{rec,\eta} > \tau_{rec,\Psi}$$

The variation with molecular weight of these recovery times is consistent with the linear viscoelastic result, $M^{3.4}$ (Figure 3.27).

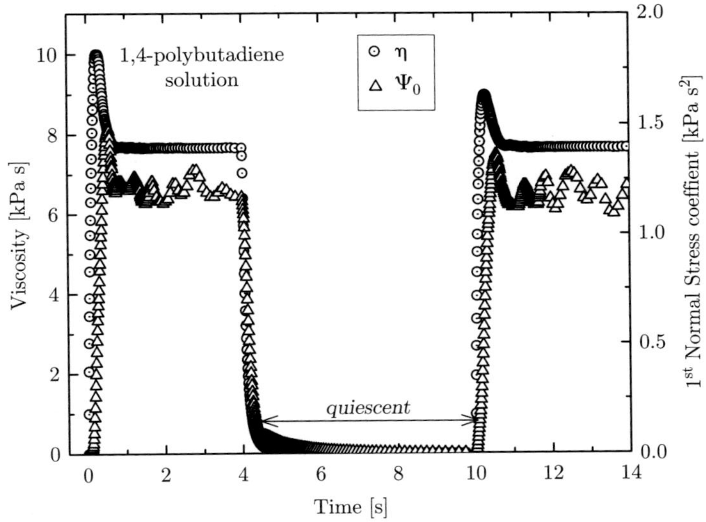

Figure 3.25 Interrupted shear flow data for an entangled PBD solution (M_w = 89.6 kg/mol; 52% polymer in oil by volume). Both the shear viscosity and the first normal stress coefficient exhibit an overshoot, the magnitude of which for subsequent flow depends on the duration of the interval between shearing. The arrows indicate the duration of the recovery periods, which varied from 1 to 30 s. The shear rate was 20 s^{-1}. Data from ref. [115].

In Table 3.3 these recovery times are compared to the linear relaxation time calculated as (*cf.* eqns (3.2) and (3.4))

$$\tau_\eta = \eta_0 J_s^0 = \frac{\sum_p g_p \tau_p^2}{\left(\sum_p g_p \tau_p\right)^2} \tag{3.25}$$

with results included for several polymers [125,126,128,129]. Especially for shear rates not much larger than the linear relaxation time, the stress overshoot recovery time can be more than 50-fold longer than τ_η, with $\tau_{rec,G'}$ even larger. Both quantities represent the time required for re-entanglement after non-linear shearing, whereas the linear relaxation time is a measure of the time for an entangled chain to diffuse by Brownian motion a distance roughly equal to its coil size. The expectation is that disentangled chains would exhibit faster motion; however, the recovery timescale involves an approach to the fully entangled state. This is a stochastic process, which evidently is not complete prior to motion over length scales significantly greater than the molecular size. Moreover, compared to the low strain rate regime of linear viscoelasticity, large shear rates effect structural changes; thus, additional mechanisms may be required to describe the ensuing relaxation process. The retarded nature of the recovery process can be exploited to obtain more processible materials. Disentangling flow produces a transient "shear-modified" material, having both lower viscosity and lower elasticity. This can be exploited to reduce the die (or extrudate) swell during polymer processing [131–139]. Freeze-drying of dilute polymer solutions has also been investigated as a route to achieving a disentangled state [140,141].

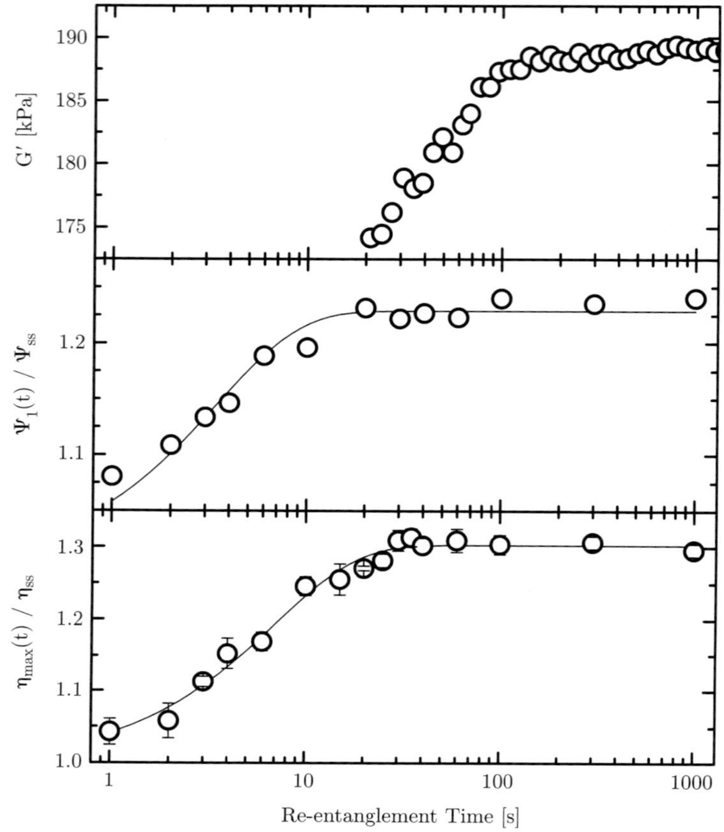

Figure 3.26 (bottom) Overshoot in transient viscosity, normalized by the steady-state viscosity, as a function of the quiescent time between shear flow. Line is fit eqn (3.23). (middle) Same plot for overshoot of first normal stress coefficient, normalized by its steady-state value. Line is fit to eqn (3.24). (top) Change in the storage modulus after cessation of shear flow. Data from ref. [127].

3.3 Practical aspects of diffusion

Intermolecular diffusion is an integral aspect of many industrial operations. Here, we consider two processes that rely on polymer diffusion: polymer dissolution by solvents and the fusion of macroscopic quantities of polymer.

3.3.1 Polymer dissolution

The dissolving of a polymer to form a solution is relevant to many technologies, including microlithography (pattern formation by removal of material from a resist [142]), phase inversion in membrane science (solvent/non-solvent exchange is used to produce microfilters or controlled-release platforms [143–145]), recycling of plastics [146,147], controlled release in which swelling of a polymer enhances diffusion of the solute [148], and even tissue regeneration (in which the polymer scaffold dissolves away as living cells grow and assemble [149,150]).

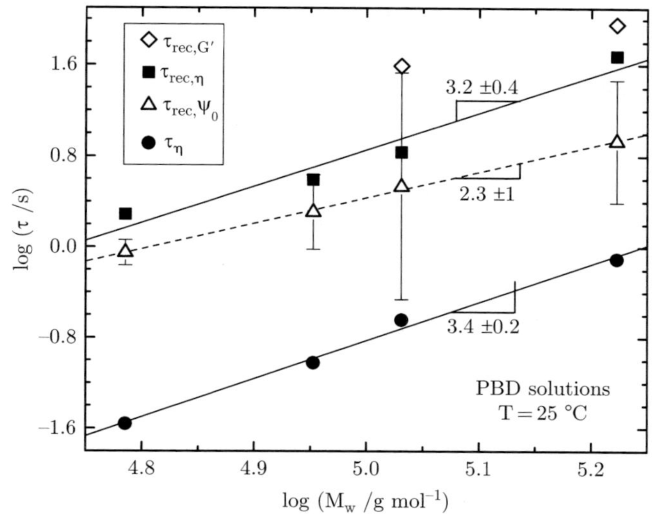

Figure 3.27 Linear relaxation time from eqn (3.25) (filled circles), recovery time of the overshoot in the first normal stress (open triangles), recovery time of the overshoot in the transient shear stress (filled squares), and time for the plateau modulus to return to its equilibrium value (open diamonds) of the PBD solutions in Figure 3.24. Within the experimental error, all timescales have the same M dependence as η_0 for the neat PBD. Data from refs. [115,127].

Whereas amorphous materials composed of small molecules can dissolve virtually instantaneously, governed only by the rapidity of solvent and solute diffusion, for polymers there is the additional requirement that the dissolving chains disentangle. For glassy polymers this can be a complex process, involving formation of an interfacial zone of swollen polymer between the solvent and solid polymer [151–153]. If a glassy polymer is immersed in solvent in the form of particles so small that disengagement of chains from the gel and their diffusion through the interfacial zone are fast, dissolution is determined by the diffusion constant of the polymer; that is, the dissolution time equals the terminal (reptation) time. This means that while higher

Table 3.3 Recovery times (s).

polymer	$c^{\#}$	M_w (kg/mol)	M_w/M_n	$M_w c/M_e$ *	τ_η (s)	Shear rate (s^{-1})	$\tau_{rec,\eta}$ (s)	$\tau_{rec,\eta}/\tau_\eta$	ref.
PBD	0.52	61.0	1.03	17	0.0274	$2/\tau_\eta$	1.95	71.1	115,127
PBD	0.52	89.6	1.04	25	0.0950	$2/\tau_\eta$	3.89	40.9	115,127
PBD	0.52	107.4	1.03	30	0.228	$2/\tau_\eta$	6.92	30.4	115,127
PBD	0.52	166.9	1.04	47	0.790	$2/\tau_\eta$	47.9	60.6	115,127
HDPE	neat	185	6.7	161	31	$3/\tau_\eta$	385	13	128
PS	0.06	8420	1.2	28	24.1	$7/\tau_\eta$	20	0.83	129
PIB	0.03	2800	broad	9	48.3	$10/\tau_\eta$	63	1.3	125
PBD	0.08	350	<1.05	15	1.07	$16/\tau_\eta$	0.95	0.89	126

#polymer volume fraction; *entanglements per chain

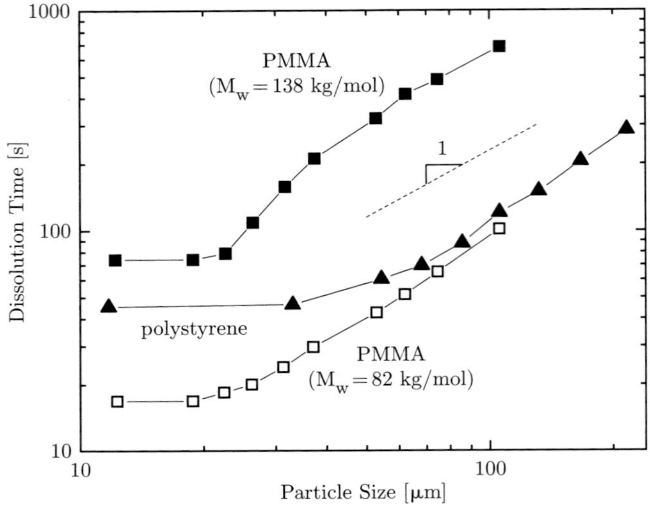

Figure 3.28 Time required to dissolve polystyrene in cyclohexane at 35 °C (triangles) and polymethylmethacrylate in benzene at 30 °C (squares) as a function of the size of the particles. The dashed line indicates the proportionality of the dissolution rate to particle size for large particles; below a certain size, the dissolution rate becomes constant as chain disentanglement becomes the rate-determining step. The stirring speed, which accelerates dissolution, was 230 rpm for the PS and 150 rpm for the PMMA. Data from ref. [154].

molecular weight polymers will dissolve much slower, the dissolution rate becomes independent of the particle size (Figure 3.28 [154]). Thus, there is no advantage to expending effort to reduce the particle size below a certain limit. This critical size, which depends on polymer, solvent, and stirring conditions, represents a competition between particle swelling and release of the chains by diffusion. An approximate relation is that the critical particle size varies as the square root of the product of the terminal relaxation time and the diffusion constant for swelling of the polymer by the solvent [155]. For large glassy particles, the dissolution time increases proportionally to the particle radius (Figure 3.28). It also varies roughly with M, which can lead to a systematic increase in the molecular weight of the undissolved material [156].

3.3.2 Adhesion and coalescence

Chain diffusion is an important mechanism for melding of separated quantities of polymer, a procedure common to many technologies. Chain interpenetration often governs the adhesion of laminates and the healing of fracture surfaces. The size of the dispersed domains in a phase-separated polymer blend is affected by their coalescence, which requires polymer diffusion. Polymerization reactions are sometimes controlled by the rate of diffusion of reactants at the end of polymer chains. Manufacture of latex products, formed by dipping a form into a fluid suspension of rubber particles, relies on coacervation and melding of the particles prior to vulcanization; the extent of chain interdiffusion governs particle coalescence and thus the barrier properties of latex products such as gloves and condoms [157].

When two surfaces of a material are brought into mutual, intimate contact, chains will interdiffuse between the layers. The wetting of the surfaces depends on the surface roughness, material viscosity, etc., and can be enhanced by larger contact pressures [158,159]. The subsequent interdiffusion requires a chain end to be oriented towards the interface; however, the configurational transitions to bring this about are much faster than chain diffusion, making the latter the rate-determining step under the usual experimental conditions. After a time period sufficient to allow diffusion over distances equal to the coil size, the interfacial region assumes the microscopic structure of the bulk; that is, the laminate becomes a homogeneous sheet. This interdigitation has the usual behavior for Brownian motion, with the timescale for disappearance of the interface expected to vary as the ratio of the chain coil size to its diffusion constant; i.e. proportional to the square-root of time. For linear chains the reptation prediction is $\tau_d \sim M^3$, although experimentally the situation is confused [160,161].

From eqn (3.9) a linear chain of typical molecular weight has a coil size in the range 10 to 20 nm. At room temperature rubbers have self-diffusion constants on the order of 0.4 to 4 nm^2/s [162–164]. From these values and using eqn (1.46), cohesion of the interfaces will require time periods from seconds to several minutes, depending on the rubber. This is in accord with experimental data, shown in Figure 3.29 for several rubbers [161–164]. In some cases the resistance to separation is invariant to contact time because interdiffusion is rapid compared to the timescale of the test. For other materials the autoadhesive force builds up over time.

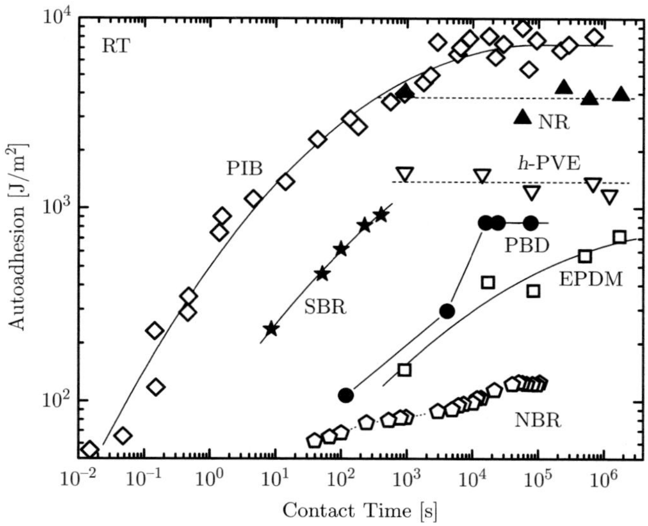

Figure 3.29 Energy to separate sheets of rubber from itself as a function of their contact time. The increase in tack over time reflects progressive diffusion across the interface. Natural rubber (filled triangles) and hydrogenated polyvinylethylene (open inverted triangles) attain a value of peel adhesion equal to their cohesive strength prior to the first measurement. For polyisobutylene (open diamonds; these data are a master curve), 1,4-polybutadiene (filled circles), and acrylonitrile-butadiene copolymer (open pentagons), a constant peel adhesion is observed at longer contact times. For styrene-butadiene copolymer (filled stars) and ethylene-propylene-diene terpolymer (open squares), the longest contact times are insufficient for complete interdiffusion. For these measurements different peel rates were used, which affects the absolute level of adhesion; the particular molecular weight of the polymers used for these tests affects both the adhesion energy and the kinetics. The ordinate values for SBR and NBR were shifted for clarity. Data from refs. [161–164].

During this interval, the interfacial zone is smaller than the coil size, and the mechanics of debonding can be complicated; for example, exhibiting a large dependence on separation rate [165,166]. The ability of rubbery materials to rapidly autoadhere is a necessary condition for "tack," the strong adhesion of an uncured rubber to itself. Tack is important in the assembly of the components of elastomer products, notably tires. In addition to forming a cohered layer by interdiffusion, the interface must be strong enough to resist subsequent failure by cohesive tearing. For this reason natural rubber has superior tack, due to a combination of rapid bond formation and high cohesive strength. The latter is a result of the capacity of NR for strain-induced crystallization. Without such mechanisms, cohesive strength in rubber requires high molecular weight, which sacrifices rapid bond formation.

When the two layers are not the same polymer, spontaneous interdiffusion occurs only if the pair is thermodynamically miscible (Chapter 7). This can serve as a test of miscibility, useful, for example, when the glass-transition temperatures are too close to be used to assess the phase morphology. Figure 3.30 [167] illustrates the strong adhesion developing due to mutual interdiffusion of two thermodynamically miscible polymers. When coalescence of the surfaces is essentially complete, the mode of failure changes from adhesive (delamination) to cohesive (tearing with substantial surface disruption). When two polymers have a negative mixing enthalpy, their diffusion constants will have a weaker M dependence, since the enthalpic interaction, which adds to the normal entropic driving force, is proportional to the number of repeat units [168].

Strength development at the interface of "welded" polymers is of industrial interest. For rubbers this is equivalent to the problem of tack (Figure 3.31). However, for plastics ($T_g >$ room temperature), the situation is more complex, because often complete interdiffusion ("healing") of the interface is never achieved. Analysis of the kinetics of adhesion development of plastics requires understanding how the bond strength is related to the extent

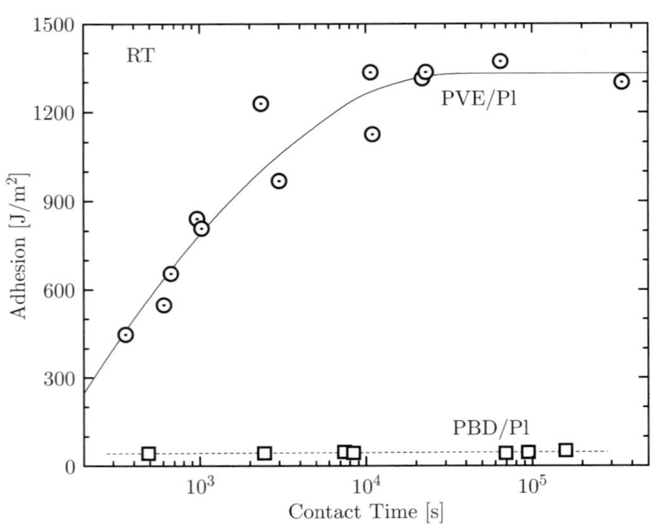

Figure 3.30 Peel adhesion as a function of contact time for two miscible polymers (circles; PVE and PI) and two immiscible one (squares; PBD and PI). For the former, the mode of failure changes from adhesive to cohesive as the interfacial region approaches the size of the chain coils. Data from ref. [167].

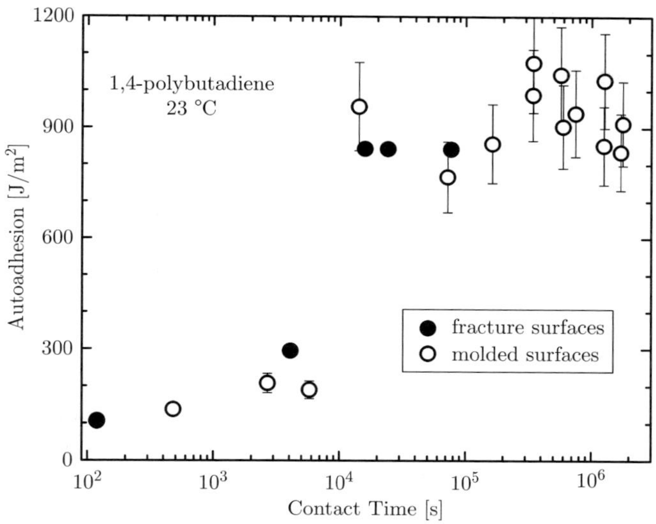

Figure 3.31 Peel energy of PBD sheets plied together for 300 s at 12 kPa pressure and then left in contact for the indicated time period. The initial sheets were molded at room temperature to have a flat surface (filled symbols) or produced by fracturing a single sheet (open symbols). The kinetics of the development of interfacial strength are equivalent for the two cases. Data from ref. [163].

of chain bridging. Similar issues arise when rubbery materials are delaminated at high rates [165]. Surprisingly, experimental data often show a debonding energy for polymers below T_g that follows the same $\sqrt{t}$ dependence of rubbery ($T > T_g$) chain diffusion [164,169,170], although this depends on the failure mechanism [171]. Various models have been developed to predict the interfacial strength as a function of the number of bridging chains and the extent of their penetration [164,169,172–174].

During mixing of immiscible polymers, strong flows[10] are used to reduce the size of the phase-separated domains. Eventually the phase morphology becomes constant, reflecting a balance between breakup of the dispersed domains and their flow-induced coalescence. The latter results from the collision and fusion of particles to form larger coalesced species; this process is depicted in Figure 3.32. Although simple liquid particles readily coalescence during flow [175,176], the brevity of a "collision" when the shear rate, $\dot{\gamma}$, is large (particle contact time $\sim \dot{\gamma}^{-1} < 1$ s) might be expected to limit fusion of polymeric particles. Nevertheless, during mixing of polymers, coalescence of the dispersed phase has been found to be rapid and extensive. For example, very brief mixing of a blend of PBD in high molecular weight polychloroprene led to coalescence of virtually all the PBD particles [177]. The particle-size distribution, however, did not change, once a steady-state condition, representing the dynamic competition between flow-induced fracture and coalescence of particles, was achieved.

[10] The ability of a flow field to disperse (breakup) suspended particles or phases is determined by the amount of stretching relative to its vorticity. For dispersive mixing pure straining flow is the strongest, rotational flow the weakest, and shear flows are intermediate in effectiveness.

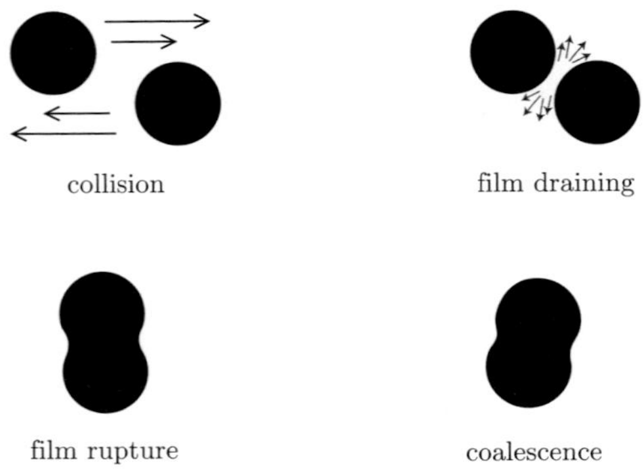

Figure 3.32 Depiction of shear-induced coalescence of dispersed fluid droplets. A collision leads to fusion if the contact time is long enough for removal of the intervening fluid layer and for sufficient interdiffusion that adhesion resists separation by the flow field.

3.4.5 Gelation

When rubbers are crosslinked, the degree of branching increases continually, eventually leading to formation of a network. Networks can also be formed from the reaction of small molecules or oligomers, although the obtained network chains typically have much lower molecular weights. The "gel point" (percolation threshold) refers to the extent of the reaction associated with formation of a branched structure having macroscopic dimensions; prior to the gel point, the sample is entirely soluble. Since gelation is a transition from a liquid to solid, existence of a gel is associated with an infinite shear viscosity and a non-zero equilibrium modulus. However, the incipient gel is mechanically very weak and these characteristic values, $\eta = \infty$ and $G_{t=\infty} > 0$, are only approached in the limit; thus, determinations of the gel point using these criteria rely on uncertain extrapolations (Figure 3.33 [178]).

If the network is formed from small molecules or unentangled chains, for which the elastic modulus is negligible, the increase in the storage modulus can be used to follow the development of a gel. Assuming the strands of the gel conform to classical rubber elasticity (eqn (4.33)), percolation theory leads to an expression for the equilibrium modulus in terms of the fraction of gelled polymer, f_{gel} [179]

$$\lim_{\omega \to 0} G' \sim f_{gel}^2 / M_{sol} \tag{3.26}$$

where M_{sol} is the molecular weight of the remaining, soluble polymer. The behavior is illustrated for a polyester synthesized by a condensation reaction in Figure 3.34 [179,180]. Dielectric spectroscopy can be used if the reactants are small molecules, since initially the molecular motions are all non-cooperative, and there is only one peak in the spectrum. The formation of macromolecular structures gives rise to a splitting of the single dielectric peak into a slow α-relaxation and a higher-frequency JG secondary process (Figure 3.35 [181]). The splitting occurs prior to the gel point, upon formation of large structures having significant degrees of freedom.

For traditional crosslinking of rubber, in which a network is formed from high polymer, eqn (3.26) does not apply, since the precursor entangled chains contribute mechanical stress. To a good approximation, however, the modulus is proportional to the gel fraction (Figure 3.36

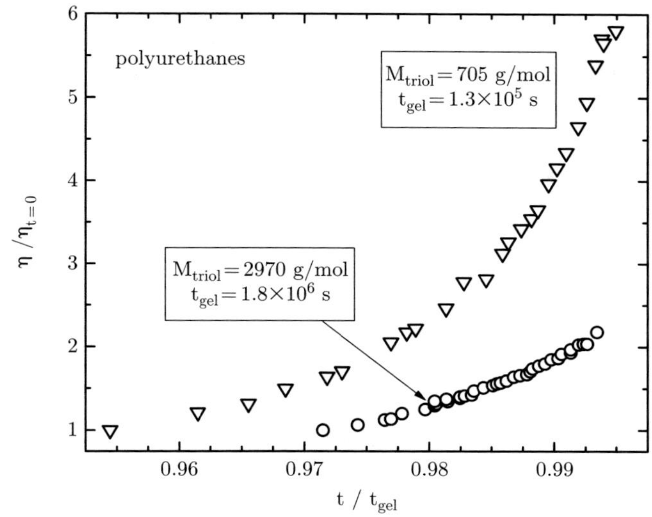

Figure 3.33 Viscosity normalized by the initially measured value during approach to the gel point for two polyurethanes formed by reaction of stoichiometric amounts of polyoxypropylene triol and an isocyanate. M_n of the precursor triols are as indicated. At $t = t_{gel}$, η would become infinite. Data from ref. [178].

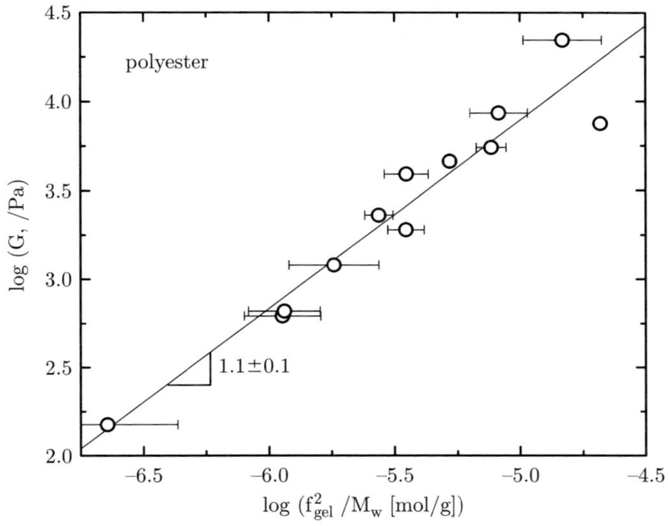

Figure 3.34 Low-frequency storage modulus as a function of the ratio of the gel fraction to the molecular weight of the soluble fraction, for polyester formed by condensation reaction. The line has a slope of two, consistent with eqn (3.26). Data from ref. [180].

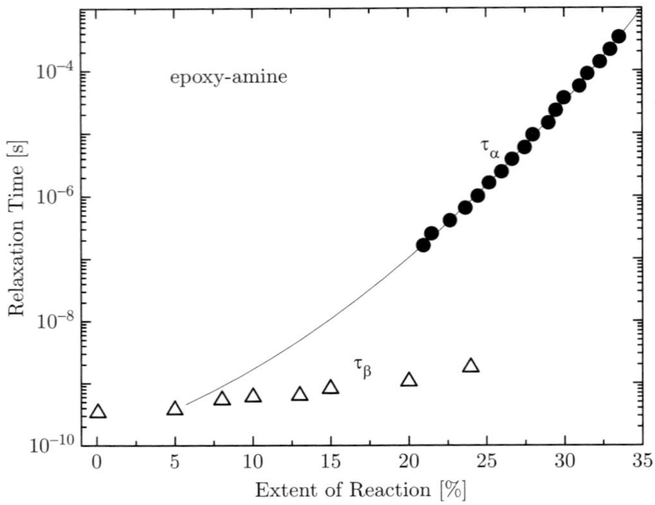

Figure 3.35 Structural relaxation times (filled circles) and JG secondary relaxation times (open triangles) measured during the reaction of stoichiometric amounts of diglycidyl ether of bisphenol A with diethylenetetriamine. Data from ref. [181].

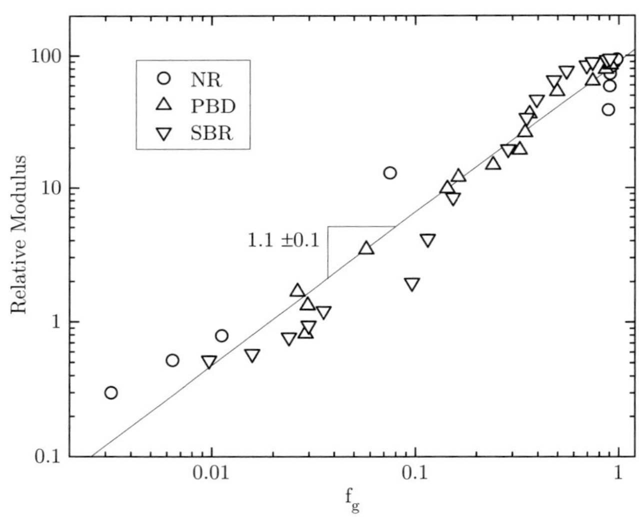

Figure 3.36 Shear modulus normalized by its long-time limiting value *versus* the gel content for three rubbers sulfur-vulcanized using N-*t*-butylbenzothiazole-2-sulfenamide (TBBS) as an accelerator. The line is the least squares fit to all points. Data from ref. [182].

[182]), although measurement of the latter is difficult and imprecise. Note that many commercial rubbers contain low levels of gel prior to vulcanization.

If crosslinking is accompanied by vitrification, as for example with epoxy resins and some polyurethanes and polyureas, the dynamic response becomes complicated, and analysis in

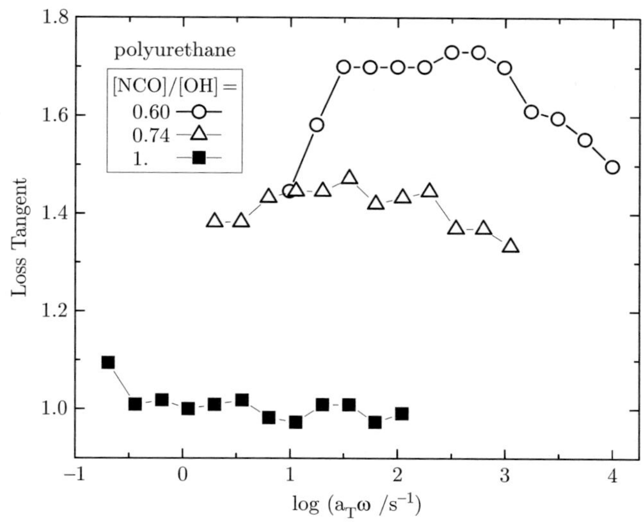

Figure 3.37 Mechanical loss tangent for polyurethanes at their gel point, with the indicated ratio of isocyanate to alcohol. Stoichiometric amount (solid squares) of the reactants yields the flattest curve. Data from ref. [184].

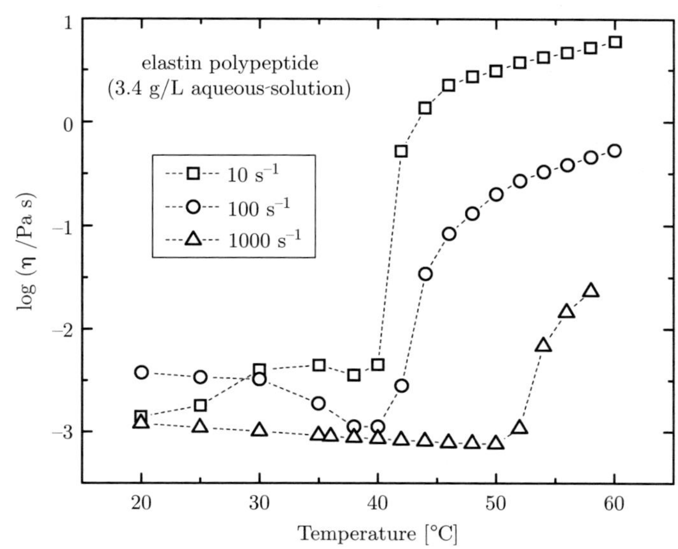

Figure 3.38 Change in the shear viscosity with temperature due to post-coacervation self-assembly of elastin peptides to form crosslinked structures. The network is partially ruptured by the flow. Shear thinning is most evident after gellation commences. Data from ref. [185].

terms of the state of the network is difficult. However, measurements can be made at temperatures above T_g (although of course this accelerates the reaction) and the mechanical response used to detect the onset of network formation. Well above T_g, the storage and loss moduli become nearly equal at the gel point [183,184]. This means that a frequency-independent loss tangent having a value near unity can be used to detect the onset of a network; however, the method is only approximate (Figure 3.37).

The self-assembly of proteins into fibrillar, elastomeric networks (Chapter 9) is another gellation process. Changes in temperature or pH cause aqueous protein solutions to phase separate (a process called "coacervation"), forming a protein-rich phase in which crosslinking occurs to yield a bioelastomer. This crosslinking can be followed rheologically, for example from the marked viscosity increase above the coascervation temperature (Figure 3.38 [185]). After gellation the dynamic properties also become essentially frequency independent. Another example of gellation in biology is the clotting of blood (Chapter 9).

References

1. P.E. Rouse, *J. Chem. Phys.* **21**, 1272 (1953).
2. B.H. Zimm, *J. Chem. Phys.* **24**, 269 (1956).
3. F. Bueche, *J. Chem. Phys.* **22**, 603 (1954).
4. J.J. Hermans, *Ann. Rev. Phys. Chem.* **8**, 179 (1957).
5. R.B. Bird, C.F. Curtiss, R.C. Armstrong, and O. Hassager, *Dynamics of Polymeric Liquids*, Vol. 2, 2nd edition, Wiley, New York (1987).
6. W. Paul and G.D. Smith, *Rep. Prog. Phys.* **67**, 1117 (2004).
7. P.J. Flory, *J. Chem. Phys.* **17**, 303 (1949).
8. R.G. Kirste, W.A. Kruse, and K. Ibel, *Polymer* **16**, 120 (1975).
9. R.W. Richards, A. Maconnachie, and G. Allen, *Polymer* **19**, 266 (1978).
10. F.S. Bates, S.B. Dierker, and G.D. Wignall, *Macromolecules* **19**, 1938 (1986).
11. P.G. de Gennes, *Macromolecules* **13**, 1069 (1980).
12. S. Onogi, T. Masuda, and K. Kitagawa, *Macromolecules* **3**, 109 (1970).
13. R.H. Colby, L.J. Fetters, and W.W. Graessley, *Macromolecules* **20**, 2226 (1987).
14. Y. Imanishi, K. Adachi, and T. Kotaka, *J. Chem. Phys.* **89**, 7585 (1988)
15. K. Adachi, H. Yoshida, F. Fukui, and T. Kotaka, *Macromolecules* **23**, 3138 (1990)
16. C. Riedel, A. Alegria, P. Tordjeman, and J. Colmenero, *Macromolecules* **42**, 8492 (2009).
17. H. Odani, N. Nemoto, and M. Kurata, *Bull. Inst. Chem. Res.* **50**, 117 (1972).
18. W.M. Prest, *J. Polym. Sci. A2* **8**, 1897 (1970).
19. D.J. Plazek and V.M. O'Rourke, *J. Polym. Sci. B2* **9**, 209 (1971).
20. N.J. Mills and A. Nevin, *J. Polym. Sci. A2* **9**, 267 (1971).
21. N. Nemoto, M. Moriwaki, H. Odani, and M. Kurata, *Macromolecules* **4**, 215 (1971).
22. N. Nemoto, H. Odani, and M. Kurata, *Macromolecules* **5**, 531 (1972).
23. J. Roovers, *Macromolecules* **8**, 1359 (1985).
24. J. Roovers and P.M. Toporowski, *Rubber Chem. Technol.* **63**, 734 (1990).
25. J. van Seggern, S. Klotz, and H.J. Cantow, *Macromolecules* **24**, 3300 (1991).

26. D.S. Pearson, L.J. Fetters, W.W. Graessley, G.V. Strate, and E. van Meerwall, *Macromolecules* **27**, 711 (1994).
27. J.S. Ham, *J. Chem. Phys.* **26**, 625 (1957).
28. M. Antonietti, J. Coutandin, and H. Sillescu, *Macromolecules* **19**, 793 (1986).
29. P.F. Green and E.J. Kramer, *Macromolecules* **19**, 1108 (1986).
30. H. Watanabe, *Macromol. Rapid Commun.* **22**, 127 (2001).
31. N. Hadjichristidis, X. Zhongde, L.J. Fetters, and J. Roovers, *J. Polym. Sci. Polym. Phys. Ed.* **20**, 743 (1982).
32. T.T. Perkins, D.E. Smith, and S. Chu, *Science* **264**, 819 (1994).
33. R.M. Robertson and D.E. Smith, *Phys. Rev. Lett.* **99,** 126001 (2007).
34. P.G. De Gennes, *J. Chem. Phys.* **55**, 572 (1971).
35. L. Leger and P.G. de Gennes, *Ann. Rev. Phys. Chem.* **33**, 49 (1981).
36. R. de L. Kronig, *J. Opt. Soc. Am.* **12**, 547 (1926); H.A. Kramers, *Atti del Congresso Internazionale dei Fisici* **2**, 545 (1927).
37. R.S. Marvin and J. Oser, *J. Res. Natl. Bur. Stand.* **66B**, 171 (1962); *ibid.* **67B**, 87 (1963).
38. V.R. Raju, E.V. Menezes, G. Marin, and W.W. Graeesley, *Macromolecules* **14**, 1668 (1981).
39. R.G. Larson, T. Sridhar, L.G. Leal, G.H. McKinley, A.E. Likhtman, and T.C.B. McLeish, *J. Rheol.* **47**, 809 (2003).
40. S. Wang, S. Q. Wang, A. Halasa, and W. L. Hsu, *Macromolecules* **36**, 5355 (2003).
41. C. Liu, J. He, R. Keunings, and C. Bailly, *Macromolecules* **39**, 3093 (2006).
42. F. Brochard and P.G. De Gennes, M*acromolecules* **10**, 1157 (1977).
43. R.H. Colby, M. Rubinstein, and J.L. Viovy, *Macromolecules* **25**, 996 (1992).
44. G. Ronca, *J. Chem. Phys.* **79**, 1031 (1983).
45. Y.-H. Lin, *Macromolecules* **20**, 3080 (1987).
46. T.A. Kavassalis and J. Noolandi, *Macromolecules* **22**, 2709 (1989).
47. L.J. Fetters, D.J. Lohse, and W.W. Graessley, *J. Polym. Sci. Polym. Phys. Ed.* **3**, 1023 (1999).
48. W.W. Graessley and S.F. Edwards, *Polymer* **22**, 1329 (1981).
49. N. Heymans, *Macromolecules* **33**, 4226 (2000).
50. L.J. Fetters, D.J. Lohse, C.A. Garcia-Francok, P. Brant and D. Richter, *Macromolecules* **35**, 10096 (2002).
51. D. Richter, B. Farago, R. Butera, L.J. Fetters, J.S. Juang, and B. Ewen, *Macromolecules* **26**, 795 (1993).
52. L.J. Fetters, D.J. Lohse, D. Richter, T.A. Witten, and A. Zirkel, *Macromolecules* **27**, 4639 (1994).
53. R. Everaers, S.K. Sukumaran, G.S. Grest, C. Svaneborg, A. Sivasubramanian, and K. Kremer, *Science* **303**, 823 (2004).
54. W.W. Graessley, *Polymeric Liquids and Networks: Dynamics and Rheology*, Taylor & Francis, London (2008).
55. L.J. Fetters, D.J. Lohse, S.T. Milner, and W.W. Graessley, *Macromolecules* **32,** 6847 (1999).
56. M. Doi and S.F. Edwards, *The Theory of Polymer Dynamics*, Oxford University Press, New York (1986).
57. T.C.B. McLeish, *Adv. Phys.* **51**, 1379 (2002).
58. H. Watanabe, *Prog. Polym. Sci.* **24**, 1253 (1999).
59. T.P. Lodge, N.A. Rotstein, and S. Prager, *Adv. Chem. Phys.* **79**, 1 (1999).

60. T.P. Lodge, *Phys. Rev. Lett.* **83**, 3218 (1999).
61. S.Q. Wang, *J. Polym. Sci. Polym. Phys. Ed.* **41**, 1589 (2003).
62. J. Klein, *Macromolecules* **19**, 105 (1986).
63. M. Daoud and P.G. deGennes, *J. Polym. Sci. Polym. Phys. Ed.* **17**, 1971 (1979).
64. M. Doi, *J. Polym. Sci. Polym. Phys. Ed.* **21**, 667 (1983).
65. S.T. Milner and T.C.B. McLeish, *Phys. Rev. Lett.* **81**, 725 (1998).
66. E.D. von Meerwall, N. Dirama, and W.L. Mattice, *Macromolecules* **40**, 3970 (2007).
67. V. N. Pokrovskii, *J. Exp. Theor. Phys.* **106**, 604 (2008).
68. C.Y. Liu, R. Keunings, and C. Bailly, *Phys. Rev. Lett.* **97**, 246001 (2006).
69. C.A. Bero and C.M. Roland, *Macromolecules* **29**, 1562 (1996).
70. C.M. Roland and C.A. Bero, *Macromolecules* **29**, 7521 (1996).
71. H. Watanabe, O. Urakaws, H. Yamada, and M.L. Yao, *Macromolecules* **29**, 755 (1996).
72. B.T. Poh, K. Adachi, and T. Kotaka, *Macromolecules* **29**, 6317 (1996).
73. M.J. Schroeder and C.M. Roland, *Macromolecules* **32**, 2000 (1999).
74. J. Colmenero, F. Alvarez, and A. Arbe, *Phys. Rev. E* **65**, 041804 (2002).
75. A. Arbe, J. Colmenero, B. Farago, M. Monkenbusch, U. Buchenau, and D. Richter, *Chem. Phys.* **292**, 295 (2003).
76. B. Frick, G. Dasseh, A. Cailliaux, and C. Alba-Simionesco, *Chem. Phys.* **292**, 311 (2003).
77. G. Tsolou, V.A. Harmandaris, and V. Mavrantzas, *J. Non-Newt. Fluid Mech.* **152**, 184 (2008).
78. M. Zamponi, A. Wischnewski, M. Monkenbusch, L. Willner, and D. Richter, P. Falus, B. Farago, and M.G. Guenza, *J. Phys. Chem. B* **112**, 16220 (2008).
79. D. Richter, R. Butera, L.J. Fetters, J.S. Huang, B. Farago, and B. Ewen, *Macromolecules* **25**, 6156 (1992).
80. M. Narkis, I.L. Hopkins, and A.V. Tobolsky, *Polym. Eng. Sci.* **10**, 66 (1970).
81. W.F. Knoff, I.L. Hopkins, and A.V. Tobolsky, *Macromolecules.* **4**, 750 (1971).
82. K.L. Ngai and D.J. Plazek, *J. Polym. Sci. Polym. Phys. Ed.* **23**, 2159 (1985); *ibid.* **24**, 619 (1986).
83. R.W. Rendell, K.L. Ngai, and G.B. McKenna, *Macromolecules* **20**, 2250 (1987).
84. P.G. Santangelo, K.L. Ngai, and C.M. Roland, *Polymer* **39**, 681 (1998).
85. R. Zorn, G.B. McKenna, L. Willner, and D. Richter, *Macromolecules* **28**, 8552 (1995).
86. C.G. Robertson and C.M. Rademacher, *Macromolecules* **37**, 10009 (2004).
87. J.L. White, *Rubber Chem. Technol.* **42**, 257 (1969).
88. P.G. Santangelo and C.M. Roland, *J. Non-Cryst. Solids* **235**, 709 (1998).
89. D.A. Tomalia, H.A Baker, J.R. Dewald, M. Hall, G. Kallos, S. Martin, J. Roeck, J. Ryder, and P. Smith, *Polym. J.* **17**, 117 (1985).
90. C.J. Hawker, *Adv. Polym. Sci.* **147**, 113 (1999).
91. B. Klajnert and M. Bryszewska, *Acta Biochim. Polon.* **48**, 199 (2001).
92. V. Gajbhiye, V.K. Palanirajan, R.K. Tekade, and N.K. Jain, *J. Pharm. Pharmacol.* **61**, 989 (2009).
93. G.V. Nazarov, S.E. Galan, E.V. Nazarova, N.N. Karkishchenko, M.M. Muradov, and V.A. Stepanov, *Pharm. Chem. J.* **43**, 163 (2009).
94. P.J. Farrington, C.J. Hawker, J.M.J. Fréchet, and M.E. Mackay, *Macromolecules* **31**, 5043 (1998).
95. S. Namba, Y. Tsukahara, K. Kaeriyama, K. Okamoto, and M. Takahashi, *Polymer* **41**, 5165 (2000).

96. J.E. Puskas, *Polymer News* **27**, 325 (2002).
97. E.A. Foreman, J.E. Puskas, and G. Kaszas, *J. Polym. Sci. Polym. Chem. Ed.* **45**, 5847 (2007).
98. G.B. McKenna, B.J. Hostetter, N. Hadjichristidis, L.J. Fetters, and D.J. Plazek, *Macromolecules* **23**, 1834 (1989).
99. S.F. Tead, E.J. Kramer, G. Hadziloannou, M. Antonietti, H. Sillescu, P. Lutz, and C. Strazielle, *Macromolecules* **25**, 3942 (1992).
100. M. Antonietti, T. Pakula, and W. Bremser, *Macromolecules* **28**, 4227 (1995).
101. W. Funke, O. Okay, and B. Joos-Muller, *Adv. Polym. Sci.* **136**, 139 (1998).
102. C.M. Roland, P.G. Santangelo, M. Antonietti, and M. Neese, *Macromolecules* **32**, 2283 (1999).
103. V.C. Long, G.C. Berry, and L.M. Hobbs, *Polymer* **5**, 517 (1964).
104. G. Kraus and J.T. Gruver, *J. Polym. Sci. A* **3**, 105 (1965).
105. W.W. Graessley, *Acct. Chem. Res.* **10**, 332 (1977).
106. L.J. Fetters, A.D. Kiss, D.S. Pearson, G.F. Quack, and F.J. Vitus, *Macromolecules* **26**, 647 (1993).
107. G. Grest, L.J. Fetters, and J.S. Huang, *Adv. Chem. Phys.* **94**, 67 (1996).
108. K.L. Ngai and C.M. Roland, *J. Polym. Sci. Polym. Phys. Ed.* **35**, 2503 (1997).
109. G.C. Berry and T.G. Fox, *Adv. Polym. Sci.* **5**, 261 (1968).
110. C.G. Robertson, C.M. Roland, C. Paulo, and J. E. Puskas, *J. Rheol.* **45**, 759 (2001).
111. P.G. Santangelo, C.M. Roland, and J.E. Puskas, *Macromolecules* **32**, 1972 (1999).
112. C.M. Roland and P.G. Santangelo, *J. Non-Cryst. Solids* **307**, 835 (2002).
113. J.M. Carella, J.T. Gotro, and W.W. Graessley, *Macromolecules* **19**, 659 (1986).
114. W.W. Graessley, *Macromolecules* **15**, 1164 (1982).
115. C.G. Robertson, S. Warren, D.J. Plazek, and C.M. Roland, *Macromolecules* **37**, 10018 (2004).
116. W.W. Graessley, *J. Chem. Phys.* **43**, 2696 (1965).
117. G.V. Vinogradov and I.M. Belkin, *J. Polym. Sci. A* **3**, 917 (1965).
118. H. Batzer, M.H. Wagner, and J. Meissner, *Makromol. Chem.* **181**, 1533 (1980).
119. Y. Isono, T. Kamohara, A. Takano, and T. Kase, *Rheol. Acta* **35**, 245 (1997).
120. J.P. Oberhauser, L.G. Leal, and D.W. Mead, *J. Polym. Sci. Polym. Phys. Ed.* **36**, 265 (1998).
121. K. Osaki, T. Inoue, and T. Isomura, *J. Polym. Sci. Polym. Phys. Ed.* **38**, 2043 (2000); K. Osaki, T. Inoue, and T. Uematsu, *J. Polym. Sci. Polym. Phys. Ed.* **38**, 3271 (2000).
122. J.D. Moore, S.T. Cui, H.D. Cochran, and P.T. Cummings, *Phys. Rev. E* **60**, 6956 (1999).
123. P.G. Santangelo and C.M. Roland, *J. Rheol.* **45**, 583 (2001).
124. J.D. Huppler, I.F. MacDonald, E. Ashare, T.W. Spriggs, R.B. Bird, and L.A. Holmes, *J. Rheol.* **11**, 181 (1967).
125. R.A. Stratton and A.F. Butcher, *J. Polym. Sci. Polym. Phys. Ed.* **11**, 1747 (1973).
126. E.V. Menezes and W.W. Graessley, *J. Polym. Sci. Polym. Phys. Ed.* **20**, 1817 (1982).
127. C.M. Roland and C.G. Robertson, *Rubber Chem. Technol.* **79**, 267 (2006).
128. J.M. Dealy and W.K.-W. Tsang, *J. Appl. Polym. Sci.* **26**, 1149 (1981).
129. J. Sanchez-Reyes and L.A. Archer, *J. Rheol.* **47**, 469 (2003).
130. Y.Z. Xu, D. Dekee and C.F.C.M. Fong, *J. Appl. Polym. Sci.* **55**, 779 (1995).
131. M. Yamaguchi and C.G. Gogos, *Adv. Polym. Testing* **20**, 261 (2001).
132. A. Rudin and H.P. Schreiber, *Polym. Eng. Sci.* **23**, 422 (1983).

133. P.J.R. Leblans and C. Bastiaansen, *Macromolecules* **22**, 3312 (1989).
134. H.A. Pohl and C.G. Gogos, *J. Appl. Polym. Sci.* **5**, 67 (1961).
135. M. Rokudai, *J. Appl. Polym. Sci.* **23**, 463 (1979).
136. H.P. Schreiber, A. Rudin, and E.G. Bagley, *J. Appl. Polym. Sci.* **9**, 887 (1965).
137. A. Ram and L. Izrailov *J. Appl. Polym. Sci.* **31**, 85 (1986).
138. R. Schertzer, A. Rudin, and H. P. Schreiber, *J. Appl. Polym. Sci.* **31**, 809 (1986).
139. P. Attane, J.M. Pierrard, and G. Turrel, *J. Non-Newt. Fluid Mech.* **18**, 295 (1985).
140. A.M. Kotliar, R. Kumar, and R.A. Back, *J. Polym. Sci. Polym. Phys. Ed.* **28**, 1033 (1990).
141. K.J. McGrath, C.M. Roland, and R.G. Weiss, *Macromolecules* **26**, 6127 (1993).
142. J.D. Marty and M. Mauzac, *Adv. Polym. Sci.* **172**, 1 (2005).
143. R.E. Kesting and A.K. Fritzsche, *Polymeric Gas Separation Membranes* Wiley, New York (1993).
144. P.D. Graham, K.J. Brodbeck, and A.J. McHugh, *J. Control. Release* **58**, 233 (1999).
145. D. Ma and A.J. McHugh, *J. Membrane Sci.* **298**, 156 (2007).
146. M. Garcia, G. Duque, I. Gracia, A. deLucas, and J. Rodriguez, *J. Mater. Cycles Waste Manag.* **11**, 2 (2009).
147. D.S. Achilias, E. Antonakou, C. Roupakias, P. Megalokonomos, and A. Lappas, *Global Nest J.* **11**, 114 (2008).
148. Controlled-Release Technology, *Pharmaceutical Applications*, (P.I. Lee and W.R. Good, eds.,) ACS Symposium Series **348**, American Chemical Society, Washington, D.C. (1987).
149. R. Langer and J.P. Vacanti, *Science* **260**, 920 (1993).
150. L.G. Griffith and G. Naughton, *Science* **295**, 1009 (2002).
151. K. Ueberreiter in *Diffusion in Polymers*, (Crank J., Park G.S., eds.), Academic Press, New York (1968).
152. B.A. Miller-Chou and J.L. Koenig, *Prog. Polym. Sci.* **28**, 1223 (2003).
153. B. Narasimhan and N.A. Peppas, *Macromolecules* **29**, 3283 (1996).
154. I. Devotta, V.D. Ambeskar, A.B. Mandhare, and R.A. Mashelkar, *Chem. Eng. Sci.* **49**, 645 (1994).
155. F. Brochard and P.G. deGennes, *Physicochem. Hydrodyn.* **4**, 313 (1983).
156. O. Pekcan and S. Ugur, *Polymer* **43**, 1937 (2002).
157. C.M. Roland, *Rubber World* **208**, 15 (1993).
158. G.R. Hamed, *Rubber Chem. Technol.* **54**, 576 (1981).
159. G.R. Hamed and C.H. Shieh, *Rubber Chem. Technol.* **55**, 1469 (1982).
160. H.R. Brown, *Ann. Rev. Mater. Sci.* **21**, 463 (1991).
161. R.G. Stacer and H.L. Schreuder-Stacer, *Int. J. Fracture* **39**, 201 (1989).
162. J.D. Skewis, *Rubber Chem. Technol.* **39**, 217 (1966).
163. C.M. Roland and G.G.A. Bohm, *Macromolecules* **18**, 1310 (1985).
164. L. Bothe and G. Rehage, *Rubber Chem. Technol.* **55**, 1308 (1982).
165. G.R. Hamed and C.H. Shieh, *Rubber Chem Technol.* **58**, 1038 (1985).
166. M. Mikrut, J.W.M. Noordermeer, and G. Verbeek, *J. Adhes.* **85**, 413 (2009).
167. C.M. Roland, *Macromolecules* **20**, 2557 (1987).
168. P.G. deGennes, *C.R. Acad. Sci. Series II* **292**, 1505 (1981).
169. R.P. Wool, *Rubber Chem. Technol.* **57**, 307 (1984).

170. K. Jud, H.H. Kausch, and J.G. Williams, *J. Mater. Sci.* **16**, 204 (1981).
171. R.P. Wool, B.L. Yuan, and O.J. McGarel, *Polym. Eng. Sci.* **29**, 1340 (1989).
172. L.L. Berger and E.J. Kramer, *Macromolecules* **20**, 1980 (1987).
173. D. Adolf, M. Tirrell and S. Prager, *J. Polym. Sci. Polym. Phys. Ed.* **23**, 413 (1985).
174. P.G. deGennes, *C.R. Acad. Sci. Series B* **291**, 219 (1980).
175. E. Barnea and J. Mizrahi, *Trans. Inst. Chem. Eng.* **53**, 75 (1975).
176. R.S. Allen and S.G. Mason, *J. Colloid Sci.* **17**, 383 (1962)
177. C.M. Roland and G.G.A. Bohm, *J. Polym. Sci. Pt. B-Polym. Phys.* **22**, 79 (1984).
178. M. Adam, M. Delsanti, and D. Durand, *Macromolecules* **18**, 2285 (1985).
179. R.H. Colby, R.H. Gilmor, and M. Rubinstein, *Phys. Rev. E* **48**, 3712 (1993).
180. R.H. Colby, M. Rubinstein, R.H. Gilmor, and T.H. Mourey, *Macromolecules* **25**, 7180 (1992).
181. B.D. Fitz and J. Mijovic, *Polym. Adv. Technol.* **9**, 721 (1998).
182. P Zhang, F. Zhao, Y. Yuan, X. Shi, and S. Zhao, *Polymer* **51**, 257 (2010).
183. F. Chambon and H.H. Winter, *J. Rheology* **31**, 683 (1987).
184. H.H. Winter, P. Morganelli, and F. Chambon, *Macromolecules* **21**, 532 (1988).
185. J.T. Cirulis and F.W. Keeley, *J. Rheol.* **53**, 1215 (2009).

4 Networks

Rubber can be deformed to a substantial degree, which gives rise to deviations from stress–strain linearity. An elastic non-linear response (strain energy depends only on the initial and final states) is referred to as hyperelasticity. There are two approaches to analyzing the large-strain behavior of rubbery networks, phenomenological and based on statistical or molecular models. The former describe the deformation using strain-energy functions drawn from continuum mechanics, with simplifying restrictions yielding tractable constitutive relations. Molecular models use statistical mechanics to compute the entropy from consideration of the topology of the chains and its modification by strain; the intramolecular energy contribution to the elastic energy (e.g., the energy difference of *gauche* and *trans* backbone bonds) may also be included in the analysis. Early rubber elasticity models[1] ignored intermolecular interactions, whereas later developments included the effect of entanglements or steric constraints on the mechanical stress.

Although the focus of rubber elasticity theories is the chain configurations, the segmental and chain dynamics are governed by the same intramolecular and intermolecular potentials and correlations. Thus, insights into the viscoelastic response can be gleaned from analysis of equilibrium mechanical properties. These parallels between dynamics and the equilibrium behavior are brought out by the coupling model of relaxation [1,2], introduced in Chapter 2. For this reason and because of its focus on experimental results, we use the model to interpret the relaxation properties in terms of rubber elasticity concepts.

Since network properties *per se* involve primarily long length scales, they depend only weakly on the chemical structure. Hence, all elastomers exhibit essentially the same mechanical behavior up through moderate strains, whatever polymer comprises the network. This axiom implies that attempts to improve the mechanical properties will have limited success. Failure properties such as strength and fatigue life exhibit a maximum versus crosslink density Figure 4.1 [3]. In order to minimize creep, the practically useful range of crosslinking falls past this maximum; thus, elastomer failure properties vary inversely with the stiffness: Increasing crosslink densities yield higher modulus but lower strength and fatigue life. A given rubber compound represents a compromise between stiffness and strength; both properties cannot be optimized using conventional approaches. To circumvent this "fundamental" limitation

[1] Early molecular theories of rubber hyperelasticity were called kinetic theories, in correspondence to the behavior of ideal gases (the molecules of which have neither size nor mutual interactions). The elastic forces of an ideal rubber are caused by thermal motion of the network chains attempting to restore the higher entropy, isotropic state; thus, analogous to ideal gases, ideal rubber elasticity is a statistical effect. In real materials there is also an energy contribution (negative or positive) due to the difference in energy between conformer of the repeat units.

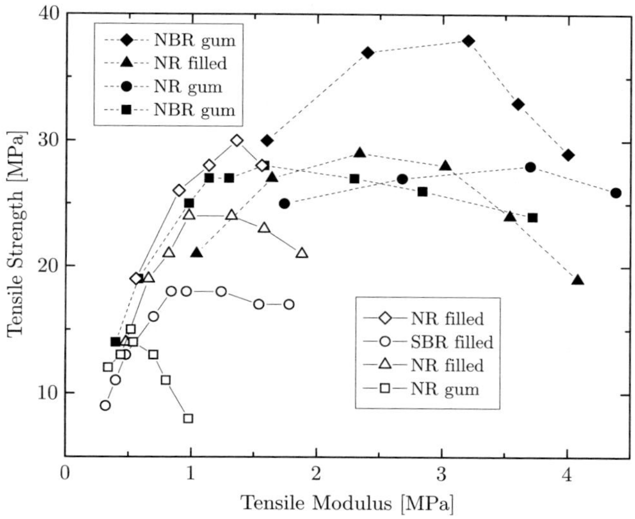

Figure 4.1 Tensile strength *versus* modulus for elastomers with covalent (hollow symbols) and ionic (solid symbols) crosslinks. Failure properties go through a maximum as a function of the degree of crosslinking. Data from ref. [3].

requires unorthodox structures or morphologies. Five examples of alternative networks—interpenetrating, double, bimodal, heterogeneous, and deswollen—are discussed at the conclusion of the chapter.

4.1 Phenomenological elasticity models

Phenomenological treatments [4–6] use mathematics to describe the mechanical potential without attempting to explain its molecular origin. The relevant potential depends on the experimental conditions: Gibbs free energy for isothermal, isobaric processes; Helmholtz free energy for isothermal, isochoric conditions; and enthalpic energy for adiabatic, isochoric conditions. The deformation is assumed to be reversible, so only the elastic strain energy, W, is derived, with the principal stresses[2] obtained as the partial derivatives of W for the relevant conditions (constant T, P, etc.). A further simplification is that the strain is uniform, without localized phenomenon such as necking. The analysis is often carried out in terms of the strain invariants, which are scalar combinations of the strain components, rather than the strains themselves. Strain invariants are independent of the axes used to define the sample geometry, enabling calculations for inhomogeneous deformations without explicit consideration of the principal directions (during a homogenous deformation, parallel lines remain parallel). Rotation of the body corresponds to a change in sign of two of the three principal stretch ratios, and in order for the strain energy to be independent of rotation, the strain

[2] Principal stresses are aligned with the coordinate system such that cross-terms are zero; that is, the stress tensor is a diagonal matrix.

invariants must be even functions of the strain. The simplest invariants are then symmetric functions of the squares of the stretch ratios[3]

$$I_1 = \lambda_1^2 + \lambda_2^2 + \lambda_3^2, I_2 = \lambda_1^2\lambda_2^2 + \lambda_1^2\lambda_3^2 + \lambda_2^2\lambda_3^2, I_3 = \lambda_1^2\lambda_2^2\lambda_3^2 \tag{4.1}$$

where the subscripts denote one of the three rectilinear coordinates. The elastic strain energy depends only on the current strain and since the body is isotropic, W is assumed to be symmetrical with the strain components; thus,

$$W = f(I_1, I_2, I_3) \tag{4.2}$$

The strain energy of any isotropic material can be expressed as a function of these three strain invariants. Since the bulk modulus of rubber is typically 3 orders of magnitude larger than the shear modulus, the assumption of incompressibility entails negligible error; hence, I_3 is taken to be unity, reducing eqn (4.2) to

$$W = f(I_1, I_2) \tag{4.3}$$

The corresponding principal stress differences are given by the derivatives of W [7]

$$\sigma_1\lambda_1 - \sigma_2\lambda_2 = 2(\lambda_1^2 - \lambda_2^2)\left(\frac{\partial W}{\partial I_1} + \lambda_3^2\frac{\partial W}{\partial I_2}\right) \tag{4.4}$$

$$\sigma_1\lambda_1 - \sigma_3\lambda_3 = 2(\lambda_1^2 - \lambda_3^2)\left(\frac{\partial W}{\partial I_1} + \lambda_2^2\frac{\partial W}{\partial I_2}\right) \tag{4.5}$$

and

$$\sigma_2\lambda_2 - \sigma_3\lambda_3 = 2(\lambda_2^2 - \lambda_3^2)\left(\frac{\partial W}{\partial I_1} + \lambda_1^2\frac{\partial W}{\partial I_2}\right) \tag{4.6}$$

Or in terms of the true stresses and derivatives with respect to the stretch ratios

$$\hat{\sigma}_1 - \hat{\sigma}_2 = \lambda_1\frac{\partial W}{\partial \lambda_1} - \lambda_2\frac{\partial W}{\partial \lambda_2}, \ldots \tag{4.7}$$

From these equations experimental stresses can be calculated for any given mode of strain. The relevant expressions are listed in Table 4.1 for the primary strain modes. The problem is to identify the form of the function f in eqn (4.3).

A general expression for the strain energy that goes to zero in the absence of strain is

$$W = \sum_{i=0,j=0}^{\infty} C_{ij}(I_1 - 3)^i(I_2 - 3)^j \quad (C_{00} = 0) \tag{4.8}$$

[3] There are various measures of strain, or relative displacement of points in a material. For normal strains the stretch (or extension) ratio, λ, which is the ratio of the final and initial dimensions parallel to the load, equals the engineering (or Cauchy) strain plus one. The true (or Hencky) strain is the natural logarithm of the stretch ratio, and is equal to the engineering strain for infinitesimal increments. Confusingly, while the Cauchy strain is the engineering strain, the Cauchy stress is the true stress, not the engineering stress. Normal strains act perpendicular to a body to stretch or compress it, whereas shear strains act parallel to a surface, inducing distortion.

Table 4.1 Stretch–stress relations for an incompressible material.

Deformation mode	Principal stretch ratios	Engineering stress	Equivalent strain
Simple shear	na	$\sigma = 2\left(\frac{\partial W}{\partial I_1} + \frac{\partial W}{\partial I_2}\right)(\lambda - \lambda^{-1})$	Tension plus rotation
Pure shear	$\lambda_1 = \lambda$ $\lambda_2 = 1$ $\lambda_3 = \lambda^{-1}$	$\sigma_1 = 2\left(\frac{\partial W}{\partial I_1} + \frac{\partial W}{\partial I_2}\right)(\lambda - \lambda^{-3})$	Plane strain compression
Uniaxial strain	$\lambda_1 = \lambda$ $\lambda_2 = \lambda_3 = \lambda^{-1/2}$	$\sigma = 2\left(\frac{\partial W}{\partial I_1} + \lambda^{-1}\frac{\partial W}{\partial I_2}\right)(\lambda - \lambda^{-2})$	Tension or compression
Biaxial strain	for equal biaxial: $\lambda_1 = \lambda_2 = \lambda$ $\lambda_3 = \lambda^{-1/2}$	$\sigma_1 = 2\left(\frac{\partial W}{\partial I_1} + \frac{\partial W}{\partial I_2}\lambda_2^2\right)(\lambda_1 - \lambda_1^3\lambda_2^2)$ $\sigma_2 = 2\left(\frac{\partial W}{\partial I_1} + \frac{\partial W}{\partial I_2}\lambda_1^2\right)(\lambda_2 - \lambda_1^2\lambda_2^3)$	Inflation

The first term in this series, $i = 1$ and $j = 0$, is identical to the classical rubber elasticity result (eqn (4.32) below). The other linear term, $i = 0$ and $j = 1$, cannot alone describe actual data. Summing both these terms yields the well-known Mooney–Rivlin equation [4,5]

$$W = C_{10}(I_1 - 3) + C_{01}(I_2 - 3) \tag{4.9}$$

The linearity in the invariants simplifies application to experimental measurements. For uniaxial strain the engineering stress along the stretch direction becomes

$$\sigma = (C_{10} - C_{01}/\lambda)(\lambda - \lambda^{-2}) \tag{4.10}$$

the other two principal stresses being zero. For shear strains the Mooney–Rivlin stress is

$$\sigma = 2(C_{10} - C_{01})\gamma \tag{4.11}$$

In Figure 4.2 data measured for an NR network subjected to tension and compression are plotted in the Mooney–Rivlin form of the reduced stress, $\sigma/(\lambda - \lambda^{-2})$, versus the reciprocal of the stretch ratio. Equation (4.10) predicts a straight line; however, even neglecting the upturn at high extensions, the slopes for tension and compression differ significantly. This is well known—different modes of deformation yield different values of the Mooney–Rivlin elastic constants; thus, fitting eqn (4.9) to the tension data yields a different set of elastic constants than fitting to the compression results. An expedient is to set $C_{01} = 0$ for shear while using it as an adjustable parameter for the tensile data. While this improves the fit, it lacks any basis. The Mooney–Rivlin equation is not a valid constitutive equation and cannot be used to predict stresses for other modes from measurements in a given mode (usually tension). The Mooney–Rivlin equation also fails at higher strains, at which there is an upturn in the stress as the network chains approach their ultimate deformability (due to non-Gaussian effects discussed below). This upturn is seen Figure 4.2.

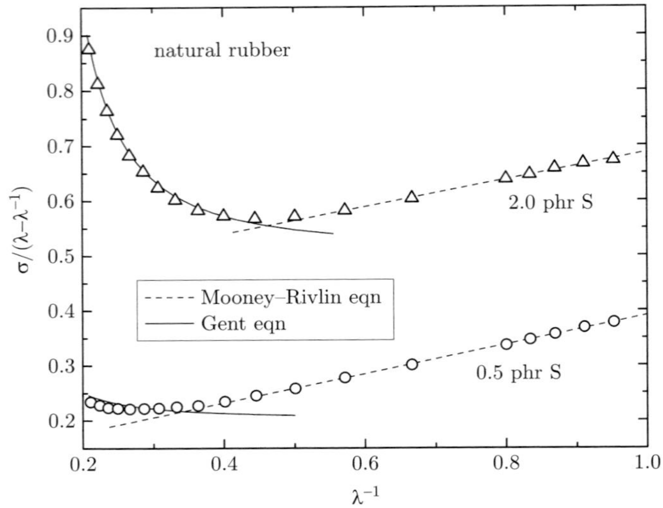

Figure 4.2 Mooney–Rivlin plot of reduced engineering stress *versus* reciprocal stretch ratio for NR networks with two degrees of crosslinking. The dashed line is a linear fit to the higher strains; that is, eqn (4.10) with $C_{10} = 0.063$ and.132 MPa for 0.5 and 2.0 phr sulfur levels, respectively. The solid line is the fit to eqn (4.26) using $C_{10} = 0.103$ MPa and $I_m = 128$ for the lower crosslink density, and $C_{10} = 0.263$ MPa and $I_m = 52$ for the higher. Data from ref. [25].

An obvious extension, found in finite analysis software, is the addition of more terms in eqn (4.8). Two popular forms for fitting experimental results for various modes of strain are a three-term expression [8]

$$W = C_{10}(I_1 - 3) + C_{01}(I_2 - 3) - C_{02}(I_2 - 3)^2 \tag{4.12}$$

or the higher-order cubic equation [9,10]

$$W = C_{10}(I_1 - 3) + C_{01}(I_2 - 3) + C_{20}(I_1 - 3)^2 + C_{11}(I_1 - 3)(I_2 - 3) + C_{30}(I_1 - 3)^3 \tag{4.13}$$

Vahapoglu and Karadeniz [11] provide an extensive bibliography of various efforts to approximate eqn (4.8) to obtain useful but not overly cumbersome expressions for the strain energy. While higher-order terms improve the accuracy of fits to experimental data, their extrapolation can lead to larger errors than the simpler Mooney–Rivlin equation.

Ogden [12] proposed an alternative form in terms of the stretch ratios rather than the invariants, which also includes the odd terms omitted in the Mooney–Rivlin formulation

$$W = \sum_i^M \frac{a_i}{b_i}(\lambda_1^{b_i} + \lambda_2^{b_i} + \lambda_3^{b_i} - 3) \tag{4.14}$$

The exponents b_i are not necessarily integers or even positive; rather, their values are chosen to best describe the mechanical response of a particular material. The Ogden equation simplifies the analysis since, unlike eqn (4.8), all terms have the same form. From eqn (4.7) the principal stress differences are

$$\hat{\sigma}_1 - \hat{\sigma}_2 = \sum_i^M a_i(\lambda_1^{b_i} - \lambda_2^{b_i}) \tag{4.15}$$

which for uniaxial strain gives

$$\sigma = \sum_i^M a_i(\lambda^{b_i-1} - \lambda^{-\frac{b_i}{2}-1}) \tag{4.16}$$

and for shear strain

$$\sigma = \sum_i^M a_i(\lambda^{b_i-1} - \lambda^{-b_i-1}) \tag{4.17}$$

The number of terms, M, depends on the span of the experimental data and the desired accuracy of the fitting. For a general description, three terms involving six adjustable parameters are usually necessary, although for finite-element modeling involving only two modes, such as axial and torsional deformations, $M = 2$ may suffice [12–14].

There are numerous variations on eqn (4.8) that invoke *ad hoc* forms for the functional dependence of the strain energy. Valanis and Landel [15], extending the earlier work of Carmichael and Holdaway [16], expressed W as the sum of three terms, one for each of the three principal strains, all having the same functional form

$$W = w(\lambda_1) + w(\lambda_2) + w(\lambda_3) \tag{4.18}$$

This separability is consistent with statistical analyses of Gaussian chains that assume fluctuations in orthogonal directions are independent. Different functions have been used for the w [17,18], and with certain restrictions eqn (4.3) can be expressed in the Valanis–Landel form [8,19]. Derivatives of eqn (4.18) yield the principal stresses

$$\hat{\sigma}_1 - \hat{\sigma}_2 = \lambda_1 \frac{dw}{d\lambda_1} - \lambda_2 \frac{dw}{d\lambda_2}, \ldots \tag{4.19}$$

which makes clear the advantage of the separability of the Valanis–Landel form. For example, if experiments are carried out at constant λ_2, such as pure shear ($\lambda_1 = \lambda_3^{-1}; \lambda_2 = 0$), the variation of w with λ_1 is obtained; thus, specification of the Valanis–Landel equation requires only measurements of a single function of just one variable. Figure 4.3 shows stresses for a NR network measured at various λ_1 for different, fixed constant values of λ_2; that is, for pure shear and various biaxial deformations ($\sigma_1 \geq \sigma_2$; $\sigma_3 = 0$). According to eqn (4.19), the principal stresses for each strain mode differ only by a constant, and accordingly, when the quantity $(dw/d\lambda_2)\lambda_2$ is added to the experimental stresses, the data collapse to a single curve. Since the Ogden equation has the Valanis–Landel form, it can describe the data in Figure 4.3, as shown by the fit of eqn (4.15) using

$$\lambda \frac{dw}{d\lambda} - c_{\lambda_2} = 0.69(\lambda^{1.3} - 1) + 0.01(\lambda^{4.0} - 1) - 0.0122(\lambda^{-2.0} - 1) \tag{4.20}$$

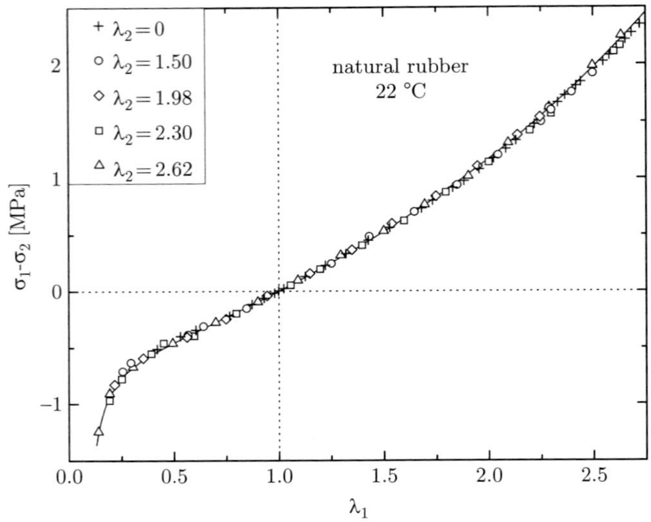

Figure 4.3 Stresses measured for an NR network as a function of λ_1. The biaxial strain data for each value of λ_2 were superposed onto the pure shear results ($\lambda_2 = 0$) by addition of a constant. The solid line is the fit of the Ogden equation (eqn (4.20)).

with the constant c_{λ_2}, reflecting the contribution of the λ_2 term, adjusted to superpose the curves.

The above constitutive relations poorly describe stresses over the entire range from low strains up through deformations at which finite extensibility of the chains becomes significant [20]. Yeoh [21] accounted for the high strain elastic behavior by assuming that $\partial W/\partial I_2 = 0$. While not strictly correct, $\partial W/\partial I_2$ is small for filled rubbers [22,23], and the upturn in the stress occurs at lower strains than in the absence of filler. With this assumption eqn (4.13) becomes [21]

$$W = C_{10}(I_1 - 3) + C_{20}(I_1 - 3)^2 + C_{30}(I_1 - 3)^3 \tag{4.21}$$

The strain energy thus depends only on the value of I_1 for any particular mode of strain. The uniaxial stress is

$$\sigma = [2C_{10} + 4C_{20}(I_1 - 3) + 6C_{30}(I_1 - 3)^2](\lambda - \lambda^{-2}) \tag{4.22}$$

and the shear stress

$$\sigma = [2C_{10} + 4C_{20}(I_1 - 3) + 6C_{30}(I_1 - 3)^2]\gamma \tag{4.23}$$

Unlike the Mooney–Rivlin equation, the shear modulus, $G = \sigma/\gamma$, in the Yeoh model varies with strain. Figure 4.4 displays the uniaxial tension data from Figure 4.2 plotted versus the strain invariant I_1–3. It can be seen that eqn (4.21) does a better job than the Mooney–Rivlin equation in accounting for the results at higher strains; however, the fits at lower strains are poor. This failing is especially apparent in a conventional plot of engineering stress versus strain (Figure 4.5), in which the low strain data are not suppressed.

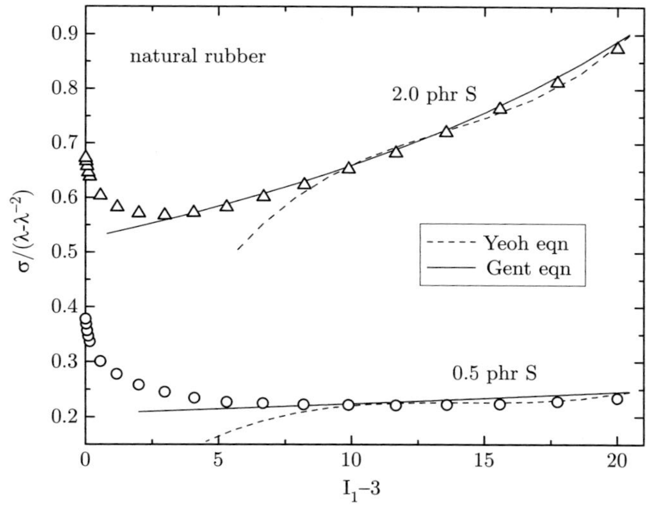

Figure 4.4 Data from Figure 4.2 plotted *versus* I_1–3. The dashed line is the fit of eqn (4.22) and the solid line the fit of the eqn (4.26).

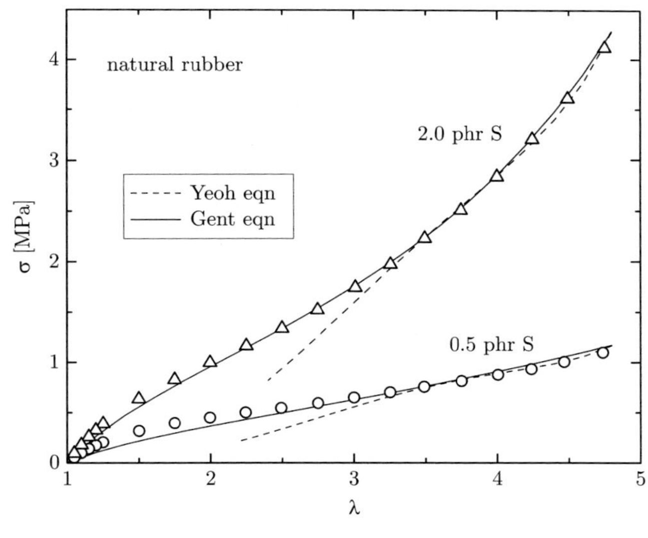

Figure 4.5 Data from Figure 4.2 plotted as the engineering stress *versus* extension ratio. The dashed line is the fit to eqn (4.22) and the solid line the fit to eqn. (4.26).

Gent [24] proposed a simple, empirical equation, in which the strain energy similarly depends only on the first strain invariant according to

$$W = -C_{10}(I_m - 3)\ln\left(1 - \frac{I_1 - 3}{I_m - 3}\right) \tag{4.24}$$

However, the deformation measure I_1 is limited to a maximum value, I_m, representing the effect of finite chain extensibility

$$I_1 \leq I_m \tag{4.25}$$

For uniaxial strain the engineering stress in the Gent model is

$$\sigma = 2C_{10}\left(1 - \frac{I_1 - 3}{I_m - 3}\right)^{-1}(\lambda - \lambda^{-2}) \tag{4.26}$$

It can be seen in Figure 4.2, Figure 4.4, and Figure 4.5 that eqn (4.26) is quite accurate at moderate to high strains. Even the deviation at low strains is small, given the limitation of only two adjustable parameters. In these figures, $I_m = 128$ and 52 for the NR networks crosslinked with 0.5 and 2.0 phr[4] sulfur, respectively. Although in practice it functions as a fitting parameter, from its definition I_m must be related inversely to the shear modulus, since both depend (oppositely) on the network chain length [25,26]. Variations on this idea of introducing limits in the constitutive equation to account for finite chain extensibility have been proposed [25,27–30], including relaxing the incompressibility requirement [31]; however, these lack the economy of eqn (4.24).

4.2 Chain models

As discussed in Chapter 1, the force to distend a long (*ca.* 100 backbone bonds), flexible chain is directly proportional to the displacement, until the *gauche* conformers become depleted with approach to full extension. The end-to-end distance of the chain has a Gaussian distribution, which is unmodified by intermolecular interactions as long as the latter do not depend on chain configuration. (An exception would be liquid-crystalline polymers, in which intermolecular forces have a strong dependence on chain orientation.) Calculation of the stress–strain relationship is derived classically from two extremes of how a chain can be envisioned to respond to deformation, the affine [6,32,33] and phantom [6,34,35] models. Both ignore any explicit consideration of interactions between chains including their uncrossability; the difference between the two models is how individual chains react to a macroscopic deformation. In the affine model the network junctions displace in proportion to the applied strain; that is, the ratio of the chain length before and after straining is the same as the ratio of the macroscopic dimensions. This corresponds to dominant interchain interactions that are independent of chain configuration. The force for a single chain was given by eqn (1.26) in Chapter 1

$$f_e = \frac{3k_B T}{\langle r^2 \rangle_0} r \tag{4.27}$$

[4] The concentration of an ingredient in a rubber formulation is usually expressed as "parts per hundred rubber" (phr), which is its weight divided by the weight of the polymer. The polymer content varies from 90% (rubber bands and gum tubing) to less than 20% (pencil erasers) of the mass of the compound.

where the subscript zero signifies the undeformed state. The strain energy is

$$W_{chain} = \frac{3k_B T}{2\langle r^2 \rangle_0}(r^2 - \langle r^2 \rangle_0) \tag{4.28}$$

Since the chains react independently, the total energy is obtained by summing over all N chains per unit volume

$$W_{affine} = \frac{3Nk_B T}{2}\left(\frac{\langle r^2 \rangle}{\langle r^2 \rangle_0} - 1\right) \tag{4.29}$$

where $\langle r^2 \rangle = \sum r^2/N_c$. In rectilinear coordinates

$$\langle r^2 \rangle = \langle x^2 \rangle + \langle y^2 \rangle + \langle z^2 \rangle \tag{4.30}$$

Since the network is isotropic, for an affine displacement

$$\frac{\langle r^2 \rangle}{\langle r^2 \rangle_0} = (\langle \lambda_1^2 \rangle + \langle \lambda_2^2 \rangle + \langle \lambda_3^2 \rangle)/3 \tag{4.31}$$

and the strain energy is

$$W_{affine} = \frac{Nk_B T}{2}(I_1 - 3) \tag{4.32}$$

This result is identical to eqn (4.8) with the elastic constant C_{10} identified as the number density of network chains times their thermal energy. For shear strains

$$\sigma_{affine} = Nk_B T\gamma \tag{4.33}$$

so that $Nk_B T$ equals the shear modulus. Assuming incompressibility, the engineering stress for uniaxial strain is

$$\sigma_{affine} = Nk_B T(\lambda - \lambda^{-2}) \tag{4.34}$$

The alternative to the affine model is to consider Gaussian chains that interact only through their crosslinks, or network junctions. The chain segments can pass through one another and occupy the same volume simultaneously (that is, there is no excluded volume); hence, the network strands are referred to as phantom chains. While junctions at the surface of a phantom network deform affinely, all other crosslink sites fluctuate around their average position. This configurational freedom reduces the entropy, lowering the elastic modulus from the affine prediction. The strain energy for the phantom network is

$$W_{phantom} = [1 - 2/\hat{f}\,]\frac{Nk_B T}{2}(I_1 - 3) \tag{4.35}$$

Both classical models, and indeed all molecular theories of rubber elasticity, predict the strain energy and the elastic modulus to be directly proportional to the number of network chains, N, and thus to the crosslink density. The phantom chain equations differ from eqns (4.32) to (4.34) of the affine model only by the factor in brackets containing the junction functionality, $\hat{f}$. For bonding of just one pair of chains, the crosslink junctions are tetrafunctional, and the strain energy and elastic stresses for a phantom network are one-half that of an affine network having the equivalent number of crosslinks.

Elasticity models only consider chains that contribute to the mechanical stress, neglecting "elastically inactive" chains, such as those that are dangling, form a loop, or have $\hat{f}=2$. This leads to the use of cycle rank, $\hat{R}$, to characterize the network topology [36]. The cycle rank, or number of independent cyclic paths, is equal to one-half the number of elastically effective network chains, N_{eff}. Equation (4.35) for the strain energy of phantom network can be rewritten

$$W_{phantom} = \frac{\hat{R} k_B T}{2}(I_1 - 3) \tag{4.36}$$

For the common case of $\hat{f} = 4$

$$N_{eff} = N(1 - 2M_c/M_n) \tag{4.37}$$

where M_n refers to the number average molecular weight prior to crosslinking and M_c is the molecular weight between crosslinks

$$M_c = \frac{\rho}{N} \tag{4.38}$$

Equation (4.37) has the same form as eqn (1.16) from Chapter 1, describing the effect of chain ends on the glass-transition temperature. Absent chain ends ($M_n \gg M_c$), the number density of network junctions is given by

$$\mu = \frac{2N}{\phi} \tag{4.39}$$

In common with phenomenological approaches to network elasticity, the classical models suffer from neglect of the limited deformability of the chains. This can be accounted for by invoking non-Gaussian statistics to describe the chains at higher strains. For a single chain this means extending the series expansion of the inverse Langevin function, eqn. (1.19) in Chapter 1, beyond the first term. This function accounts for the reduction in configurations, and thus entropy, of the chain as it becomes highly stretched [6,37]. For a network of chains, however, the problem becomes enormously complex, since the strain energy depends on the stretch, orientation, and spatial distribution of each chain. One approach is to develop a full-network model, which is then solved either for selected modes of deformation [38,39] or only approximately [40]. More progress has been made by representing the bulk elastomer as a simplified network of a few chains deforming in an assumed fashion relative to the bulk strain. Examples include the 3-chain [41], 4-chain [33,42], and 8-chain models [43]. The latter, known as the Arruda–Boyce model, employs the inverse Langevin function, with the strain energy expressed as a series in the first strain invariant

$$W = G\left[\frac{1}{2}(I_1 - 3) + \frac{1}{20\,\hat{N}}(I_1^2 - 9) + \frac{11}{1050\hat{N}^2}(I_1^3 - 27) + \frac{19}{7000\hat{N}^3}(I_1^4 - 81) + \frac{519}{673750\hat{N}^4}(I_1^5 - 243) + \ldots\right] \tag{4.40}$$

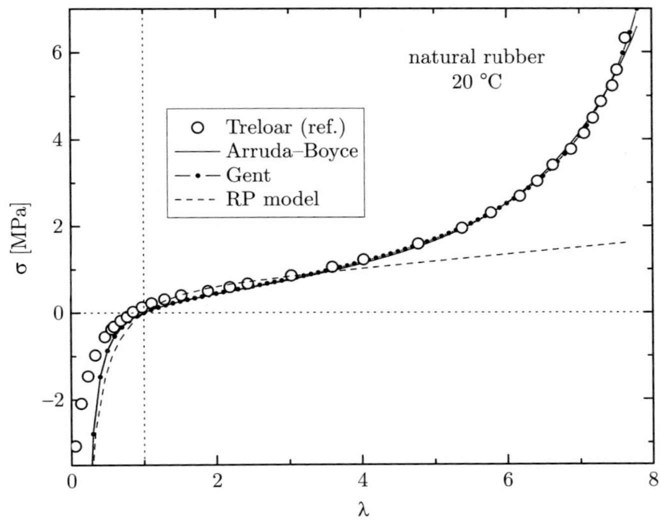

Figure 4.6 Engineering stresses for two NR networks in compression and tension [44]. The solid curve is the fit of eqn (4.41) with $G = 0.23$ MPa and $\hat{N} = 19.6$ and the (nearly coincident) dash-dotted line is the fit of eqn (4.26) with $C_{10} = 0.12$ MPa and $I_m = 83$. Also shown (dashed line) is the fit of eqn (4.57) to the low strain tensile data, which gives poor accuracy for high extensions and in compression.

Here, G is the shear modulus and the parameter $\hat{N}$ is related to the finite extensibility of the chains. The engineering stress for this model is

$$\sigma = 2G(\lambda - \lambda^{-2})\left[\frac{1}{2} + \frac{1}{10\,\hat{N}}(I_1) + \frac{33}{1050\hat{N}^2}(I_1^2) + \frac{76}{7000\hat{N}^3}(I_1^3) + \frac{2595}{673750\hat{N}^4}(I_1^4) + \ldots\right] \quad (4.41)$$

In Figure 4.6 the stress data of Treloar [44] for a vulcanized NR measured in uniaxial tension and compression are shown, along with the fit of eqn (4.40) to the tension data. Although the tension data are well described to high strains, using the same fit values of G and $\hat{N}$ gives a poor description of the compression results. This is a failing that molecular models have in common with phenomenological models.

There is a significant amount of literature comparing various elasticity models to experimental results [20,45–51]. An evaluation of twenty different approaches in terms of their ability to represent high strain data encompassing a range of deformation modes found the phenomenological Gent model (eqn (4.24)) [24] and the Arruda–Boyce chain model [43] to be the best among those with only two adjustable parameters [46]. These two models give very similar results [52], as shown in Figure 4.6.

4.3 Constraint models

The crosslink sites in a real network are embedded in a high concentration of neighboring chain segments. At typical crosslink densities, spatially neighboring junctions are not topological neighbors; that is, the volume existing between a directly connected pair of junctions will contain many other junctions. The dense packing of chains (sometimes referred to as entanglements, although these local constraints are not necessarily manifested equivalently to the long-range topological interactions discussed in Chapter 3) imposes limitations on the segment fluctuations. Neutron spin echo studies provide direct evidence that crosslink motions in networks are less than phantom network predictions [53,54]. The effects of these intermolecular interactions are ignored in the phantom chain model (even the uncrossability of chains), but overestimated in the affine model. A more accurate treatment of the mutual effect of neighboring chains has to consider the manner in which the interactions modify the stress in a strain-dependent manner. Constraint models of rubber elasticity have the general form

$$\sigma = \sigma_{phantom}[1 + f_{cnstr}(\lambda)] \tag{4.42}$$

in which the function f_{cnstr} reflects steric hindrances to Brownian motion of the network. From eqn (4.35)

$$\sigma_{phantom} = [1 - 2/\hat{f}\,]Nk_B T\gamma \tag{4.43}$$

for shear strains, and

$$\sigma_{phantom} = [1 - 2/\hat{f}\,]Nk_B T(\lambda - \lambda^{-2}) \tag{4.44}$$

for uniaxial strains.

The original constraint model of Flory [55] assumed the effect of constraints was to restrict fluctuations specifically of the network junctions. These constraints operate within a spatial domain, which deforms affinely with the macroscopic strain. The engineering stress for uniaxial strain in the constrained-junction (CJ) model is given by [55]

$$\sigma_{CJ} = \sigma_{phantom}\left[1 + \frac{2}{\hat{f} - 2}\frac{\lambda K(\lambda^2) - \lambda^{-2}K(\lambda^{-1})}{\lambda - \lambda^{-2}}\right] \tag{4.45}$$

where incompressibility is assumed. The function $K(\lambda^2)$ is defined in terms of functions of the strain

$$K(x) = \frac{B_1\dot{B}_1}{B_1 + 1} + \frac{B_2\dot{B}_2}{B_2 + 1} \tag{4.46}$$

where

$$B_1(x) = \frac{\kappa(x^2 - 1)}{(x^2 + \kappa)^2} \tag{4.47}$$

$$\dot{B}_1(x) = \frac{\partial B_1}{\partial x^2} = B_1\left[\frac{1}{x - 1} - \frac{2}{x + \kappa}\right] \tag{4.48}$$

$$B_2(x) = \frac{x^2}{\kappa}B_1(x^2) \tag{4.49}$$

and

$$\dot{B}_2(x) = \frac{\partial B_2}{\partial x^2} = \frac{1}{\kappa}[x^2 \dot{B}_1 + B_1)] \tag{4.50}$$

The parameter κ is a measure of the severity of the constraints. Equation (4.45) reduces to the phantom and affine models respectively for $\kappa = 0$ and ∞.

Subsequently, Flory and Erman [56] extended the CJ model, introducing an additional parameter, ξ (≥ 0), to allow for departures from uniformly affine distortion of the domain of constraints. A large value of ξ implies that the constraints are alleviated by extension of the network more rapidly than if the domains deformed affinely. In this extended constrained junction (ECJ) model, $K(\lambda^2)$ is still given by eqn (4.46), but the functions $B_1(\lambda^2)$ and $B_2(\lambda^2)$ become

$$B_1(x) = \frac{\kappa^2[x^2 - 1 - x^2\xi(x-1)]}{[x^2 + \kappa + \kappa x^2(x-1)]^2} \tag{4.51}$$

and

$$B_2(x^2) = x^2[\kappa^{-1} + \xi(x-1)]B_1 \tag{4.52}$$

with the corresponding derivatives for $\dot{B}_1$ and $\dot{B}_2$. Substitution of these quantities into eqn (4.45) yields the engineering stress for the ECJ model, with the original CJ model recovered for $\xi = 0$. In Figure 4.7 the ECJ model is compared to elastic mechanical results for NR at two different crosslink densities [57]. The model is fit to the tension data and, as seen, deviates strongly from the compression results.

In both the CJ and ECJ models, the constraints exert their effect directly on the network junctions. Alternatively, the chain's center of gravity can be the locus of the constraints [58,59], or more generally, the constraints can act all along the entire chain, as in the continuously constrained chain (CC) model [60]. The uniaxial engineering stress for the CC model is

$$\sigma_{cc} = \sigma_{phantom}\left[1 + \frac{\hat{f}}{\hat{f} - 2}\int_0^1 \Omega(z)\frac{\lambda K(\lambda^2) - \lambda^{-2}K(\lambda^{-1})}{\lambda - \lambda^{-2}}\,dz\right] \tag{4.53}$$

where $K(\lambda^2)$ is again given by eqn (4.46). $\Omega(z)$ is a distribution function describing the effectiveness of the constraints at each point along the chain ($\Omega(z) = 1$ if the strength of the constraints is uniform) and z, which varies from 0 to 1, is the relative distance of a point on the chain from the junction site. So in the CC model the constraint parameter in eqns (4.47)–(4.50) is not constant, but varies along the chain according to

$$\kappa(z) = \bar{\kappa}\left[1 + \frac{(\hat{f} - 2)^2 z(1-z)}{\hat{f} - 1}\right] \tag{4.54}$$

where $\bar{\kappa}$ is a constant, adjusted to fit experimental data. As seen in Figure 4.7, the CC model is slightly better than the ECJ model in describing the NR data; however, the problem remains that both tension and compression results cannot be simultaneously fit with good accuracy.

An idea of long-standing attraction is that the elastic response of a rubbery network reflects the same entanglement interactions that give rise to the plateau in the dynamic modulus of high molecular weight polymers (Chapter 3). There have been various attempts to connect the

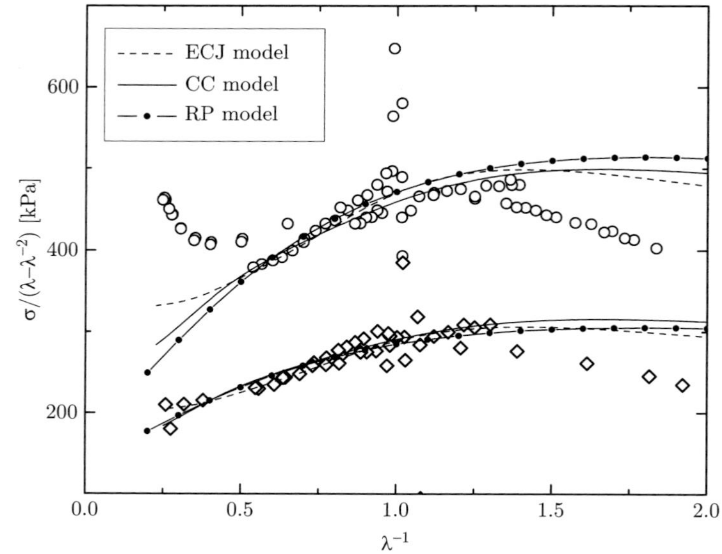

Figure 4.7 Reduced stresses for two NR networks in compression and tension [57]. The dashed curves are the fit of eqns (4.45), (4.51), and (4.52), the solid line the fits of eqn (4.53), and the dashed-dotted line fits to eqn (4.57), all to the tensile data. Note that experimental inaccuracies are magnified by plotting the ratio of the engineering stress and the strain function $\lambda - \lambda^{-2}$.

C_{01} term of the Mooney–Rivlin equation ((4.10)) to the concentration of entanglements. Possin and Graessley [61] included this effect by expressing the stress as the sum of contributions from the phantom network and from the entanglements, with the latter proportional to the plateau modulus of the polymer in its precursor melt state

$$\sigma = \sigma_{phantom} + G_N f_N \tag{4.55}$$

The "trapping factor" f_N represents the fraction of entanglements with paths covalently bonded to the network [62]. This expression does not address the non-linearity of the mechanical behavior.

Rubinstein and Panyukov [63,64] also proposed identifying the constraints with the chain entanglements, however with $f_N = 1$. Depending on the assumptions made about how these entanglements restrict fluctuations of the network chains and how this varies with strain, the engineering stress in the RP model is either [63]

$$\sigma = \sigma_{phantom} + G_N \frac{\lambda - \lambda^{-2}}{\lambda - \lambda^{1/2} + 1} \tag{4.56}$$

or [64]

$$\sigma = \sigma_{phantom} + G_N \frac{\lambda - \lambda^{-2}}{0.74\lambda + 0.61\lambda^{-1/2} - 0.35} \tag{4.57}$$

Table 4.2 Network chain densities from Figure 4.7.

Determination	N (mole/m^3)	
	1.0 phr dicumyl peroxide	2.0 phr dicumyl peroxide
crosslinking chemistry*	66	136
ECJ model (eqns (4.45), (4.51), and (4.52))	162	264
CC model (eqn (4.53))	130	206
RP model (eqn (4.57))	218	264

* assuming complete reaction of the peroxide crosslinker with $\hat{f} = 4$ [57].

Although in principle G_N is the plateau modulus of the uncrosslinked polymer, in practice it serves as an adjustable parameter. As shown in Figure 4.7 the RP gives results quite close to the CC model, and thus shares its failure to describe simultaneously results from more than one mode of deformation. And since they are based on modifications of the phantom chain approach, the constraint models do not account for the upturn in the stress at high strains, as illustrated in Figure 4.6.

In Table 4.2 are collected the results from Figure 4.7, comparing the number of network strands deduced from the crosslinking reaction to the value of N obtained by application of the various models. The latter are all substantially higher, indicating that $\hat{f}$ may be greater than the value of four used in the calculation. The usual assumption is that peroxide crosslinking of natural rubber yields a stable tertiary radical that terminates by combination to form tetra-functional junctions [65].

Equations (4.56) and (4.57) embody ideas drawn from slip-link models of rubber elasticity [66–68]. Heinrich *et al.* [69,70] developed a related model that introduced constraint release effects that, as discussed in Chapter 1 describing the rheology of entangled polymer liquids, reduce the severity of the entanglement constraints, yielding more phantom-like behavior.

4.4 Role of molecular motions in the elastic response

The picture that emerges from rubber-elasticity theories is one of chains frustrated by intermolecular constraints in their effort to achieve all the configurations available to an isolated chain; thus, the elastic response reflects the effect the network structure exerts on the chain dynamics. Since the topology and motion of chain molecules are both governed by the same intramolecular and intermolecular potentials and correlations, there have been various efforts to analyze rubber elasticity from study of the chain dynamics [71–78]. The underlying assumption is the usual one of FDT – equilibrium fluctuations of the network segments and junctions are related to the linear relaxation behavior (see Chapter 1).

The coupling model of relaxation [2] addresses the effect on the local dynamics of the same constraints restricting the network configurations. According to this model, at short times before unbalanced forces and torques have attained sufficient magnitude, the chains exhibit exponential time dependence

$$C_{ph}(t) \sim \exp[-(t/\tau_{ph})],\, t < t_c \tag{4.58}$$

with τ_{ph} the junction relaxation time. (Equation (4.58) applies at all times for phantom chains, since they experience no intermolecular constraints.) After some temperature-insensitive timescale, t_c, the average relaxation rate of the segments becomes slowed down by intermolecular cooperativity. From experimental data for amorphous polymers [76,77], t_c is found to have a magnitude between 10^{-12} and 10^{-11} s. For times longer than t_c, which fall within the usual, experimentally accessible realm, the dynamics of the network chains are slowed by their interactions with neighboring chains, causing the simple exponential decay to become stretched (*cf.* eqn (1.7))

$$C_{cnstr}(t) = [-(t/\tau_\alpha)^{\beta_K}],\, t > t_c \tag{4.59}$$

In the model the important metric of intermolecular constraints is the coupling parameter, $n = 1 - \beta_K$, so that the Kolhrausch exponent varies inversely with the strength of these interactions. At $t = t_c$ eqns (4.58) and (4.59) can be set equal to one another, whereby the correlation time at longer times is given by

$$\tau_\alpha = [\beta_K t_c^{\beta_K - 1}\tau_0]^{1/\beta_K} = [(1-n)t_c^{-n}\tau_0]^{1/(1-n)} \tag{4.60}$$

This relation (*cf.* eqn (2.18)) makes clear the strong, non-linear slowing down of the dynamics at longer time, since $\tau_\alpha \gg \tau_0$. In the model the onset of intermolecularly constrained relaxation is accompanied by an increase of the activation energy, E_a, which from eqn (4.60) is related to E_a^0, the microscopic conformational energy barrier to segment motion, as

$$E_a = E_a^0/(1-n) \tag{4.61}$$

E_a is the value prevailing at longer times when constraints are dominant, although a consequence of these constraints is deviation from Arrhenius behavior

$$\tau_\alpha = \tau_\infty \exp[-E_a(T)/RT] \tag{4.62}$$

This means that the observed E_a varies with temperature. (Note that the inverse correlation of fragility and the Kohlrausch β_K described in Chapter 2 follows from eqn (4.61)). Equation (4.61) is the same relation that leads to the coupling model prediction of a breakdown in time-temperature superpositioning in the glass-transition zone, due to the stronger coupling (larger n) of the segmental modes in comparison to the global chain modes (Chapter 6). There is an analogous relationship to eqn (4.60) between the respective pre-exponential factors, $\tau_{0,\infty}$ and τ_∞, for uncoupled and coupled relaxation.

The expectation is that the severity of the constraints on the junctions, giving rise to departures of the elastic response from phantom-like behavior, is evident directly in the junction dynamics. This means that experimental variables that affect the parameter κ in the constraint models (which characterizes the domain of the constraints on the fluctuations) will similarly affect the coupling parameter n describing intermolecular cooperativity of the junction dynamics (see Table 4.3). This is illustrated with NMR spin-lattice relaxation measurements [79] on PTHF networks having various molecular weights between crosslinks. The network junctions were tris(4-isocyanatophenyl) thiophosphate, so that the NMR experiment measured the motion of the junctions. From spin-lattice relaxation times, spectral density functions, corresponding to the Fourier transforms of the $C(t)$ in eqn (4.59), were

Table 4.3 Comparison of constraint models of elasticity and coupling model of relaxation.

Network alteration	effect on constraints	κ^*	β_K	E_a^*/E_a	$\tau_\infty^*/\tau_\infty$
larger N	smaller	smaller	larger	smaller	smaller
larger $\hat{f}$	larger	larger	smaller	larger	larger
larger λ	smaller	—	larger	smaller	smaller
diluent	smaller	smaller	larger	smaller	smaller

*eqns (4.45)–(4.54)

obtained. The obtained coupling parameters are displayed in Figure 4.8, showing that the junctions in the more crosslinked PTHF experience stronger intermolecular constraints (larger n) [77,79]. A similar trend in the parameter κ of the constraint elasticity models is observed [80].

As predicted by eqn (4.61), the apparent activation energy increases with n (Figure 4.9). The product $(1-n)E_a$ is constant, equal to the activation energy in the absence of cooperative dynamics, $E_a^0 = 26$ kJ/mol. This value is of the same magnitude as the conformational energy barrier of the polyether backbone. It is unaffected by intermolecular interactions, reflecting only the intramolecular barriers to changes in rotameric state, as well as the static, mean-field background potential. Also in Figure 4.9 are data representing the most crosslinked network with added diluent. The motion of small molecules is very rapid on the timescale of the chain dynamics, so the diluent alleviates the intersegmental constraints, decreasing both the coupling parameter and E_a^*.

The motion of the junctions followed by the NMR experiment is, of course, coupled to the motion of the chains connecting the junctions. The differences between the CJ and CC models are artifacts of the respective assumptions of these approaches. Thus, the dynamics of

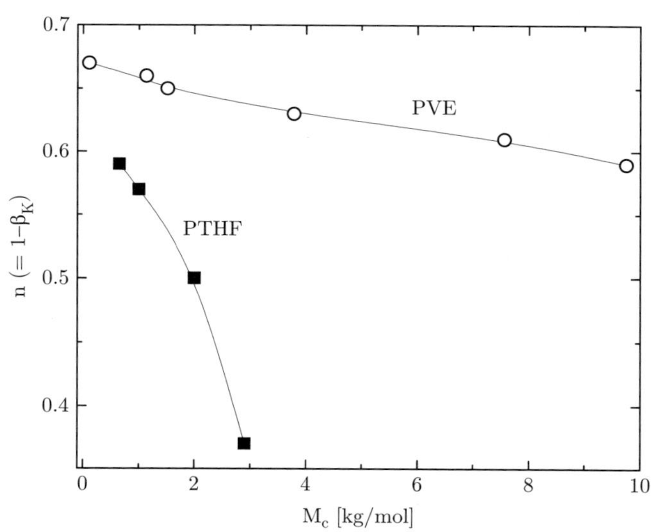

Figure 4.8 Coupling parameter *versus* molecular weight between crosslink junctions for polytetrahydrofuran (solid symbols) [77,79] and polyvinylethylene (open symbols) [78] networks.

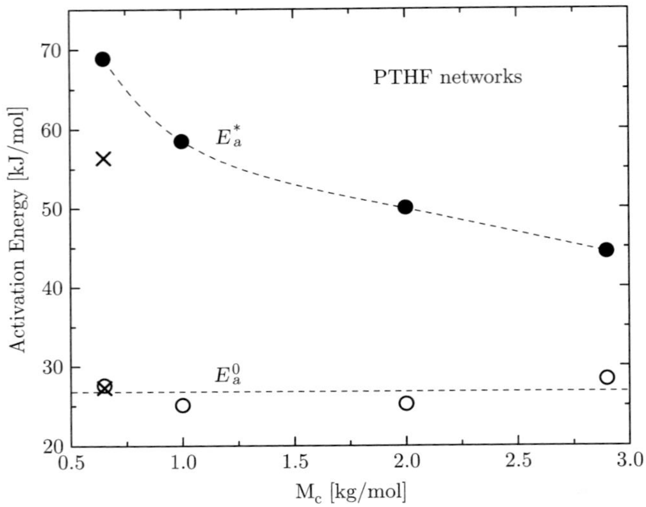

Figure 4.9 Apparent activation energy (filled symbols) measured by ^{31}P NMR [79] and non-cooperative activation from eqn (4.61) (hollow symbols) using the coupling parameters in Figure 4.8 [77], as a function of the molecular weight between crosslinks. The crosses are data for the diluted network.

networks measured by a technique such as dielectric or mechanical spectroscopy, which reflects the motion of all segments rather than specifically the junctions, can be analyzed in a similar manner. However, in a randomly crosslinked polymer the chain segments have different mobilities depending on their proximity to a junction. Segments closer (topologically or spatially) to a crosslink will have their motions more retarded, especially for large junction functionalities, wherein the confluence of many strands imposes more severe steric constraints. For this reason the segmental relaxation dispersion (peak in the dielectric or mechanical loss) will be inhomogeneously broadened, whereby its shape cannot be interpreted directly in terms of intermolecular cooperativity (*viz.*, eqn (4.59) with $n = 1 - \beta_K$).

This broadening is evident in dielectric spectra for a series of polyvinylethylene (PVE or 1,2-polybutadiene) networks (Figure 4.10), which have large $\hat{f}$ (due to the free-radical crosslinking). The peaks deviate strongly from the stretched exponential form, and the coupling parameter cannot be obtained directly from fitting the spectra to eqn. (4.59). However, since E_a^0 must be the same for each PVE sample (since it represents the intrinsic barrier to conformational transitions of the isolated chain), n can be determined as the value that satisfies eqn (4.61). (In molecular dynamic simulations this heterogeneous broadening is circumvented, since the dynamics of the chain end-to-end vector can be determined directly. Simulation results for rubber networks confirm that both the coupling parameter and the relaxation time increase systematically with increasing degree of crosslinking [81].)

Determining the coupling parameter from E_a is opposite to the procedure used to analyze the NMR data on PTHF networks. Since $\tau_\alpha(T)$ is non-Arrhenius, rather than the apparent activation energy *per se*, the fragility, or T_g-normalized activation energy (Chapter 2), is used. The results are shown in Figure 4.11. After removal of the effect of intermolecular cooperativity by assuming a value for the coupling parameter, the calculated activation energy and relaxation times for segmental relaxation are essentially independent of crosslink density.

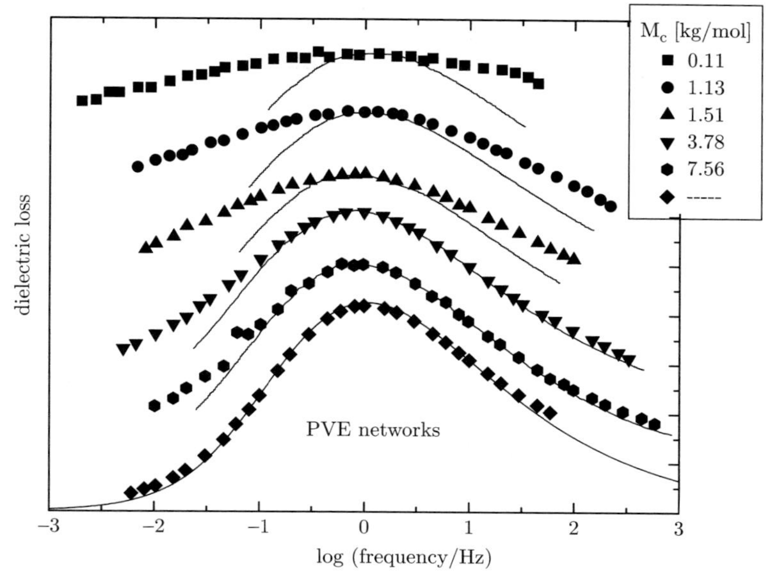

Figure 4.10 Peak in the dielectric loss for PVE networks of increasing crosslinking from bottom (uncrosslinked) to top; curves have been vertically displaced for clarity. The solid lines are calculated using the coupling parameters determined from applying eqn (4.61) to the fragility curves in Figure 4.11. Data from ref. [78].

Using the values deduced for n, eqn (4.59) is plotted in Figure 4.10. The experimental peak is much broader than the deduced Kohlrausch function because of the inhomogeneous broadening arising from the distribution of environments in the random crosslinked PVE. The assumption that the non-cooperative activation energy, E_a^0, is independent of crosslink density can be evaluated directly from dielectric relaxation measurements on the PVE networks [82]. The Johari–Goldstein secondary peak is present in the spectra, with the obtained JG relaxation times showing an Arrhenius temperature dependence with an activation energy E_a^{JG} (Figure 4.12). According to the coupling model, these τ_{JG} can be indentified with τ_0 in eqn (4.60); thus, $E_a^0 = E_a^{JG}$. From the parallelism of the slopes in Figure 4.12, it is seen that the measured E_a^{JG} are indeed independent of crosslinking and equal to the value for linear PVE. However, the value of $E_a^{JG} = 44$ kJ/mol [82] is much larger than that deduced from the analysis in Figure 4.11. The discrepancy comes from the assumption that τ_0 is constant. As seen in Figure 4.12, τ_{JG} $(= \tau_0)$ becomes smaller with crosslinking, opposite to the effect on τ_α. This is an interesting anomaly, seen also during physical aging of glassy PVE [83].

4.5 Alternative network structures

In developing an elastomeric compound, the particular polymer selected for the formulation has minimal effect on the network properties *per se*. That is, while the chemical structure of a rubber affects various aspects of performance (e.g., chemical and heat resistance, friction, adhesion, gas permeability, biocompatibility, etc.), the mechanical behavior up through

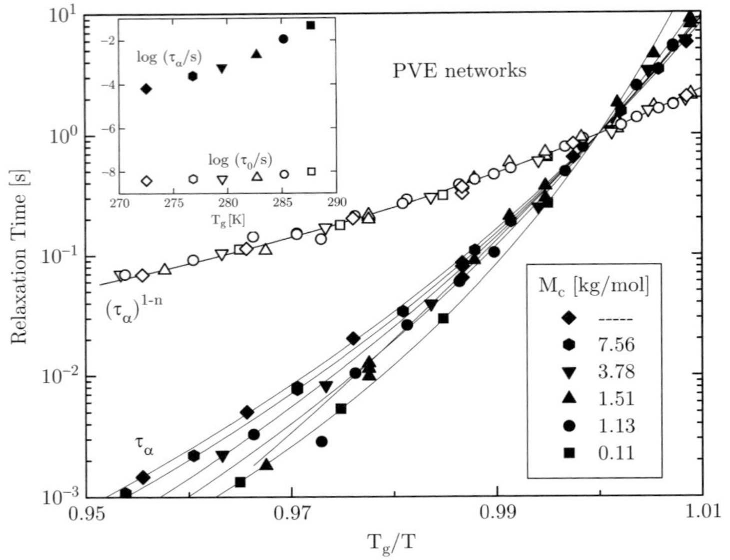

Figure 4.11 T_g-normalized Arrhenius plots of the dielectric relaxation times as measured (solid symbols) and after raising to the power of β_K (open symbols; see eqn (4.60)). The inset displays the relaxation times for each network (identified by T_g) at $T = T_g$. Data from ref. [78].

moderate deformations is essentially the same for all flexible-chain polymers. Even the fracture and fatigue properties, in the absence of strain-induced crystallization, are not especially dependent on the polymer. Of course, the chemical structure has an indirect effect, through its influence on the crosslinking reaction, but circumventing the usual compromise between stiffness and strength requires unconventional network architectures. One method is to incorporate ionic crosslinks, which dissociate under stress. Elastomeric "ionomers" incorporate an ionizable monomer or have an inorganic salt grafted pendant to the main chain. The result is a crosslink density that effectively decreases with the extent of deformation, as the junctions progressively dissociate during large deformation to yield substantial toughness (Figure 4.1) [3,84,85]. However, this same mechanical lability confers large unrecovered strain (permanent set) and susceptibility to degradation. In these respects ionically crosslinked rubbers are similar to vulcanizates with polysulfidic bonds, although an additional drawback of the former is poor processability, due to the limited control over when the ionic bonds form.

Several approaches to improving the mechanical properties of elastomers by alternative network structures are discussed below.

4.5.1 Interpenetrating polymer networks

An interpenetrating polymer network (IPN) refers to a blend ostensibly consisting of cocontinuous, interlocking networks (catenanes) of the constituents [86–88]. This cocontinuity is achieved by kinetic retention of an initially homogenous mixture of small molecules during subsequent polymerization and crosslinking, the formation of a network precluding phase

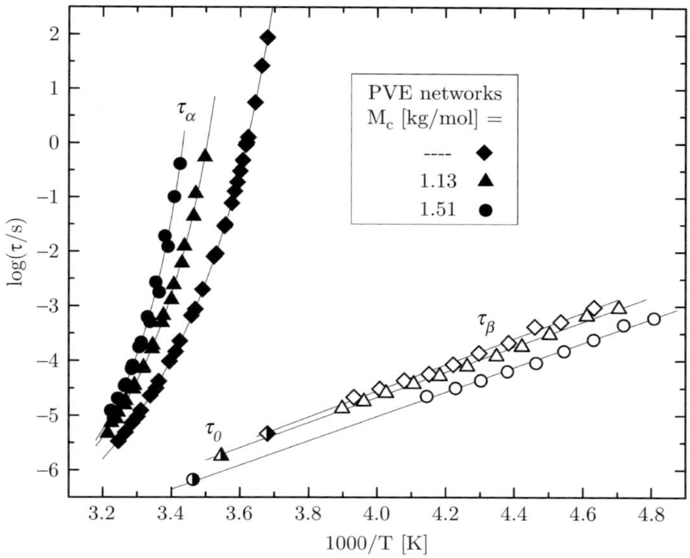

Figure 4.12 Relaxation times for the local segmental (filled symbols) and JG secondary (open symbols) processes in linear (diamonds) and crosslinked PVE. The half-filled symbols denote the value of τ_0, obtained by extrapolating $\tau_\beta(T)$ to the temperature at which for $\tau_\alpha = 3.5 \times 10^4$ s for each sample. Data from ref. [82].

segregation of the immiscible polymers [89,90]. The objective is a micro-heterogeneous morphology stabilized by physical interlocking; however, segmental interpenetration of the two network structures may not actually be obtained, so that in practice the designation IPN connotes the method of preparation, rather that the actual blend morphology.

IPNs can be made simultaneously or sequentially. A simultaneous IPN is formed from the polymerization and crosslinking of premixed monomers or linear prepolymers. During network formation the tendency for phase separation is promoted by the increasing molar mass of the constituents (which reduces the combinatorial entropy) but opposed by their concomitantly slower dynamics. In a sequential IPN [91–93], one network is swollen in monomers of the other component, the latter then rapidly polymerized and crosslinked, often using radiation curing. Phase separation is more extensive for sequential IPNs, but cocontinuity can still be achieved. "Semi-IPN" is used in reference to an IPN in which one component remains uncrosslinked [94,95]. The T_g of an IPN is expected to be intermediate to the neat component T_g values, as shown in Figure 4.13 for an IPN [96] and Figure 4.14 for a semi-IPN [97]. (Note in these figures the component glass transitions are better resolved in the loss tangent than in the loss modulus. This is generally the case; however, there is some ambiguity because the exact position of the loss tangent is not solely a function of T_g; it also is affected by the rubbery modulus; see Figures 10.22 and 10.23.) Even though the constituents are incompatible, small phase domains are obtained, homogenizing the properties. The glass-transition temperature becomes size dependent for domain sizes on the order of or less than the cooperativity length scale, so that a change in the component T_g does not necessarily indicate mixing on the segmental level. (See section 2.3.2 for a discussion of dynamic correlation length scales,

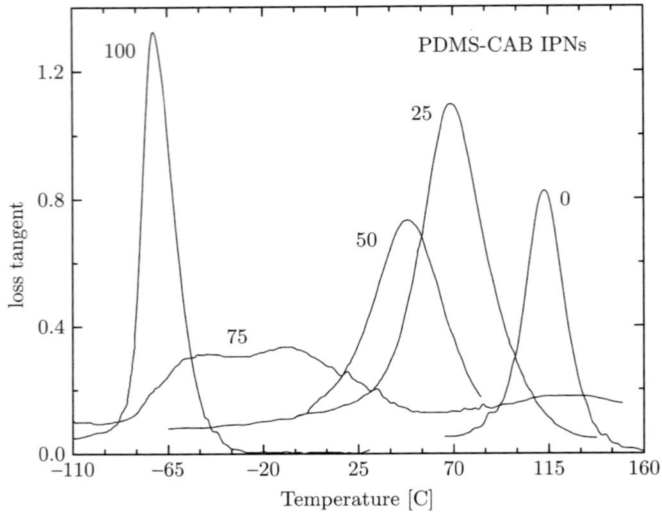

Figure 4.13 Loss tangent for polydimethylsiloxane – acetate butyrate cellulose IPNs having the indicated weight % of PDMS. Data from ref. [96]

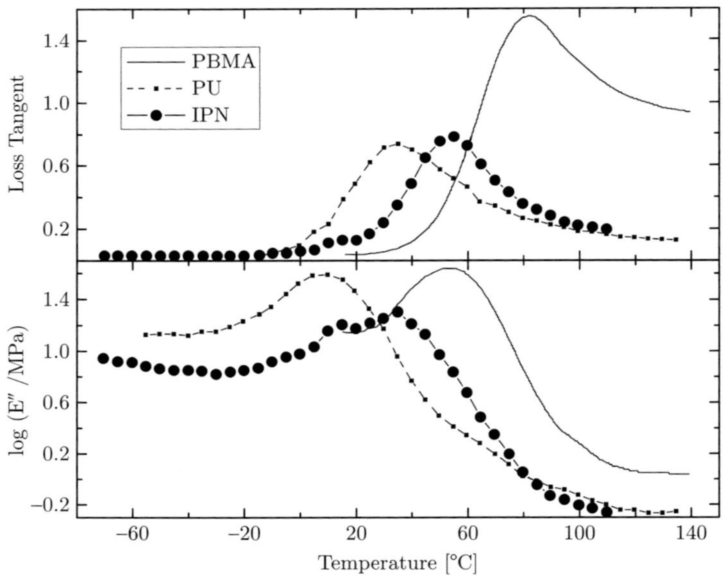

Figure 4.14 Loss tangent for linear polybutylmethacrylate, polyurethane network, and their semi- IPN. Data from ref. [97].

which can exceed the size of the chain.) A distribution of phase sizes gives rise to a very broad glass transition dispersion, and for this reason IPNs have attracted interest as materials for acoustic damping and vibration isolation [98,99]. Determinations of the degree of homogeneity and interpenetration of an IPN from its glass transition behavior can be ambiguous when the component T_g values are close.

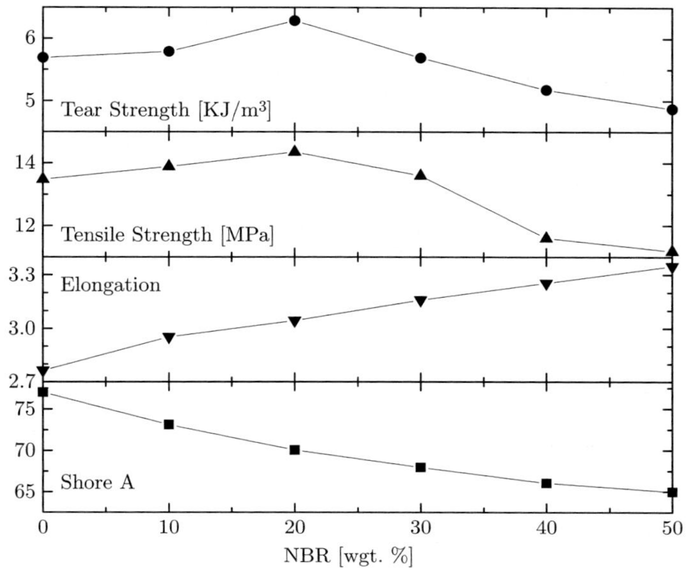

Figure 4.15 Properties of an IPN based on nitrile rubber and poly(vinylidene fluoride-co-hexafluoroproylene). With increasing NBR content, the material becomes more compliant; however, the tensile and tear strengths exhibit a maximum. Data from ref. [103].

The premise of IPN formation is the absence of interference among the polymerization and crosslinking reactions, although such interferences can be unavoidable. Grafting between the components may occur, with consequences for both the morphology (such as the inhibition of phase separation) and properties [86,100–103]. While from the structure of IPNs it is expected that mechanical performance will be additive with regard to the component properties, interlocked networks can confer greater mechanical integrity than a completely phase-separated blend morphology. This potential for better properties is illustrated in Figure 4.15 [102] for an IPN of a fluoroelastomer and nitrile rubber, which exhibits a maximum in failure properties as a function of composition.

Phase homogeneity in IPNs can be greatly enhanced if a cosolvent is used; e.g., a popular type of IPN are hydrogels. These are dilute aqueous IPNs, which react to changes in pH, temperature, or the presence of specific chemicals. Their biocompatibility, high water content, and ability to respond to stimuli have led to use of IPN hydrogels as biomaterials for tissue engineering [104–106] and for controlled drug delivery [107–110]. These materials, sometime confusingly referred to as double-network hydrogels, can exhibit enhanced performance [111–114]. For example, Figure 4.16 [115] shows the remarkable increase in compressive strength of an IPN hydrogel compared to the corresponding single network hydrogel; however, the components of this material have very different crosslink densities, and this difference is an essential element in the property improvements (see Sections 4.5.3 and 4.5.4).

4.5.2 Double networks

A special type of IPN is a double-network rubber, in which the same chain segments belong to the two networks, both of which are oriented. A double network is formed by two crosslinking

Figure 4.16 Compression of (left) a poly(2-acrylamido-2-methylpropanesulfonic acid) hydrogel and (right) an IPN hydrogel based on poly(2-acrylamido-2-methylpropanesulfonic acid) and polyacrylamide, showing the greater resistance to fracture of the IPN. From ref. [115] with permission.

processes, the second while the elastomer is in a deformed state (usually tension). Since the properties of an elastomeric network depend not only on the crosslink density, but also on the distribution and orientation of the chains, the properties of double networks can be quite different from those of the corresponding single network at equal degrees of crosslinking. Double networks are a means to increase both the stiffness and strength of rubber. They can arise spontaneously via chain scission [116,117], strain-induced crystallization [118,119], due to the presence of reinforcing fillers [120], or as described in Chapter 8 by crosslinking liquid-crystalline elastomers while in their ordered state. The effect of aging on stress relaxation ("chemical stress relaxation") [121] and of trapped entanglements on the elastic modulus [122,123] both involve the contribution of inadvertent double networks.

Analysis of double networks relies on the independent network hypothesis [124–127], which assumes that the behaviors of the component networks are uncoupled, so that the mechanical response is the sum of the individual contributions. The equilibrium configuration of each network corresponds to that existing during network formation, so the strain energy of a double network is

$$W_{DN}(\lambda) = W_1(\lambda) + W_2(\lambda/\lambda_X) \tag{4.63}$$

in which the subscripts refer to the first and second network, and λ_X is the stretch ratio during the second crosslinking (the initial network being formed in the unstressed, isotropic state). The strain is referenced to the undeformed network; hence, $\lambda = \lambda_1$ and $\lambda_2 = \lambda/\lambda_X$. The residual strain at zero stress of the double network represents the state for which the forces from the initial network, which is subsequently extended, are balanced by those from the second network, which is compressed. Although from eqn (4.63) the stress of a deformed double network is just the sum of the stresses from the component networks, modeling of the mechanical behavior is limited by the poor ability of elasticity theories to simultaneously describe tensile and compressive strains (Figure 4.6 and Figure 4.7).

The orientation of the network chains gives rise to anisotropic mechanical properties [128–133]. Generally, the modulus increases parallel to the curing strain, with negligible changes in the transverse direction. This modulus enhancement is illustrated in Figure 4.17 [134] for a series of elastomers having the same total crosslink density, but varying λ_X. For double networks reinforced with carbon black, lower electrical conductivity [135] and suppression of the Payne effect (Chapter 5) [136] have been observed, both consistent with less filler aggregation due to the strain imposed during curing.

Although the experimental results are scattered [128–133], a double-network structure affects the strength of the elastomer only weakly. Given their higher modulus, this means that double networks can be used to circumvent the usual compromise between stiffness and strength. Moreover, double networks of strain-crystallizing rubbers having substantially larger fatigue lifetimes (Figure 4.18 [129,137]). This appears to result from retention of crystallinity through the minimum of the strain cycle [137], analogous to the greater fatigue life of crystallizing rubbers subjected to non-relaxing strain cycles [138].

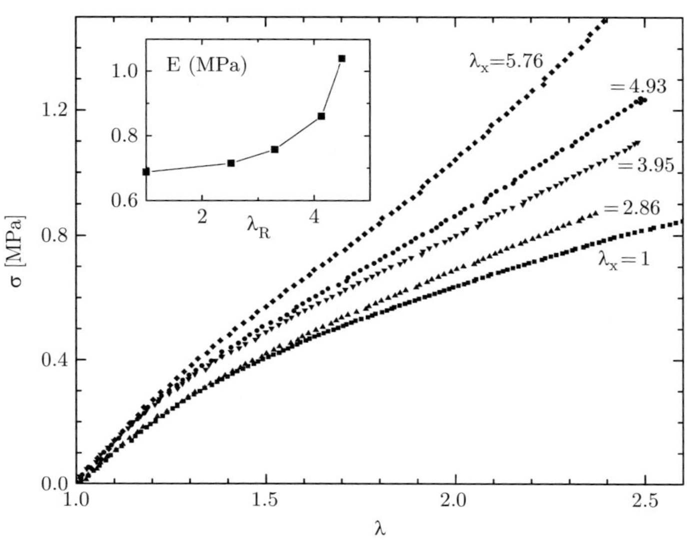

Figure 4.17 Engineering stress as a function of stretch ratio for double networks having the indicated strain during the second crosslinking. The crosslink density was fixed at 102 mol/m^3. The inset shows the stress at 100% strain. Data from ref. [134].

As seen in Figure 4.19 (also Figure 5.5), double networks are birefringent absent external stress [132,134], contrary to the stress optical law (eqn. (5.14)). This stress-free birefringence is consistent with constraint models of rubber elasticity, which predict that the ratio of the birefringence to the true stress varies with strain. For example, the CC model gives [139]

$$\Delta n_{cc} = \Delta n_{phantom}\left[1 + \frac{\hat{f}}{\hat{f}-2}\int_0^1 \Omega(z)\frac{B_1(\lambda) - B_1(\lambda_\perp) + b_{cc}[B_2(\lambda) - B2(\lambda_\perp)]}{\lambda^2 - \lambda^{-1}}\,dz\right] \tag{4.64}$$

where $\Delta n_{phantom}$ is a constant proportional to the true stress. The parameter b_{cc}, which has a value between 0 and 1, accounts for the contribution of the constraints to the birefringence. Since the effect of constraints is strain dependent, the apparent stress optical coefficient varies with strain, as confirmed by measurements on silicone [140] and polybutadiene [141] rubbers. Thus, even if the stress and birefringence of the components networks are both additive quantities, double networks will exhibit non-zero Δn in the stress-free state [134].

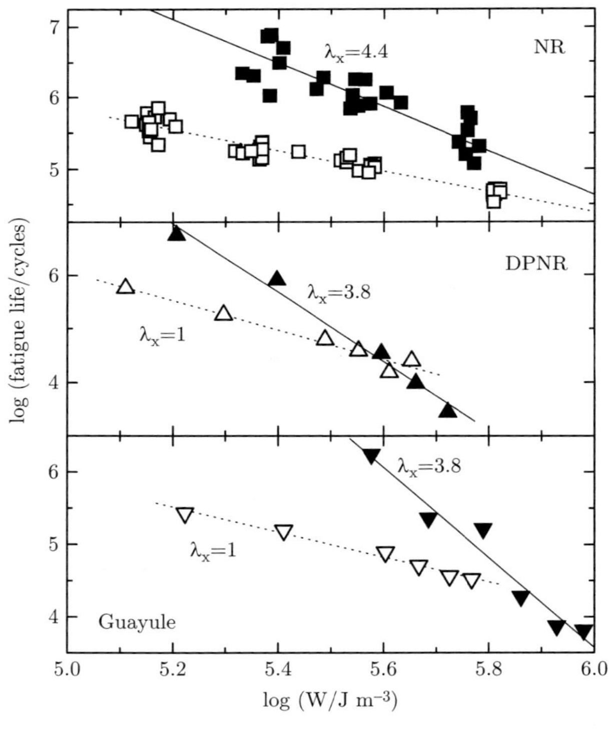

Figure 4.18 Double-logarithmic plots of the mechanical fatigue life *versus* recoverable strain energy for single (hollow symbols) and double (solid symbols) networks of natural rubber (squares), deproteinized natural rubber (triangles), and guayule rubber (inverted triangles). Measurements were for uniaxial extension parallel to the curing strain. Data from refs. [129, 137].

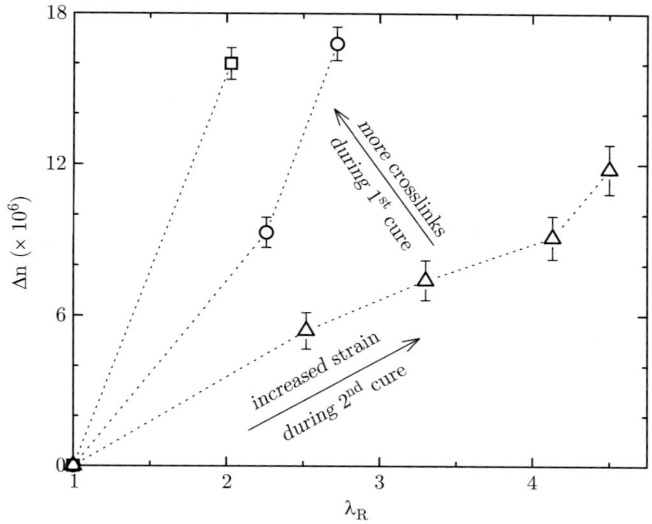

Figure 4.19 Birefringence measured at zero stress for various double networks, identified according to their residual stretch ratio, all having the same total crosslink density (= 102 mol/m^3). The latter, and hence Δn, increases with increase in either the proportion of crosslinks in the first network or the strain during formation of the second network. Data from ref. [134].

4.5.3 Bimodal networks

A bimodal network is an elastomer in which a portion of the chains between crosslink junctions are very short and the remaining are very long. Model bimodal networks are prepared by end-linking a mixture of low and high M_w precursor chains [142]. Scattering experiments show the inhomogeneity of bimodal networks on the size scale of the network mesh [143], with some indication in PDMS networks of segregation of the short and long chains [144]. Clustering of the short and long chains has also been seen in simulations of bimodal networks [145]. However, the respective orientations of the long and short chains are not very different, as found in infrared dichroism measurements on deuterium-labeled networks of poly(tetrahydrofuran) (PTHF, also known as polytetramethylene oxide); see Figure 4.20 [146]. In bimodal networks with a preponderance of short chains, the mesh size of the network was found to be essentially the same as for unimodal networks of the same short chains [147]. This is not the case for bimodal networks composed primarily of long chains, in which this correlation length relates to the entanglement spacing [147].

The appeal of bimodal networks is that they can exhibit substantially higher toughness and strength (Figure 4.21) [148,149]. Such results are contrary to the early ideas of rubber elasticity, according to which failure is due to rupture of the shortest network chains [150], implying the short chains in a bimodal network would weaken the elastomer. The improved properties require that the short chains be low in molecular weight (typically a few hundred g/mol) and lower by as much as a factor of 100 than the molecular weight of the longer chains [151]. The data in Figure 4.21 exhibit maxima, indicating that the number density of the short chains should be large, typically around 95 mol%, which still corresponds to a low weight fraction. Note that the converse, a small weight fraction of long chains, also improves

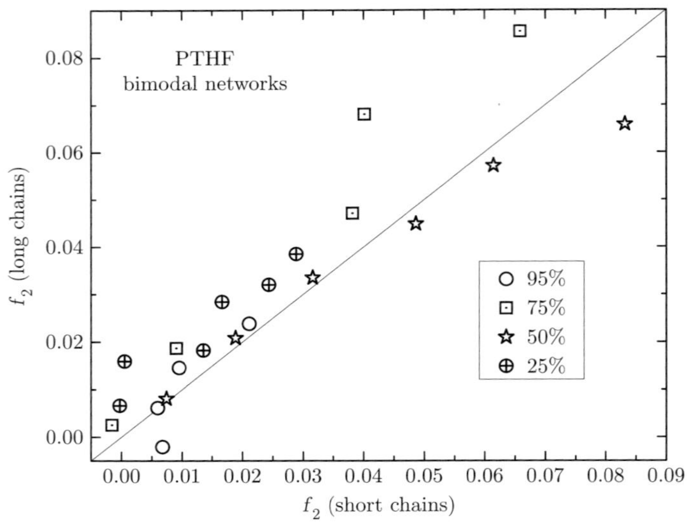

Figure 4.20 Hermans orientation function (see eqn (5.18)) measured for PTHF bimodal networks having the indicated mole percent of short chains. M_n = 11 076 and 1560 g/mol for the long and short chains, respectively. The solid line corresponds to equal orientation. Data from ref. [146].

toughness, although such materials are not elastomeric [152,153]. The improved properties of bimodal networks are ascribed to a synergy between the high modulus of the short chains and the extensibility of the longer ones [142,151]. Experiments on polyethylene oxide [154] and polytetramethylene oxide [155] indicate a greater propensity to strain crystallize, presumably because the short chains orient sooner to serve as nucleating sites. This enhanced strain crystallization would contribute to better failure properties.

4.5.4 Heterogeneous miscible networks

The components of an IPN are immiscible, with crosslinking used to minimize phase segregation; however, the morphology of IPNs is rarely homogeneous on the segmental level. Hydrogels employ water as a solvent to compatibilize the components and, as described above, enhanced properties can be achieved [111–115]. If the polymers comprising a network are thermodynamically miscible, there is no driving force for phase separation and a composition can be achieved that is uniform on the sub-nm level. Various commercial materials are based on miscible polymer blends, although a homogeneous phase morphology *per se* does not necessarily provide improved mechanical properties. The expectation is that the blend properties will be some average of those of the pure components. Thus, there is a second important requirement—a large disparity in the respective crosslink densities of the components. This disparity causes their respective contributions to the network mechanical response to differ diametrically, in this manner emulating bimodal networks and the hydrogel in Figure 4.16, but without the need for end-linking or a solvent. Unlike IPNs, the components are uniformly mixed, but there is only one network. Chains of the more crosslinked component become highly stretched upon deformation of a heterogeneous miscible network, raising the stress

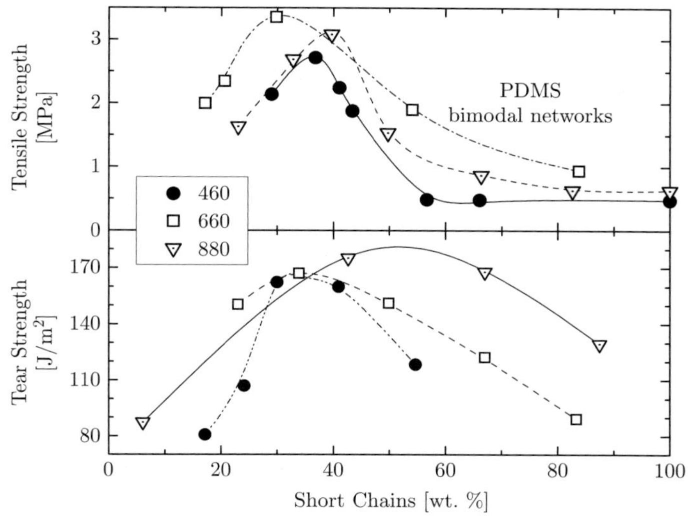

Figure 4.21 Failure properties of PDMS bimodal networks with short chains having the indicated M_n; for the long chains $M_n = 21{,}300$ g/mol. Data from refs. [148,149].

locally, while chains of the lightly crosslinked component are able to alleviate overstressing and consequent spread of rupture nuclei, which would otherwise lead to macroscopic failure.

There are two approaches to achieve a miscible mixture of a highly crosslinked polymer with one that is lightly or even uncrosslinked. One means is to use chemically identical components, except that only one has crosslink sites. Examples would be EPDM mixed with ethylene-propylene copolymer and butyl rubber mixed with polyisobutylene.

Some data illustrating the first method is shown in Figure 4.22 for EPDM (a random terpolymer of ethylene, propylene and a diene[5]) blended with a random copolymer of ethylene and propylene (EP) [156]. The blend is miscible over significant ranges of backbone composition [157]. The EP is unreactive to sulfur vulcanization, so a large disparity in crosslink density can be achieved. At constant modulus the failure properties of EPDM/EP networks exhibit a maximum at approximately equal concentration of the components.

The second approach to heterogeneous networks is blending chemically different polymers that are thermodynamically miscible and have different crosslinking reactivities. One example is the mixtures of PI and PVE. Miscibility of natural rubber and high vinyl polybutadiene was first reported by Bartenev and Kongarov [158], and subsequently investigated by various groups [159–161]. Thermodynamic miscibility is the result of a fortuitous near-equivalence of the polarizabilities (and thus van der Waals energies) of the repeat units for the two polymers [162,163]. Although infrared spectroscopy indicates an absence of specific interactions [164], the interaction parameter is negative, implying a lower critical solution temperature (LCST) [162]. The blend is attractive for hetergeneous networks because tertiary vinyl carbons do not sulfurize [65], so that only PI can be vulcanized.

[5] EPDM rubbers employ ethylidene norbornene, dicyclopentadiene, 1,4-hexadiene, vinyl norbornene, or norbornadiene for the unsaturated repeat unit at concentrations up to 12%.

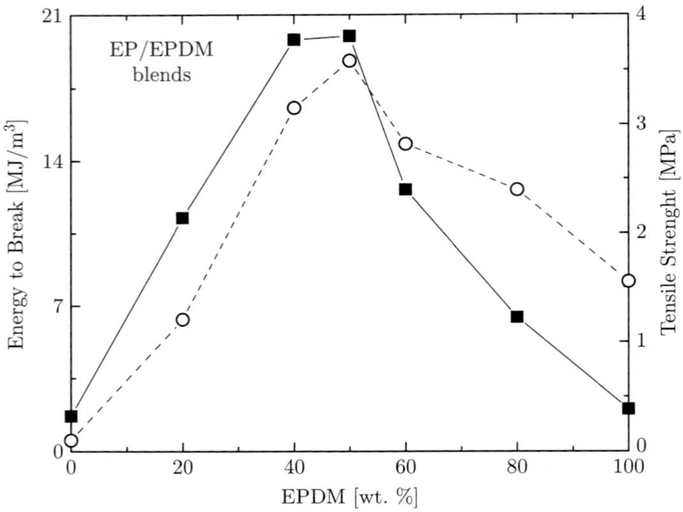

Figure 4.22 Fracture energy (filled squares) and stress at break (open circles) for vulcanized blends of EPDM with ethylene-propylene copolymer, in which the curative level was adjusted to maintain a constant crosslink density of the EPDM (Young's modulus = 1.5 ± 0.1 independent of composition). Data from ref. [156].

This selectivity of sulfur crosslinking in PI/PVE blends is illustrated in DSC curves (Figure 4.23) [165]. The gel phase is essentially PI, while the soluble fraction is the unreacted PVE. However, the chemical changes effected by the crosslinking reaction reduce the miscibility of PI with the PVE, so that some degree of heterogeneity of the phase morphology arises, governed by the rate of curing [165]. Alternatively, organic peroxide can be used to crosslink the blend. Although the PVE is more reactive than PI to free radicals, the method does not give a sufficient disparity in the respective degrees of cure. Consequently, the mechanical properties are poorer, notwithstanding the more homogeneous phase morphology (Figure 4.24) [165]. Similar to the behavior of bimodal networks, the best mechanical properties correspond to the largest disparity in crosslinking.

Another miscible blend with components having different crosslinking chemistries is 1,4-polychloroprene (PC) and epoxidized 1,4-polyisoprene (EPI). The miscibility arises from specific interaction of the chlorine with the oxirane group [166,167]. While EPI sulfur vulcanization is similar to that of polyisoprene, PC can be crosslinked using zinc and magnesium oxides without an organic accelerator. The zinc oxide produces ether crosslinks through the tertiary allylic chorine, and the magnesium oxide extends the reaction by scavenging the chlorine ion byproducts [168]. The distinct crosslink mechanisms of PC and EPI afford control of the rate and degree of network formation in their blends. It has been shown that a miscible morphology is retained after curing, although the properties obtained for the blend were only intermediate between those of the pure components (Figure 4.25) [169].

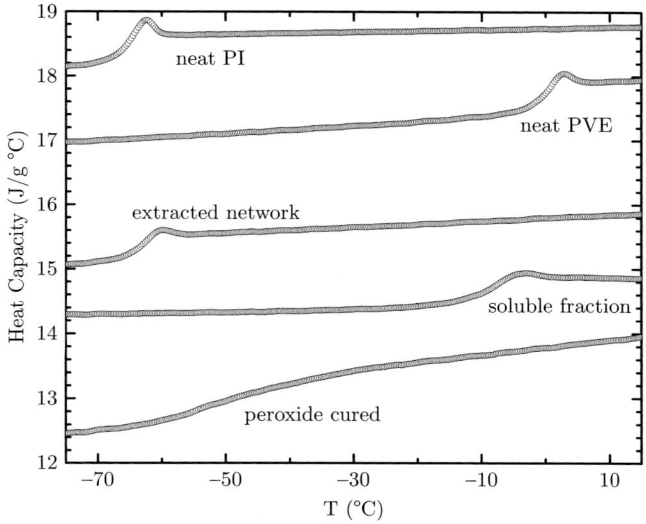

Figure 4.23 DSC curves for pure, uncrosslinked components (upper curves) and for the insoluble and soluble fractions of a sulfur-vulcanized blend of PI and PVE. The transition temperature for the insoluble fraction is equal to that of the neat PI ($T_g = -64$ °C) and for the soluble fraction to that of neat PVE ($T_g = -1$ °C), indicating that only the PI has undergone the crosslinking reaction. The lowest curve is for the peroxide-crosslinked blend, which has a homogeneous phase morphology but with similar degrees of crosslinking of the components. Data from ref. [165].

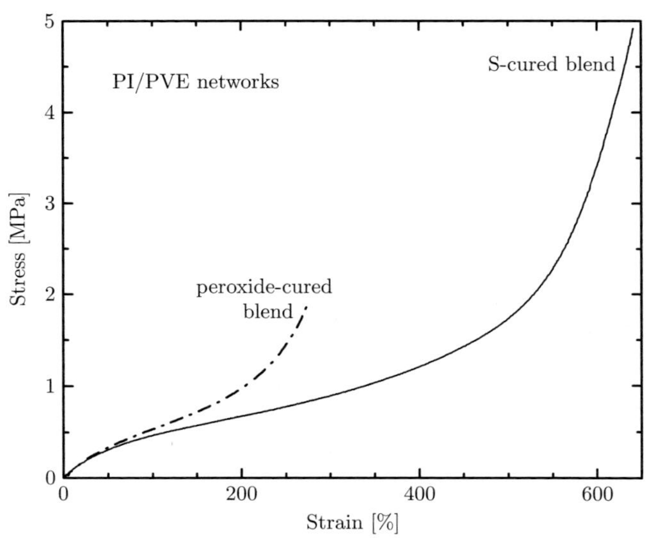

Figure 4.24 Engineering stress *versus* tensile strain for the sulfur and peroxide cured blends of equal weights of PI and PVE. The PVE is relatively unreactive to the S-vulcanization, with the consequent disparity in crosslink density of the components yielding better mechanical properties. Date from ref. [165].

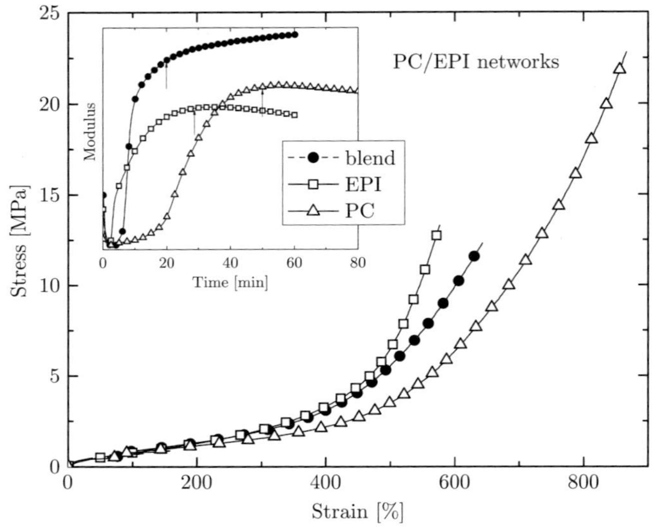

Figure 4.25 Engineering stress *versus* tensile strain for equal weight blends of polychloroprene and 50 mol% epoxidized polyisoprene. The blend mechanical properties are intermediate to those of the pure components due to the near parity of the crosslink density of the components. The inset show the cure curves for the neat components and the blend, with the arrow indicating the state of cure at which the respective stress–strain curves were measured. Data from ref. [169].

4.5.5 Deswollen networks

The stretch ratio of the chains comprising a network, λ_0, is related to the macroscopic extension ratio, λ, according to

$$\lambda_0 = \left(\frac{V}{V_x}\right)^{1/3} \lambda \tag{4.65}$$

where V_x and V are the volumes during crosslinking and subsequent deformation, respectively. Equation (4.65) indicates that deswelling of a network prepared by crosslinking diluted chains results in a compacted or "supercoiled" network [170]. This compaction of the chains is seen in radius of gyration measurements on deswollen ($V/V_x = 0.1$) polystyrene networks of different crosslink density (Figure 4.26) [171]. The behavior is analogous to networks swollen by a poor solvent, which similarly causes some collapse of the chains [172].

The highly compressed "internal state", in combination with a lowered degree of chain entanglements, causes deswollen networks to exhibit low modulus and very large failure strains [173,174]. This "superelasticity" is illustrated in Figure 4.27 comparing stress–strain curves for conventional and deswollen networks [175]. The greater packing of the collapsed chains increases the mass density (about 15% for $V_x \gg V$ in Vasiliev *et al.* [176]), which contributes to an increase in modulus. The result is a minimum in the dependence of the modulus on V_x (Figure 4.28). The more compact chain configurations in deswollen networks also suppresses the rate of thermal (unoriented) cystallization [177].

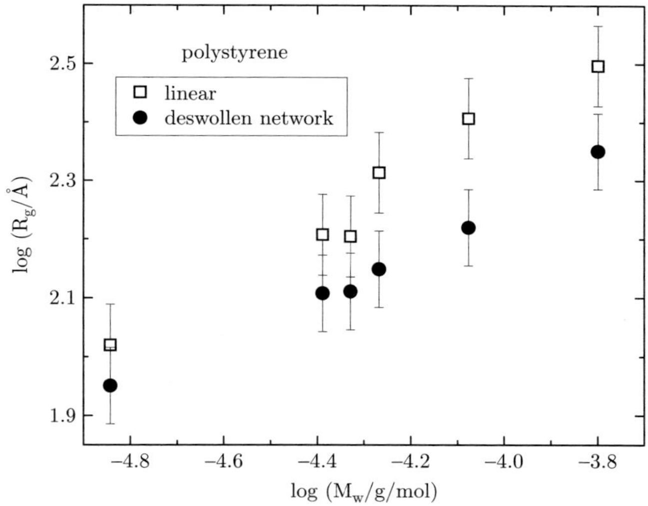

Figure 4.26 Radius of gyration of deuterated polystyrene chains dispersed in a melt (squares) and in a network crosslinked in a tenfold-swollen state and then dried (circles), as a function of the molecular weight of the respective melt or network chains. Data from ref. [171].

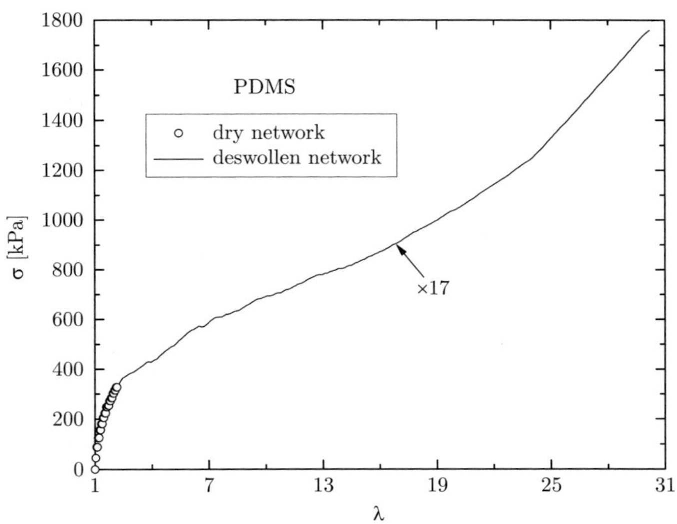

Figure 4.27 Engineering stress–strain curve for PDMS conventional network (circles) and formed in toluene solution at 25% volume fraction and tested after solvent removal (solid line). The data for the latter were multiplied by 17 to superpose the initial moduli. The stretch ratio at failure increases from 2 to about 30 for the deswollen elastomer. Data from ref. [175].

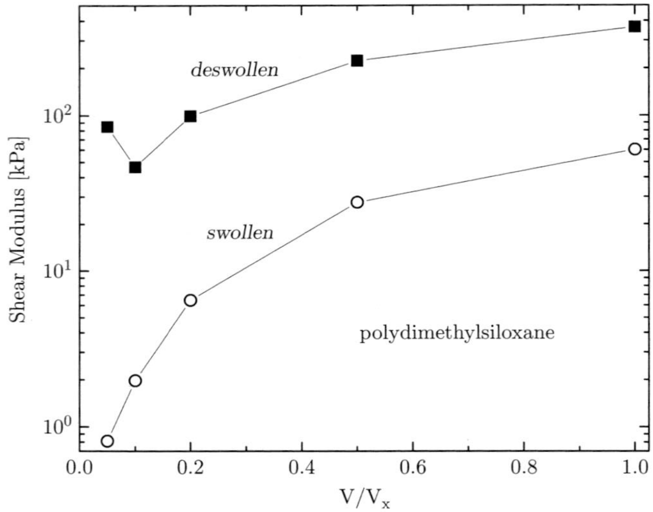

Figure 4.28 Shear modulus of PMDS crosslinked at the indicated volume fraction measured before (circles) and after (squares) deswelling (note that the ordinate scale is logarithmic). Data from ref. [176].

References

1. K.L. Ngai, A.K. Rajagopal, and S. Teitler, *J. Chem. Phys.* **88**, 5086 (1988).
2. K.L. Ngai and R.W. Rendell, *ACS Symp. Ser.* **676**, 45 (1997).
3. I.S. Choi, C.M. Roland, and LC. Bissonnette, *Rubber Chem. Technol.* **67**, 892 (1994).
4. M. Mooney, *J. Appl. Phys.* **11**, 582 (1940).
5. R.S. Rivlin, *Rubber Chem. Technol.* **65**, G51 (1992).
6. L.R.G. Treloar, *Rep. Prog. Phys.* **36**, 755 (1973).
7. R.S. Rivlin in "Rheology," Vol. 1 (F.R. EIrich, ed.) Academic Press, New York, Chapter 10 (1956).
8. R.S. Rivlin, *J. Elasticity* **73**, 291 (2003).
9. K.N. Morman and T.Y. Pan, *Rubber Chem. Technol.* **61**, 503 (1988).
10. C.L. Chow and C.H. Cundiff, *Tire Sci. Technol.* **15**, 73 (1987).
11. V. Vahapoglu and S. Karadeniz, *Rubber Chem. Technol.* **79**, 489 (2006).
12. R.W. Ogden, *Proc. Roy. Soc. A.* **326**, 565 (1972).
13. R.H. Finney and A. Kumar, *Rubber Chem. Technol.* **61**, 879 (1988).
14. O.H. Yeoh, *Rubber Chem. Technol.* **70**, 175 (1997).
15. K.C. Valanis and R.F. Landel, *J. Appl. Phys.* **38**, 2997 (1967); **11**, 1969 (1978).
16. A.J. Carmichael and H.W. Holdaway, *J. Appl. Phys.* **32**, 159 (1961).
17. R.W. Ogden, *Rubber Chem. Technol.* **59**, 361 (1986).
18. E.A. Kearsley and L.J. Zapas, *J. Rheol.* **24**, 483 (1980).
19. R.S. Rivlin, *Intl. J. Non-Linear Mech.* **41**, 141 (2006).

20. C.O. Horgan and G. Saccomandi, *Rubber Chem. Technol.* **79**, 152 (2006).
21. O.H. Yeoh, *Rubber Chem. Technol.* **63**, 792 (1990).
22. M.J. Gregory, *Plast. Rubber: Mater. Appl.* **4**, 184 (1979).
23. C.K.L. Davies, D.K. De, and A.G. Thomas, *Rubber Chem. Technol.* **67**, 716 (1994).
24. A.N. Gent, *Rubber Chem. Technol.* **69,** 59 (1996).
25. O.H. Yeoh and P.D. Fleming, *J. Polym. Sci.* **B35**, 1919 (1997).
26. C.O. Horgan and G. Saccomandi, *J. Elasticity* **68**, 167 (2002)
27. C.O Horgan and J.G. Murphy, *J. Elasticity* **86**, 101 (2007).
28. A. N. Gent, *Int. J. Nonlinear Mech.* **40**, 165 (2005).
29. K. Takamizawa and K. Hayashi, *J. Biomechanics* **20**, 7 (1987).
30. E. Pucci and G. Saccomandi, *Rubber Chem. Technol.* **75**, 839 (2002).
31. C.O. Horgan and G. Saccomandi, *J. Elasticity* **77**, 123 (2004).
32. W. Kuhn, *J. Polym. Sci.* **1**, 380 (1946).
33. P.J. Flory and J. Rehner, *J. Chem. Phys.* **11**, 512 (1943).
34. H.M. James and E. Guth, *J. Chem. Phys.* **11**, 455 (1943); **15**, 669 (1947); *J. Polym. Sci.* **4**, 153 (1949).
35. P.J. Flory, *Proc. Roy. Soc. London A* **351**, 351 (1976).
36. P.J. Flory, *Macromolecules* **15**, 99 (1982).
37. W. Kuhn and F. Grün, *Kolloid-Z.* **101**, 248 (1942).
38. L.R.G. Treloar and G. Riding, *Proc. Roy. Soc. London A* **369**, 261 (1979).
39. L. R. G. Treloar, *Trans. Faraday Soc.* **50,** 881 (1975).
40. P.D. Wu and E. van der Giessen, *J. Mech. Phys. Sol.* **41**, 427 (1993).
41. M. Wang and E. Guth, *J. Chem. Phys.* **20,** 1144 (1952).
42. L. R. G. Treloar, *Trans. Faraday Soc.* **42,** 83 (1946).
43. E.M. Arruda and M.C. Boyce, *J. Mech. Phys. Sol.* **41**, 389 (1993).
44. L.R.G. Treloar, *Trans. Faraday Soc.* **40**, 59 (1944).
45. T. Beda, *J. Polym Sci. Polym. Phys. Ed.* **45**, 1713 (2007).
46. G. Marckmann and E. Verron, *Rubber Chem. Technol.* **79**, 835 (2006).
47. L. Meunier, G. Chagnon, D. Favier, L. Orgeas, and P. Vacher, *Polym. Testing* **27**, 765 (2008).
48. M. Sasso, G. Palmieri, G. Chiappini, and D. Amodio, *Polym. Testing* **27**, 995 (2008).
49. D.J. Seibert and N. Schoche, *Rubber Chem. Technol.* **73**, 366 (2000).
50. M. Rachik, F. Schmidt, N. Reuge, Y.L. Maoult, and F. Abbe, *Polym. Eng. Sci.* **41**, 532 (2001).
51. M. Beatty, *J. Elasticity* **70**, 65 (2003).
52. M. Boyce, *Rubber Chem. Technol.* **69**, 781 (1996).
53. R. Oeser, B. Ewen, D. Richter, and B. Farago, *Phys. Rev. Lett.* **60**, 1041 (1988); D. Richter, B. Farago, B. Ewen, and R. Oeser, *Physica B* **156**, 426 (1989).
54. B. Ewen and D. Richter, *Adv. Polym. Sci.* **134**, 1 (1997).
55. P.J. Flory, *J. Chem. Phys.* **66**, 5720 (1977).
56. P.J. Flory and B. Erman, *Macromolecules* **15,** 800 (1982).
57. P.H. Mott and C.M. Roland, *Macromolecules* **29**, 6941 (1996).
58. B. Erman and L. Monnerie, *Macromolecules* **22,** 3342 (1989).

59. F. Fontaine, C. Morland, C. Loel, L. Monnerie, and B. Erman, *Macromolecules* **22**, 3348 (1989).
60. A. Kloczkowski, J.E. Mark, and B. Erman, *Macromolecules* **28**, 5089 (1995).
61. L.M. Dossin and W.W. Graessley, *Macromolecules* **12**, 123 (1979).
62. N.R. Langley, *Macromolecules* **1**, 348 (1968).
63. M. Rubinstein and S. Panyukov, *Macromolecules* **30**, 8036 (1997).
64. M. Rubinstein and S. Panyukov, *Macromolecules* **35**, 6670 (2002).
65. A.Y. Coran in *Science and Technology of Rubber*; Ehrich, F. R., ed.; Academic Press, New York (1978).
66. R.C. Ball, M. Doi, S.F. Edwards, and M. Warner, *Polymer* **22**, 101 (1981).
67. S.F. Edwards and T. Vilgis, *Polymer* **27**, 483 (1986).
68. T.A. Vilgis and B. Erman, *Macromolecules* **26**, 6657 (1993).
69. M. Kaliske and G. Heinrich, *Rubber Chem. Technol.* **72**, 602 (1999).
70. C.M. Roland, P.H. Mott, and G. Heinrich, *Comput. Theor. Polym. Sci.* **9**, 197 (1999).
71. T.A. Vilgis and F. Boue, *J. Poly. Sci.* **26**, 2291 (1988).
72. T.A. Vilgis and G. Heinrich, *Macromol. Theory Simul.* **3**, 172 (1994).
73. M. Warner, *J. Phys. C.* **14**, 4990 (1981).
74. S.F. Edwards, *Polymer J.* **17**, 271 (1985).
75. N.R. Kenkare, S.W. Smith, C.K. Hall, and S.A. Khan, *Macromolecules* **31**, 5861 (1998).
76. K.L. Ngai, C.M. Roland, and A.F. Yee, *Rubber Chem. Technol.* **66**, 817 (1993).
77. K.L. Ngai and C.M. Roland, *Macromolecules* **27**, 2454 (1994).
78. C.M. Roland, *Macromolecules* **27**, 4242 (1994).
79. J.F. Shi, L.C. Dickinson, W.J. MacKnight, and J.C.W. Chien, *Macromolecules* **26**, 5908 (1993).
80. J. Sanjuan and M.A. Llorente, *J. Polym. Sci. Polym. Phys. Ed.* **26**, 235 (1988).
81. J. Liu, D. Cao, and L. Zhang, *J. Chem. Phys.* **131**, 034903 (2009).
82. R. Casalini and C.M. Roland, *J. Polym. Sci. Pt. B-Polym. Phys.* **48**, 582 (2010).
83. R. Casalini and C.M. Roland, *J. Chem. Phys.* **131**, 114501 (2009).
84. W.J. MacKnight and R.D. Lundberg, *Rubber Chem. Technol.* **57**, 652 (1984).
85. P. Antony and S.K. De, *J. Macromol. Sci. Polym. Rev.* **C41**, 41 (2001).
86. L.H. Sperling, *Interpenetrating Polymer Networks and Related Materials*, Plenum, New York (1981).
87. *Advances in Interpenetrating Polymer Networks*, Vol. 1, (D. Klempner and K.C. Frisch eds.) Technomic, Lancaster (1989).
88. *IPNs Around the World: Science and Engineering* (S.C. Kim, L.H. Sperling, eds), Wiley, Chichester (1997).
89. N. Gupta and A.K. Srivastava, *Polym. Int.* **35**, 109 (1994).
90. B. Suthar, H.X. Xiao, D. Klempner, and K.C. Frisch, *Polym. Adv. Technol.* **7**, 221 (1996).
91. J.M.M. Dueñas and J.L.G. Ribelles, *J. Therm. Anal. Calor.* **72**, 695 (2003).
92. N.R. Manoj, R.D. Raut, P. Sivaraman, D. Ratna, and B.C. Chakraborty, *J. Appl. Polym. Sci.* **96**, 1487 (2005).
93. R.P. Burford, M.G. Markotsis, and R.B. Knott, *Physica B* **385–386**, 766 (2006).
94. T. Tamai, A. Imagawa, and Q. Tran-Cong, *Macromolecules* **27**, 7486 (1994).
95. X. Ramis, A. Cadenato, and J.M. Morancho, *Polymer* **42**, 9469 (2001).

96. F. Vidal, O. Fichet, J. Laskar, and D. Teyssié, *Polymer* **47**, 3747 (2006).
97. N.V. Babkina, Y.S. Lipatov, T.T. Alekseeva, L.A. Sorochinskaya, and Y.I. Datsyuk, *Polym. Sci. Ser. A* **50**, 798 (2008).
98. *Sound and Vibration Damping with Polymers*, ACS Symp. Ser. 424, R.D. Corsaro and L.H. Sperling, eds. (1990).
99. D.I.G. Jones, *Handbook of Viscoelastic Vibration Damping*, Wiley Interscience, New York (2001).
100. R.B. Fox, J.P. Armistead, C.M. Roland, and D.J. Moonay, *J. Appl. Polym. Sci.* **41**, 1281 (1990).
101. K.C. Frisch, D. Klempner, S. Migdal, H.L. Frisch, and H. Ghiradella, *Polym. Eng. Sci.* **14**, 76 (1974).
102. Y.M. Lee, Y.S. Yang, and L.J. Lee, *Polym. Eng. Sci.* **27**, 716 (1987).
103. C. Chunming, X. Chuanxi, Y. Jian1, and D. Lijie, *J. Wuhan Univ. Technol.—Mater. Sci. Ed.* **23**, 50 (2008).
104. D. Myung, W. Koh, J. Ko, Y. Hu, M.I. Carrasco, J. Noolandi, C.N. Ta, and C.W. Frank, *Polymer* **48**, 5376 (2007).
105. E.F. Irwin, K. Saha, M. Rosenbluth, L.J. Gamble, D.G. Castner, and K.E. Healy, *J. Biomater. Sci. Polym. Ed.* **19**, 1363 (2008).
106. Y. Liu, and M. Chan-Park, *Biomaterials* **30**, 186 (2009).
107. Z.S. Akdemir and N. Kayaman-Apohan, *Polym. Adv. Technol.* **18**, 932 (2007).
108. Q. Wang, S. Li, Z. Wang, H. Liu, and C. Li, *J. Appl. Polym. Sci.* **111**, 1417 (2008).
109. H. Ajiro, Y. Takemoto, and M. Akashi, *Chem. Lett.* **38**, 368 (2009).
110. Y. Yue, P. Wang, Q. Chen, and X.G. Liu, *J. Appl. Polym. Sci.* **112**, 2261 (2009).
111. J. Kopecek and J. Yang, *Polym. Int.* **56**, 1078 (2007).
112. D. Myung, D. Waters, M. Wiseman, P.-E. Duhamel, J. Noolandi, C.N. Ta, and C.W. Frank, *Polym. Adv. Technol.* **19**, 647 (2008).
113. L.K. Kostanski, R.X. Huang, C.D.M. Felip, and R. Ghosh, *J. Biomater. Sci. Polym. Ed.* **20**, 271 (2009).
114. H. Tsukeshiba, M. Huang, Y.-H. Na, T. Kurokawa, R. Kuwabara, Y. Tanaka, H. Furukawa, Y. Osada, and J.P. Gong, *J. Phys. Chem. B* **109**, 16304 (2005).
115. J.P. Gong, Y. Katsuyama, T. Kurokawa, and Y. Osada, *Adv. Mater.* **15**, 1155 (2003).
116. A.V. Tobolsky, Y. Takahashi, and S. Naganuma, *Polym. J.* **3**, 60 (1972).
117. K.T. Gillen, *Macromolecules* **21**, 442 (1988).
118. L. Mandelkern, D.E. Roberts, A.F. Diorio, and A.S. Posner, *J. Am. Chem. Soc.* **81**, 4148 (1959).
119. R.A.M. Hikmet, J. Lub, and P.M. Vanderbrink, *Macromolecules* **25**, 4194 (1992).
120. W.F. Reichert, D. Goritz, and E.J. Duschl, *Polymer* **34**, 1216 (1993).
121. J. Scanlan and W.F. Watson, *Trans. Faraday Soc.* **54**, 740 (1958).
122. S. Granick and J.D. Ferry, *Macromolecules* **16**, 39 (1983).
123. W. Batsberg and O. Kramer, *J. Chem. Phys.* **74**, 6507 (1981).
124. R.D. Andrews, A.V. Tobolsky, and E.E. Hanson, *J. Appl. Phys.* **17**, 352 (1946).
125. T. Termonia, *Macromolecules* **23**, 1976 (1990).
126. L.G. Baxandall and S.F. Edwards, *Macromolecules* **21**, 1763 (1988).
127. B. Meissner and L. Matejka, *Polymer* **44**, 4611 (2003).
128. P.G. Santangelo and C.M. Roland, *Rubber Chem. Technol.* **67**, 359 (1994).

129. P.G. Santangelo and C.M. Roland, *Rubber Chem. Technol.* **68**, 124 (1995).
130. G.R. Hamed and M.Y. Huang, *Rubber Chem. Technol.* **71**, 846 (1998).
131. A. Greene, K.J. Smith, and A. Ciferri, *Trans. Faraday Soc.* **61**, 2772 (1965).
132. C.M. Roland and M.L. Warzel, *Rubber Chem. Technol.* **63**, 285 (1990).
133. C.V. Marykutty, G. Mathew, and S. Thomas, *Rubber Chem. Technol.* **64**, 790 (1991).
134. P.H. Mott and C.M. Roland, *Macromolecules* **33** 4132 (2000).
135. C.M. Roland and K.L. Peng, *Rubber Chem. Technol.* **64**, 790 (1991).
136. J. Wang, G.R. Hamed, K. Umetsu, and C.M. Roland, *Rubber Chem. Technol.* **78**, 76 (2005).
137. P.G. Santangelo and C.M. Roland, *Rubber Chem. Technol.* **76**, 892 (2003).
138. G. J. Lake and P. B. Lindley, *J. Appl. Polym. Sci.* **10**, 343 (1966).
139. A. Kloczkowski, J.E. Mark, and B. Erman, *Comp. Polym Sci.* **5**, 37 (1995).
140. B. Erman and P.J. Flory, Macromolecules **16**, 1607 (1983).
141. R.S. Stein, R.J. Farris, S. Kumar, and V. Soni, *ACS Symp. Ser.* **193**, 452 (1982).
142. J. Mark, *Macromol. Symp.* **191**, 121 (2003).
143. V.K. Soni and R.S. Stein, *Macromolecules* **23**, 5257 (1990).
144. W. Wu, L.D. Coyne, L. Jong, A. Hanyu, and R.S. Stein, *Macromolecules* **23**, 351 (1990).
145. J.U. Sommer and S. Lay, *Macromolecules* **35**, 9832 (2002).
146. A. Hanyu and R.S. Stein, *Makromol. Chem. Macromol. Symp.* **45**, 189 (1991).
147. K. Urayama, T. Kawamura, Y. Hirata, and S. Kohjiya, *Polymer* **39**, 3827 (1998).
148. T.L Smith, B. Haidar, and J.L. Hedrick, *Rubber Chem. Technol.* **63**, 256 (1990).
149. J.E. Mark and M.Y. Tang, *J. Polym. Sci. Polym. Phys. Ed.* **22**, 1849 (1984).
150. F. Bueche, *Physical Properties of Polymers*, Wiley-Interscience, New York (1962).
151. J.E. Mark, *Rubber Chem. Technol.* **72**, 465 (1999).
152. G.A. Holmes and A. Letton, *Polym. Eng. Sci.* **34**, 1635 (1994).
153. N.C.B. Tan, B.J. Bauer, J. Plestil, J.D. Barnes, D. Liu, L. Matejka, K. Dusek, and W.L. Wu, *Polymer* **40**, 4603 (1999).
154. C.C. Sun and J.E. Mark, *J. Polym. Sci., Polym. Phys. Ed.* **25**, 2073 (1987).
155. A. Hanyu and R.S. Stein, *Makromol. Chem., Macromol. Symp.* **45**, 189 (1991).
156. G.S. Buckley, D.M. Fragiadakis, and C.M. Roland, unpublished.
157. H.G. Kikland and M. Van Duin, *Rubber Chem. Technol.* **76**, 495 (2003).
158. G.M. Bartenev and G.S. Kongarov, *Rubber Chem. Technol.* **36**, 668 (1963).
159. R.E. Cohen, *Adv. Chem. Series* **193**, 489 (1982).
160. C.M. Roland, *Macromolecules* **20**, 2557 (1987).
161. S. Kawahara, S. Akiyama, and A. Ueda, *Polymer J.* **21**, 221 (1989).
162. D.W. Tomlin and C.M. Roland, *Macromolecules* **25**, 2994 (1992).
163. C.A. Trask and C.M. Roland, *Macromolecules* **22**, 256 (1989).
164. C.M. Roland, *J. Polym. Sci. Polym. Phys.* **26**, 839 (1988).
165. J. Wang and C.M. Roland, *Polymer* **46**, 4160 (2005).
166. J.B. Nagode and C.M. Roland, *Polymer* **32** 505 (1991).
167. J.H Walton, J.B. Miller, and C.M. Roland, *J. Poly. Sci. Poly. Phys. Ed.* **30** 527 (1992).

168. A.C. Stevenson in *Vulcanization of Elastomers* (G. Alliger and I.J. Sjothun, eds) Reinhold Publ., New York (1964).
169. Y.T. Vu and C.M. Roland, unpublished.
170. J.J. Hermans, *J. Polym. Sci.* **59**, 191 (1962).
171. J. Bastide, in *Physics of Finely Divided Matter*, Springer Proceedings of Physics, Vol. 5, (edited by N. Boccara and M. Daoud), Springer, Berlin (1985).
172. P. Cifra and T. Bleha, *J. Polym. Sci. Polym. Phys. Ed.* **37**, 2013 (1999).
173. S.P. Obukhov, M. Rubinstein, and R.H. Colby, *Macromolecules* **27**, 3191 (1994).
174. K. Sivasailam and C. Cohen, *J. Rheol.* **44**, 897 (2000).
175. K. Urayama and S. Kohjiya, *Eur. Phys. J. B* **2**, 75 (1998).
176. V.G. Vasiliev, L.Z. Rogovina, and G.L. Slonimsky, *Polymer* **26**, 1667 (1985).
177. K. Urayama, K. Yokoyama, and S. Kohjiya, *Polymer* **41**, 3273 (2000).

5

Constitutive modeling, non-linear behavior, and the stress-optic law

The stresses accompanying material flow can be calculated using constitutive equations, which relate the displacements and mechanical loads in terms of material properties, which are independent of sample dimensions and geometry. Constitutive equations are essential for quantitative descriptions of polymer rheology and processing. Chapters 1 and 3 discussed the Rouse model for unentangled polymers and the tube model for entangled chains. There is an enormous literature devoted to constitutive models for the linear viscoelastic response of polymer solutions and entangled melts [1–4]; however, non-linear effects present additional challenges. Flow at high rates can reduce the degree of chain entanglement (Chapter 3). A broad consideration of the non-linear flow behavior of polymers is eschewed herein, with consideration limited primarily to aspects unique to elastomers and filler-reinforced rubber.

5.1 Linearity and the superposition principle

The problem of analyzing viscoelastic mechanical behavior is greatly facilitated when the material exhibits linearity, which implies that no strain interferes with the response to any prior strain. The strict definition is: (i) stress and strain remain directly proportional to one another, and (ii) time invariance of the mechanical response is observed. The stress resulting from an applied strain can then be described by the Boltzmann equation

$$\sigma(t) = \int_0^t E(t-u)\frac{d\varepsilon}{du}\,du \tag{5.1}$$

where $E(t)$ is the stress relaxation function (for tension as written) and ϵ is the uniaxial strain. Writing this equation in the form

$$\sigma(t) = \int_0^t \frac{dE(t-u)}{du}\Big(\varepsilon(t)-\varepsilon(u)\Big)\,du + E(t)\varepsilon(t) \tag{5.2}$$

makes clear the idea that the stresses are additive with $\frac{dE(t-u)}{du}du$ representing the survival probability at the present time t of a stress created at time u. Analogous expressions can be written for other modes of deformation such as shear.

For crosslinked rubber the strain can be defined in terms of the strain function suggested by the statistical theories of rubber elasticity (defined in Chapter 4)

$$\varepsilon \equiv f(\lambda) = \lambda - \lambda^{-2} \tag{5.3}$$

All polymers obey eqn (5.1) at sufficiently small strains; that is, within the bounds of linear behavior. For elastomers the limits of linearity extend over a wider range than for most materials [5]; however, adherence to eqn (5.1) is usually not evident from a single experiment [6]. Beyond the maximum strain for which strict linear viscoelasticity applies, an integral constitutive equation, known as the K-BKZ equation[1] [7,8] can be employed

$$\sigma(t) = \int_0^t [E\Big(t-u, \varepsilon(t-u)\Big)] \frac{d\varepsilon}{du}\, du \tag{5.4}$$

The application of this integral is limited by the strain dependence in the kernel. If the relaxation behavior *per se* is independent of strain, the effects of time and strain are uncoupled. Such "time invariance" permits the simplification

$$\sigma(t) = \int_0^t [E(t-u)g(\varepsilon)] \frac{d\varepsilon}{du}\, du \tag{5.5}$$

In eqn (5.5) $g(\varepsilon)$, known as a damping function, accounts for the fact that the stress on the material does not remain directly proportional to strain during the course of the deformation; thus, the product $g(\varepsilon)\varepsilon$ quantifies the non-linear elasticity, and $g(\varepsilon)$ is unity for linear behavior. Figure 5.1 and Figure 5.2 respectively show relaxation data illustrating this separability of the strain and time dependences for a synthetic NR at different shear strains and for NR under three different modes of deformation [9].

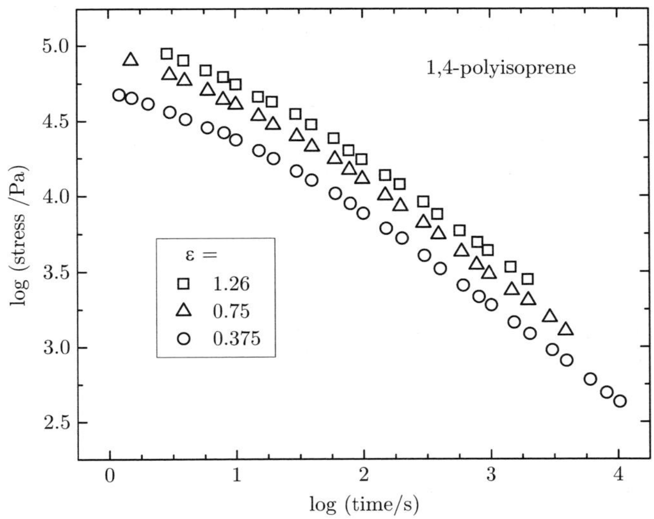

Figure 5.1 Stress relaxation of PI at the indicated shear strains [9]. Separability of time and strain is reflected in the parallel nature of the curves.

[1] Equation (5.4) is known as the K-BKZ equation after the coauthors of ref. [7] and A. Kaye, who in 1962 published equivalent ideas in an obscure note to the College of Aeronautics in Cranfield, UK.

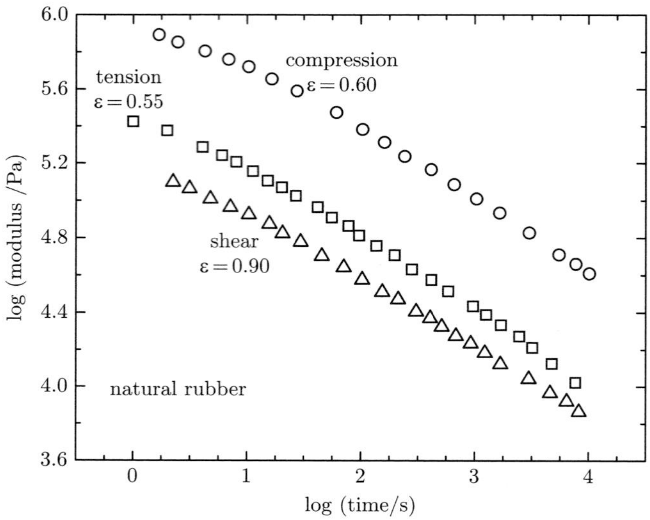

Figure 5.2 Ratio of engineering stress to strain for NR [9]. To a good approximation the stress-relaxation behavior is independent of the mode of deformation.

For a simple step strain the K-BKZ formulation gives

$$\sigma(\varepsilon, t) = E(t)g(\varepsilon)\varepsilon \tag{5.6}$$

For continuous straining the integral in eqn (5.5) is approximated as a series of steps, each described by eqn (5.6). Within the limits for which this separability is valid, the integral constitutive equation, eqn (5.5), enables prediction of the stresses for arbitrary strain histories. The calculation requires knowledge of the stress-relaxation function and the strain dependence of the modulus, both determined by separate experiments. This phenomenology has often been applied to the rheology of polymer melts, with generally good results for strain-increasing deformations [10]. It can also be extended to the normal stresses, which are usually measured in shear

$$N_1 = G(t)g(\gamma)\gamma^2 \tag{5.7}$$

where N_1 is the primary normal stress, acting perpendicular to the strain direction. Combining eqn (5.6) for shear and eqn (5.7) gives the Lodge–Meissner relation [11,12][2]

$$N_1/\sigma = \gamma \tag{5.8}$$

for stress relaxation after a step strain; that is, $\psi_1(t) = G(t)$.

The accuracy of K-BKZ predictions can be improved with the addition of correction terms to account for other mechanisms [13–15]. With the "independent alignment approximation"[3] the Doi–Edwards model of polymer dynamics yields the K-BKZ form [16]. The advantage of

[2] Much earlier, Rivlin derived an equivalent relation applicable quite generally to highly elastic materials [12].

[3] The *independent alignment approximation* of the Doi–Edwards tube model simplifies the problem of calculating the coupled motion of different parts of a polymer chain by ignoring the initial stretch–retraction of the chain following a deformation. By assuming the tube does not retract, the location of a segment within the tube does not change, so that a single segment description can be used for the chain orientation.

molecular theories is the possibility of providing a connection of the structure to the rheological properties. The Doi–Edwards damping function for shear can be approximated as

$$g(\gamma) = \left(1 + \frac{\xi}{3}\gamma^2\right)^{-1} \tag{5.9}$$

with ξ taken as an empirical parameter having a value between 0 and 0.6 [17,18]. $\xi = 0.43$ gives results approximated by an alternative expression derived from the tube model with constraint release [19].

In the absence of mechanisms such as strain crystallization or the finite extension of the chains, rubbery materials usually strain-soften; thus, generally the damping function decreases with strain. For elastomers in uniaxial extension, the form of $g(\varepsilon)$ should be consistent with the equilibrium elastic behavior (Chapter 4). An empirical equation used to describe uniaxial extension measurements is the Mooney–Rivlin relation (eqn (4.10) [20,21]

$$\sigma = f(\lambda)[c_1 + c_2/\lambda) \tag{5.10}$$

in which c_1 and c_2 are material constants and $f(\lambda)$ is given by eqn (5.3). The damping function can then be expressed as [22]

$$g(\varepsilon) = \left(1 + \frac{c_2}{c_1}\lambda^{-1}\right) \tag{5.11}$$

The Mooney–Rivlin equation was derived for elastic equilibrium but can be applied to unrelaxed mechanical stresses, with a larger value of c_2 obtained [23].

Figure 5.3 illustrates the use of eqn (5.4) to predict the stresses for a peroxide-crosslinked SBR stretched continuously at a constant velocity [24]. The modulus and stress relaxation

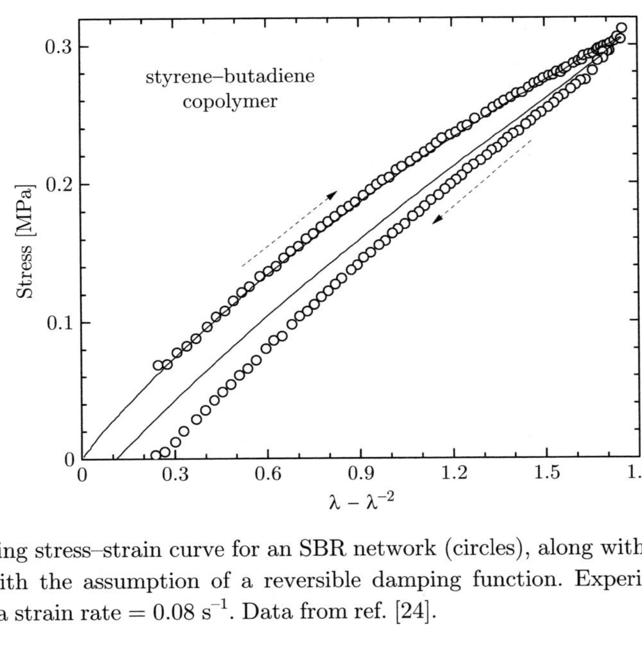

Figure 5.3 Reversing stress–strain curve for an SBR network (circles), along with the curve calculated using eqn (5.5) with the assumption of a reversible damping function. Experiments done at room temperature with a strain rate = 0.08 s^{-1}. Data from ref. [24].

function are both determined in separate experiments, then using these measured strain and rate dependences, the stress for the imposed strain history is calculated with the assumption that $g(\varepsilon)$ depends only on the instantaneous strain, a condition known as reversible damping.[4] As seen in the figure, the measured and calculated extensional stresses are equivalent.

5.2 Internal stress and optical birefringence

Constitutive models describe stresses that result from the imposition of external constraints and can be measured. However, there can be "internal stresses" within solids and very viscous liquids, which are independent of any exterior forces. These are also known as residual stresses, since a common origin is "frozen" orientation that arises during processing of a material subsequently cooled to a non-equilibrium glassy or crystalline state. Figure 5.4 illustrates internal stresses in a PI network [25], which formed when the elastomer was stretched and then quenched below its glass-transition temperature (= −65 °C). In the glassy state no stress is measured; however, upon warming through T_g, the material exhibits a "memory effect", reflecting the presence of the internal stresses. Another example of stresses is found in double-network elastomers, which have orientation stabilized by crosslinking in the strained state (Chapter 4). Since internal stresses exert no pressure on a transducer, they cannot be measured by mechanical instruments, nor can they be described using conventional constitutive equations. Optical probes sensitive to stress are used for their characterization, with the most important being birefringence of visible light.

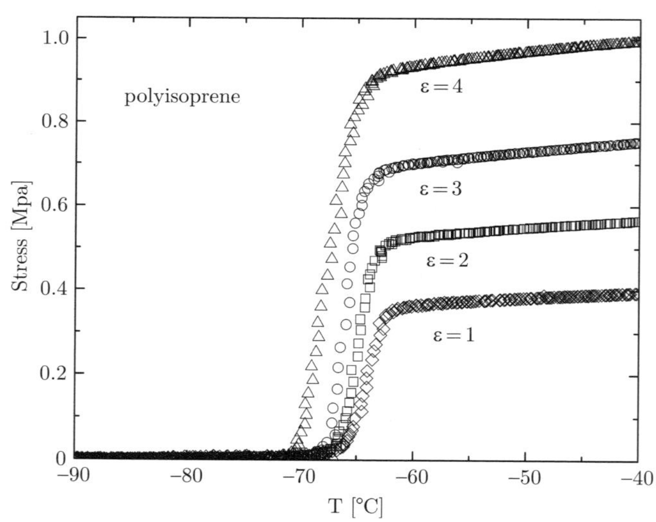

Figure 5.4 Stress measured in a crosslinked 1,4-polyisoprene during heating at 1 K/min. The sample was initially stretched at RT to the indicated tensile strain and then quenched below T_g. Data from ref. [25].

[4] Reversible damping attempts to account for any strain-induced changes in the material that do not recover on the timescale of the experiment.

5.2.1 Stress optic law and chain orientation

The effect of light on matter is to polarize the material, with the induced polarization expressible as a power series in the applied electric field

$$P = \chi_1 E + \chi_2 E^2 + \chi_3 E^3 + \ldots \tag{5.12}$$

where the electric susceptibility χ may be a tensor. Except for intense laser sources, the linear term is sufficient to describe the response[5]. The refractive index is then independent of the light intensity with the Lorentz–Lorenz equation giving

$$\frac{n^2 - 1}{n^2 + 2} = \frac{4\pi}{3} N\chi_1 \tag{5.13}$$

in which N is the number density of molecules (chain segments) having polarizability χ_1. Classically, the material is considered a continuous dielectric with certain symmetry properties. The molecular origin of the polarizability is the dispersibility of the electron cloud, and since the outer electrons are more easily displaced along their bond axis than transversely, this polarizability is inherently anisotropic; however, the average over many randomly arranged molecules yields a symmetric refractive index. When deformation induces orientation, the principal components of the refractive index, n_x, n_y, and n_z, become non-equal, causing birefringence, Δn, of light propagating through the material. Birefringence ("double refraction") refers to a difference between the refractive indices within the material for the two transverse polarizations of propagating light. This birefringence can be used to characterize the deformation according to the stress-optic law,[6] which states that the anisotropy of the refractive index is proportional to the difference in the principal stresses. For uniaxial stress in the x-direction, ($|\hat{\sigma}_x| > 0; \hat{\sigma}_y = \hat{\sigma}_z = 0$, where the caret signifies true stress[7])

$$\Delta n \equiv n_x - n_y = n_x - n_z = C\hat{\sigma}_x \tag{5.14}$$

C is the stress-optical coefficient, a material constant determined by the chemical structure of the repeat unit. Equation (5.14) has long been exploited to quantify stresses and their spatial distribution in glasses, plastics, and other transparent materials, wherein the relation is known as the photoelastic effect. The magnitude of Δn also reflects internal stresses, since it arises from the anisotropy of the polarizability and thus only requires molecular ordering, not external forces. An example is double-network elastomers (Chapter 4), which as mentioned above, are internally stressed. As shown in Figure 5.5 (see also Figure 4.19), this leads to birefringence, notwithstanding the absence of external stress or strain.

The stress-optic law has been widely applied to deformed rubber. The repeat units of chain molecules are optically anisotropic, so macroscopic orientation leads to birefringence conforming to eqn (5.14), as first reported by Kuhn and Grün [26]. The proportionality of Δn

[5] The quadratic term in eqn (5.12), which vanishes for molecules possessing an inversion center, is responsible for hyper-Raman scattering and optical rectification, while χ_3, describing events involving three simultaneous photons, gives rise to stimulated Raman scattering, coherent anti-Stokes Raman scattering, and the optical Kerr effect.

[6] The stress optic law is sometimes referred to as Brewster's Law, after the Scottish physicist who observed compression-induced birefringence of glass in the early 19th century. The subsequent discovery of the proportionality between stress and birefringence was by Maxwell. Brewster's Law also refers to an equation for the angle of incidence at which a reflected light beam will be polarized. "Brewster" is a unit of birefringence equal to 1 TPa^{-1}.

[7] The true or Cauchy stress is the force relative to the instantaneous area. For uniaxial strain, $\hat{\sigma} = [1 + \varepsilon]\,\sigma$; for shear $\hat{\sigma} = \sigma$.

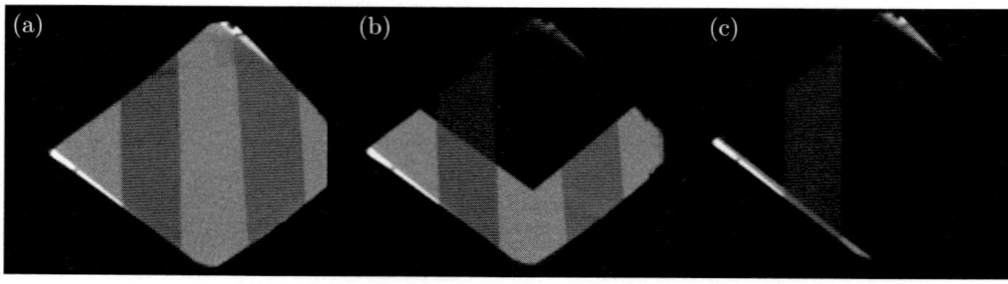

Figure 5.5 (a) Natural rubber double network (rectangular strip on left) and conventional NR elastomer (strip on right), both in a state of mechanical equilibrium ($\sigma = 0$); nevertheless, (b and c), through crossed polarizers the double network transmits light, due to its inherent orientation. Since the external stress is zero, this is a violation of the stress-optic law, eqn (5.14).

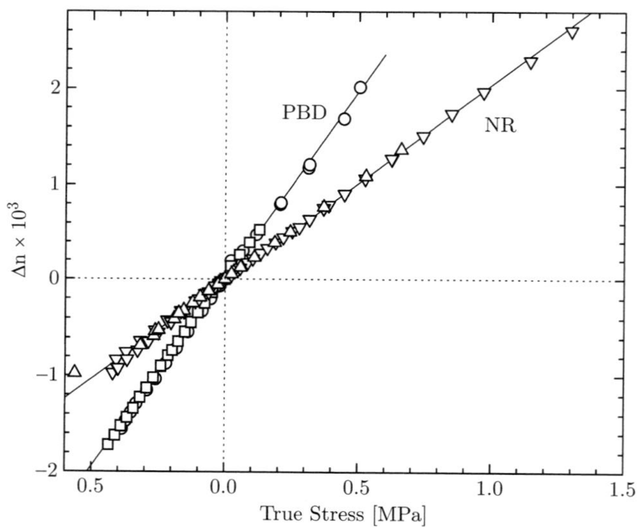

Figure 5.6 Birefringence as a function of true stress for crosslinked 1,4-polybutadiene and natural rubber subjected to tension and compression; the slopes yield 3.6 and 2.0 GPa^{-1} for the respective stress-optical coefficients. Different symbols represent different test specimens. Data from ref. [27].

and the stress of a deformed elastomer, assuming phantom network behavior (Chapter 4), gives the result

$$\Delta n = \left[\frac{2\pi\Delta\alpha(n^2+2)^2}{45kTn}\right]\sigma \tag{5.15}$$

where $\Delta\alpha$ is the optical anisotropy of the chain segment, n is the refractive index in the absence of orientation, and the bracketed factor defines the stress-optical coefficient. C is independent of crosslink density or molecular weight and can be used to determine the difference in parallel and perpendicular polarizabilities of the segment. Figure 5.6 shows the birefringence measured for two crosslinked rubbers subjected to both tension and compression [27]. A linear fit through the origin ($\varepsilon = 0$) represents the data well, in accord with eqn (5.15) for phantom networks;

however, more realistic elasticity models predict a small difference in C for different modes of deformation, so that the slope in Figure 5.6 would change on passing through the origin.

The measured birefringence represents the average value of all segments, each having a probability of alignment at a given angle to the reference direction described. This can be described by an orientational distribution function, which for uniaxial orientation of linear (axially symmetric) chains can be expressed as a series

$$f(\theta) = \sum_{j=0}^{\infty} a_j P_j(\cos\theta) \tag{5.16}$$

where $P_j(\cos\theta)$ is the jth Legendre polynomial[8] and θ is the angle the long axis of the segment makes with the reference axis. It is a property of Legendre polynomials that the weighting factor in eqn (5.16) is just the mean value times a constant factor [28]

$$f(\theta) = \sum_{j=0}^{\infty} \frac{2j+1}{2} \langle P_j(\cos\theta) \rangle P_j(\cos\theta) \tag{5.17}$$

The relevant Legendre function will depend on the experimental variable (the order of its tensor), although X-ray scattering can in principle yield the complete orientation function, enabling determination of all orders. Birefringence is related to the second Legendre polynomial

$$f_2(\theta) = \frac{\Delta n}{\Delta n_0} = \frac{3 < \cos^2\theta > -1}{2} \tag{5.18}$$

Equation (5.18) is known as the Hermans orientation function.[9] The quantity Δn_0, proportional to $\Delta\alpha$, is the magnitude of the birefringence for complete alignment, whereby $f_2(\theta) = 1$ for perfect orientation. For random orientation the average is over all angles

$$\langle \cos^2\theta \rangle = \int_0^{\pi} \cos^2\theta \left[\frac{1}{2} \sin\theta \right] d\theta = \frac{1}{3} \tag{5.19}$$

where the bracketed quantity is the weighting function for θ; thus $f_2(\theta) = 0$.

The Hermans orientation function can also be determined from the dichroic ratio, defined as the ratio of the infrared absorbance for radiation polarized parallel and perpendicular to the stretching direction [29]

$$f_2 = \frac{(R-1)(R_0+2)}{(R+2)(R_0-1)} \tag{5.20}$$

R_0 is the dichroic ratio for perfect orientation, $R_0 = 2\cot^2\varphi$, with φ the angle between the chain axis and the transition moment vector of the vibration. The Hermans orientation function can also be obtained from fluorescence polarization and from analysis of the amorphous halo in X-ray scattering.

Equations (5.15) and (5.18) can be used to determine the anisotropy of, not only the network chains in an elastomer, but also deformed or flowing polymer melts and solutions. Figure 5.7 [30] shows that conformance to the stress-optic law is maintained during stress

[8] Legendre polynomials are given by the formula $f_n = \frac{1}{2^n n!} \frac{d^n}{dx^n} [(x^2-1)^n]$, $n = 0, 1, 2, \ldots$

[9] Named for the Dutch scientist P.H. Hermans, eqn (5.18) is known in the liquid-crystal literature as the order parameter.

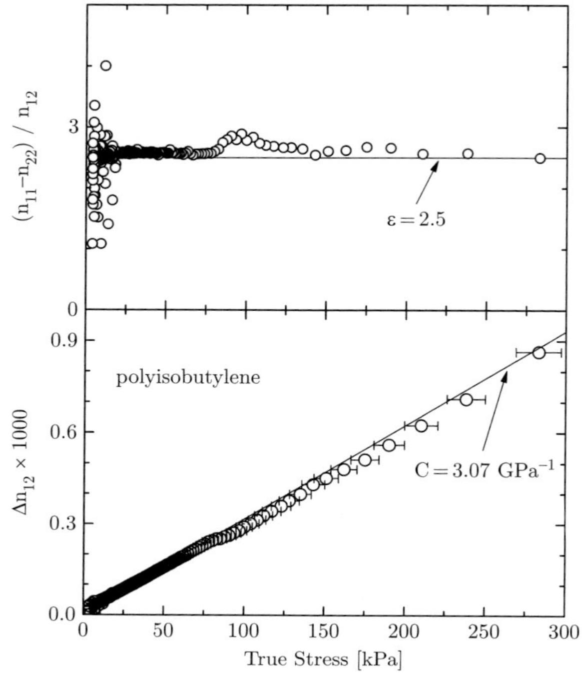

Figure 5.7 (bottom) Birefringence *versus* shear stress for uncrosslinked PIB following a step shear strain = 2.5. Solid curve is the linear fitting yielding the indicated stress-optical coefficient. (top) Ratio of refractive index components (1 is the flow direction and 2 the direction of the gradient), which is constant and equal to the shear strain, in accord with eqn (5.8). Data from ref. [30].

relaxation of an uncrosslinked PIB for both the shear and normal components. The upper half of this figure also confirms the Lodge–Meissner relation, eqn (5.8). The stress-optical behavior is independent of strain rate [31–33], although as seen in data for stretched PS melts (Figure 5.8 [33]), the stress-optical coefficient decreases in the non-linear regime. The latter effect increases in prominence for more polydisperse materials, probably related to the varying fraction of chains oriented by the applied field [34]. Since orientation of rubber also affects heat conduction [35], there is a corresponding proportionality, known as the "stress-thermal rule", between stress and the anisotropy of the thermal conductivity [36]. This in turn causes the latter to be proportional to the birefringence (Figure 5.9) [30,36].

The Hermans orientation function quantifies the mean orientation of the segments, and it is this microscopic or molecular strain that $f_2(\theta)$ connects to Δn. Thus, any changes in the molecular polarizability *per se* can cause birefringence unrelated to molecular orientation, and likewise, orientation due to internal stress will contribute to Δn. Both these situations give rise to deviations from the stress-optic law. An example is hydrodynamic drag forces [37], constituting the fluid resistance to motion. These can orient the polymer chains but do not affect the measured stress; thus, their contribution to Δn is not accounted for by the stress-optic law. Another example is when the material itself changes, such as by strain-induced crystallization. The intrinsic birefringence of crystals can be very high, which in combination with a contribution from "form birefringence" (refraction at the interface between crystalline and

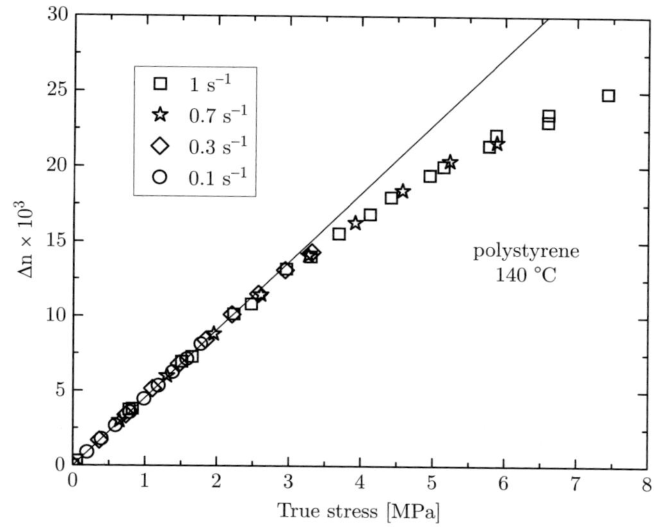

Figure 5.8 Birefringence *versus* true stress for a PS ($M_w = 206$ kg/mol) during uniaxial extension at the indicated strain rates. The data superpose with deviation from the stress-optic law observed for $\hat{\sigma}$=2.2 MPa. Data from ref. [33].

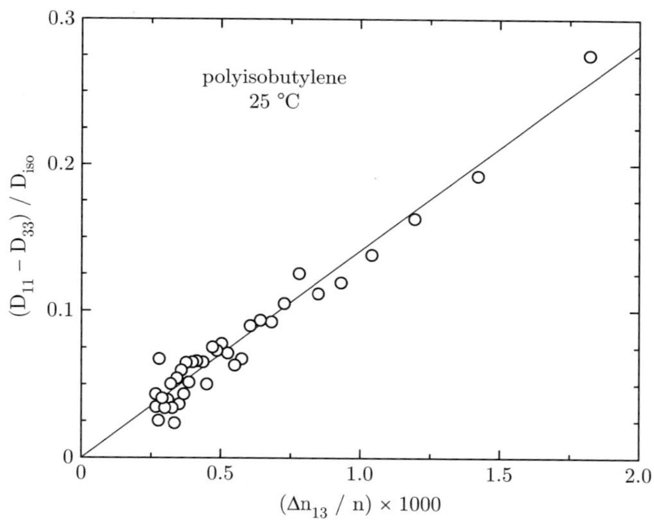

Figure 5.9 Normalized difference in thermal diffusivity *versus* normalized birefringence for polyisobutylene ($M_w = 85$ kg/mol) after a step shear strain equal to eight. The solid line through the data suggests a proportionality between these quantities. Data from ref. [36].

amorphous domains) makes measurements of Δn very sensitive to the presence of crystallinity. This is illustrated in Figure 5.10 showing Δn as a function of stretch for various natural rubber networks [38]. Departures from the stress-optic law are observed at strains for which other techniques give no indication of crystallization.

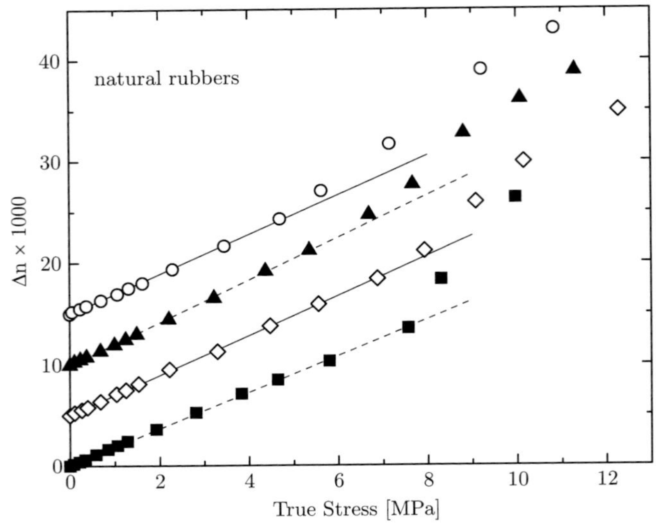

Figure 5.10 Birefringence measured at room temperature for natural-rubber elastomers, in order of increasing purity of the material: guayule NR (circles); "ribbed smoked sheet" grade of NR (triangles); "SML" grade NR (diamonds); deproteinized NR (squares). The solid lines are fits to the data at lower strains, yielding a mean $C = 1.9 \pm 0.1$ GPa^{-1}. An excess Δn appears at higher strains due to strain-induced crystallization. The data, from ref. [38], are shifted vertically for clarity.

5.2.2 Birefringence in the glass-transition zone

More interesting, but of less practical value, are departures from proportionality between σ and Δn that arise when there is a disconnect between stress and the chain orientation. One example is during deformation at frequencies high relative to the rate at which the chains alter their configuration through backbone bond rotation. In this circumstance the material responds by stretching and bending of these bonds, which generate stresses not described by eqn (5.15). Nevertheless, birefringence measured well within the glassy zone of the viscoelastic spectrum is still proportional to the stress, but with a different value of the stress-optical coefficient.

This difference in the stress-optical coefficients causes breakdown of the stress-optic law in the softening regime of the viscoelastic spectrum, where both chain orientation and bond-distortion mechanisms are operative. Experimental measurements show that Δn^* and σ^* have very different frequency-dependences, as observed in the glass-transition zone of master curves[10] for PI in Figure 5.11 [39]. To analyze this region eqn (5.14) is rewritten in complex form

$$\Delta n^* = C(\omega)\sigma^*(\omega) \tag{5.21}$$

[10] As discussed in Chapter 6 time–temperature superpositioning is invalid in the glass-transition zone because of the different temperature-dependences of the orientation and distortion mechanisms; therefore, the shape of the master curve in this region is not quantitative.

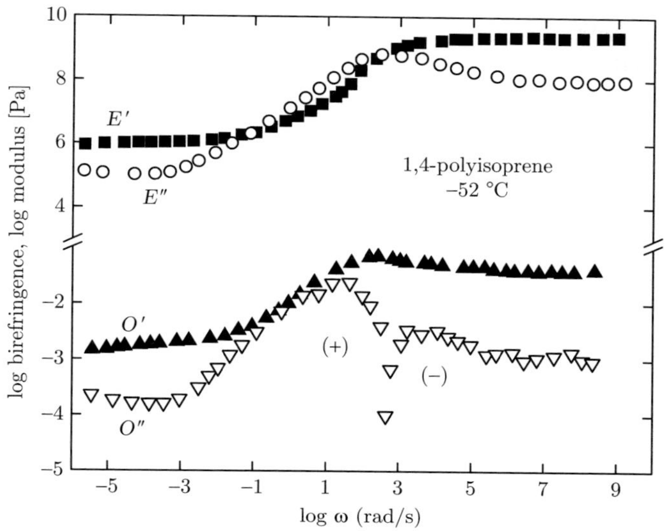

Figure 5.11 Time–temperature master curves of the dynamic moduli (upper curves) and strain optical coefficients (lower curves) for PI. There is a change in sign of $O''(\omega)$ at log $\omega = 2.5$. Data from refs. [39].

Dividing by the strain gives the strain-optical coefficient

$$O^*(\omega) = C(\omega)E^*(\omega) \tag{5.22}$$

The birefringence measured in the softening zone can be assumed to have two independent components [40–42]

$$\Delta n^*(\omega) = C_G\,\hat{\sigma}_G{*}(\omega) + C_R\,\hat{\sigma}_R{*}(\omega) \tag{5.23}$$

In these equations $\hat{\sigma}_G{*}$, $\hat{\sigma}_R{*}$ and C_G, C_R are the respective complex stresses and stress-optical coefficients for the "glassy" (chain distortion) and "rubbery" (chain orientation) components, the latter underlying rubber elasticity and the Rouse dynamics. Equation (5.23) follows from the assumption that the corresponding stresses are additive

$$\hat{\sigma}\,{*}(\omega) = \hat{\sigma}_G{*}(\omega) + \hat{\sigma}_R{*}(\omega) \tag{5.24}$$

This means that stresses within the same chain segment arise independently and are completely decoupled. The expression for the complex strain optical coefficient is then

$$O^*(\omega) = C_G E_G^*(\omega) + C_R E_R^*(\omega) \tag{5.25}$$

with E_G^* and E_R^* the respective components of the dynamic modulus

$$E^*(\omega) = E_G^*(\omega) + E_R^*(\omega) \tag{5.26}$$

To implement this approach beyond a trivial, multi-parameter curve-fitting exercise, some definition must be adopted for the stress-optical coefficients. Read [40] assumed

$$C_R = \lim_{\omega\to 0}\frac{O'(\omega)}{E'(\omega)} = \lim_{\omega\to 0}\frac{O''(\omega)}{E''(\omega)} \tag{5.27}$$

and

$$C_G = \lim_{\omega\to\infty} \frac{O'(\omega)}{E'(\omega)} \tag{5.28}$$

Inoue and coworkers [41,42] employed a different definition for the glassy coefficient

$$C_G = \lim_{\omega\to\infty} \frac{O''(\omega)}{E''(\omega)} \tag{5.29}$$

These equations have been applied to dynamic birefringence data on various polymers [40–43]. The analyses yield predictions for the frequency-dependent contributions of the high- and low-frequency components. Two questions are apparent:

5.2.2.1 Are the assumptions of this approach (eqns (5.23) through (5.29)) correct and/or verified by their application?

The only quantities not experimentally determined are the (hypothetical) components of the modulus, which from eqns (5.23) and (5.26) are given by

$$E_G^*(\omega) = \frac{O^*(\omega) - C_R E^*(\omega)}{C_G - C_R} \tag{5.30}$$

and

$$E_R^*(\omega) = \frac{O^*(\omega) - C_G E^*(\omega)}{C_R - C_G} \tag{5.31}$$

Dynamic birefringence data can be described equally well using either eqns (5.28) or (5.29), but with different values obtained for the high- and low-frequency components. This means that the approach of Read [40] and Inoue and coworkers [41–43] is ambiguous and thus not corroborated by successful fitting of experimental data.

The premise of the method is that the stresses are additive. This is consistent with the virial stress formulation, used in molecular-dynamics simulations, in which the atomic virial stresses,[11] arising from the kinetic and potential energies, are summed to obtain the (continuum) Cauchy stress [39,44]. However, simulations can also be analyzed using alternative conservation laws, for example based on summation of the local strains [45]. Experimentally, it is known that the total strain during creep flow of linear viscoelastic materials is equal to the sum of the recoverable strain and the viscous flow; in terms of the shear compliance, this strain additivity is expressed as

$$J(t) = J_r(t) + t/\eta_0 \tag{5.32}$$

where J_r is the recoverable creep compliance. The validity of eqn (5.32) is confirmed by creep-recovery experiments on polymer melts, in which the permanent deformation measured after steady state flow is equal to t/η_0 times the creep stress [46]. Equation (5.32) indicates that at least for terminal flow, stresses (elastic and viscous) are coupled rather than additive.

Analysis of birefringence in the softening zone can be carried out using the assumption that the strains are additive [47]; thus,

[11] The term "virial" refers to the configurational part of the pressure, with the virial stress given by $\sigma_{12} = V^{-1}\sum_i\left(\frac{1}{2}\sum_j(r_{1,i} - r_{1,j})F_{2,ij} - m_i v_{1,i} v_{2,i}\right)$ where the index (1,2) denotes directions in a three-dimensional rectilinear coordinate system, and the sums are over all i atoms in volume V, interacting through a force F_{ij} with each jth neighbor. The last term is the thermal kinetic energy. The virial pressure for an ideal gas is zero.

$$\varepsilon = \varepsilon_G + \varepsilon_R \tag{5.33}$$

In terms of the compliances, $D^*(\omega) = 1/E^*(\omega)$,

$$D^*(\omega) = D^*_G(\omega) + D^*_R(\omega) \tag{5.34}$$

with the stress-optic law given by

$$O^*_G(\omega)^{-1} = \frac{D^*_G(\omega)}{C_G} + \frac{D^*_R(\omega)}{C_R} \tag{5.35}$$

The assumption of strain additivity gives for the glassy component [48]

$$D^*_G(\omega) = \frac{C_R/O^*(\omega) - D^*(\omega)}{C_R/C_G - 1} \tag{5.36}$$

and for the rubber component

$$D^*_R(\omega) = \frac{C_G/O^*(\omega) - D^*(\omega)}{C_G/C_R - 1} \tag{5.37}$$

Note that C_R and C_G are no longer defined by eqns (5.27), (5.28), or (5.29); without additional assumptions, they are free parameters adjusted to give agreement between the calculated and experimental birefringence data.

5.2.2.2 Are the results of eqns (5.23) through (5.29) reasonable?

The most important aspect of such analyses is what they reveal about the molecular mechanism underlying the viscoelastic behavior. Ostensibly, the relative contribution of the chain segment reorientation and the distortional modes to the various parts of the viscoelastic spectrum can be quantified. In Figure 5.12 these contributions, as reflected in the magnitudes of rubbery and glassy dynamic moduli, are plotted versus frequency. Of course, the rubbery modes dominate at the lowest frequencies, while the glassy component governs the behavior at the highest frequencies. More interesting is the relative contribution in the softening zone in Figure 5.12 (*ca.* $-1 < \log \omega(\text{rad/s}) > 3$). The assumption of stress additivity leads to the inference that the glassy modes are the dominant relaxation mechanism throughout the transition region, including at frequencies over which the loss component of the birefringence, $O''(\omega)$, changes sign. When strain additivity is assumed, however, the relative contribution of the glassy component is significantly smaller throughout most of the softening zone. And the main contribution changes from the glassy to the rubbery component at the same frequencies for which $O''(\omega)$ changes sign. Thus, the different assumptions used in interpreting dynamic stress birefringence measurements lead to diametric conclusions.

One difficulty with results of this type is the reliance on master curves (e.g., Figure 5.11), obtained by time–temperature shifting. Although this is necessary in order to obtain the entire viscoelastic spectrum, superpositioning breaks down for polymers in the glass-transition zone, as discussed in Chapter 6. While the interpretation of dynamic birefringence can potentially yield useful insights, the interplay of forces in a dynamically correlated system such as polymer melts may be too complicated for an analysis in terms of distinct "rubbery" and "glassy" components. It remains unclear how these putative entities interact, so that rote application of the stress-optic law to behavior in the softening zone is dubious.

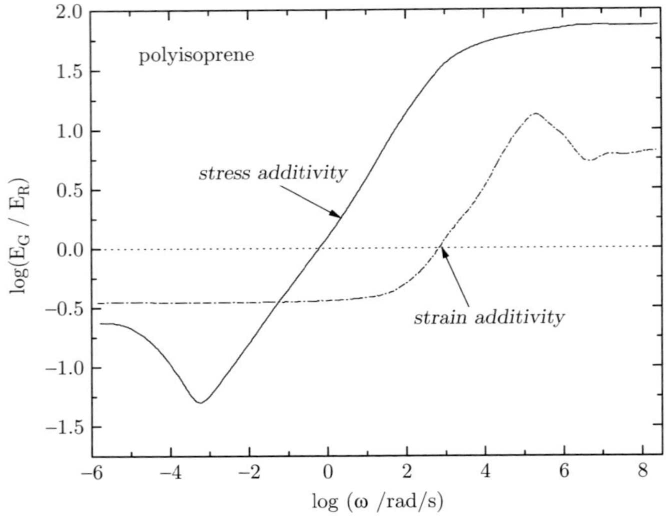

Figure 5.12 The relative contribution of the distortional and orientational components of the stress birefringence, as reflected in the magnitude of the respective glassy and rubbery moduli, determined from the data in Figure 5.11 using eqn (5.25) (solid line) or eqn (5.34) (broken line). The horizontal dotted line denotes equal contributions from the two components.

5.2.3 Orientational coupling (nematic interaction)

Since the mechanism of birefringence arises on the level of the chemical bonds, whereas rubbery stresses involve orientation of subchains involving numerous backbone bonds, another departure from proportionality between stress and birefringence can result from local ordering that does not affect the stress. In the classical depiction of rubber elasticity, network chains orient independently in response to applied stresses; however, intermolecular interactions, mainly steric effects from aligned neighboring segments, can also contribute to the local orientation. For this reason unattached polymer chains acquire a stable orientation when present in an oriented network [49–54]. Similar effects are seen in polymers with bidisperse molecular weight distributions [55–58] and in miscible blends of chemically dissimilar polymers [59,60]. This orientational coupling (or nematic interaction) in polymer melts causes the relaxation times of shorter chains to approach the value of τ_η for the longer host chains.

Orientational coupling is only locally effective, so that short probe chains in a strained network retain their isotropic dimensions [61,62] and there is no effect on the stress. However, orientational coupling contributes to optical birefringence and can also be detected by infrared dichroism, deuterium NMR by the quadrupole interaction, fluorescence polarization, and from the anisotropy of neutron scattering. Molecular-dynamics simulations of chains interacting via a Lennard-Jones 6–12 potential also reveal nematic interactions [63,64]. A metric of the strength of the orientational coupling is the ratio of the strain of the probe chain to that of the matrix [65]. Values of this coupling coefficient for different probes in various rubbery hosts are listed in Table 5.1. In mixtures of different polymers, including

Table 5.1. Coupling coefficients for various probe/host mixtures.

Materials	Comment	Coupling coefficient	Method	Reference
PMPS		1	^{2}H NMR	61
PBD	probe $M_w > 2$ kg/mol probe $M_w > 2$ kg/mol	0.9 ± 0.1 0.4 ± 0.05	birefringence/ IR dichroism	51
poly(ethylene-*co*-propylene)	M_w (kg/mol) = 53–370	0.45 ± 0.05	birefringence/ IR dichroism	55
polydiethylsiloxane (PDES)	M_w (kg/mol) = 27–16310–71	> 0	SANS	62
modified PBD	1–4% H-bonding monomer units	0.6–1.0	birefringence/ IR dichroism	58
polyisoprene	small molecule probes	> 0	^{2}H NMR	49
polyisoprene	small molecule probes	0.4–0.65	fluorescence polarization	65
PVE/PI blend	PVE M_w (kg/mol) = 23–134; PI network	0.13–0.20	IR dichroism	52
	PVE M_w (kg/mol) = 80–110; linear PI	0.30–0.35	birefringence	60
	PVE M_w (kg/mol) = 204; linear PI	0.35 ± 0.05	birefringence/ IR dichroism	59
PDMS/PDES blend	PDMS M_w (kg/mol) = 15; PDES network	> 0	^{2}H NMR	53
bead–spring model (L-J 6–12 potential)	probe chains: $N = 5$–25; host chains: $N = 5000$	0.28 ± 0.01	simulations	63
	probe chains: $N = 10$–20; host chains: $N = 20$–120	0.35 ± 0.05	simulations	64

bidisperse molecular-weight distributions, the stress-optic law breaks down because of the orientational coupling; however, it does not affect the proportionality between the stress and the total optical anisotropy from all chains, so that eqn (5.14) is maintained.

5.3 Reversing strain histories

The Boltzmann superposition approach (eqn (5.1)) treats a continuous deformation such as constant strain rate as a sequence of incremental, discrete strains separated by a small time interval during which the chains relax. Strict linearity implies the material response to each step is independent of all other responses. As discussed above, when this condition is not met, non-linearities can be incorporated into the definition of the strain, so that the stress remains proportional to a strain function that is factored separately from viscoelastic effects (eqn (5.5)). However, this method fails when the deformation changes sign (direction). Reversing strain histories in general gives rise to anomalous behavior, as illustrated here with two examples, the Mullins effect and the optical birefringence during recovery from creep strain. These anomalies reveal the need for further development of molecular-based constitutive equations for both entangled melts and rubbery networks.

5.3.1 Mullins effect

A change in the sign of the strain during deformation of a rubber, such as retraction of stretched rubber, causes stresses that are lower than the corresponding values prior to the reversal [66]. The only exception to this behavior is low strain rates yielding an elastic response without accompanying structural or chemical changes. Mechanisms for mechanical hysteresis during reversing strains include: (i) Changes in chemical structure, for example rupture of network chains ("chemical stress relaxation") [67–70] or their detachment from filler particles [71–73]. These effects are irreversible or recover only very slowly, so that the reduction in stiffness persists throughout the deformation cycle. (ii) Crystallization induced by large deformation. Orientation decreases the conformational entropy and thus the entropy penalty associated with crystal formation. Strain crystallization in turn lowers the stress by reducing the number of elastically effective network strands and diminishing their microscopic strain[12] [74–76]. The reduced back-folding (crystal stem re-entry) of crystals formed while oriented leads to more stable crystals, which may persist even after removal of the strain. (iii) The third origin of strain softening is viscoelasticity, a phenomenon common to all rubbers as described in Chapter 1. The response of the network chains is retarded due to their mutual interactions; thus, unless mechanical equilibrium is maintained through low strain rates, the recovery stresses are smaller.[13]

Although the term "Mullins softening" is commonly applied to the mechanical hysteresis of carbon-black-filled rubbers, this is an unfortunate use of the term. The effect is not specific to filler reinforcement, as first shown by Harwood and coworkers [77,78]. They found for cyclically stretched rubber that the retraction stresses were independent of carbon-black content, as long as the strain for each compound was adjusted to give constant maximum stress. The maximum strain was reduced as the filler content increased, yielding a roughly constant degree of stress softening. This comparable behavior of filled and gum rubbers is seen in plots (Figure 5.13) of the stress during cyclic straining, plotted as a function of the stress during the initial extension [77].

As discussed above, stresses calculated using K-BKZ superpositioning are generally in good agreement with measurements on elastomers for shear or tensile strains up to at least 100%. However, this success is limited to stress-increasing strain histories. When the sign of the strain changes, the energy dissipation is significantly underestimated. This is the Mullins effect, which refers specifically to *anomalously* small recovery stresses. The stresses during retraction are smaller than can be accounted for by the mechanisms enumerated above (bond breaking, crystallization, or viscoelastic relaxation) [77–79]. Thus, assessment of Mullins

[12] At mechanical equilibrium strain-induced crystallization reduces the stress in accord with Le Chatelier's principle; however, by functioning as inextensible filler particles, the crystallites usually increase the transient stress.

[13] This type of mechanical hysteresis is analogous to the enthalpy changes accompanying strain crystallization of rubber. The heat evolved during the initial stretching is greater than that absorbed during the subsequent recovery because the latter only reflects the latent heat of crystallization, while the former has a contribution from internal friction. The result is an increased temperature of strain-crystallizing rubber when cyclically strained.

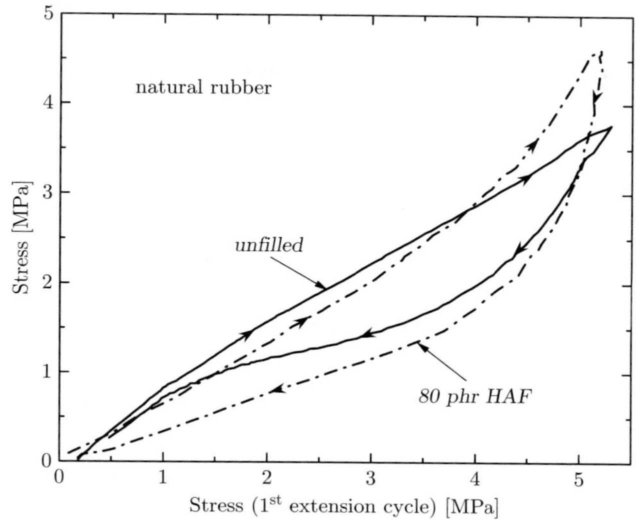

Figure 5.13 Engineering stress for crosslinked unfilled NR and NR with carbon black during the second cycle of extension–retraction plotted *versus* the stress for the initial extension. The strain rate was 0.03 s^{-1}. Data from ref. [77].

softening requires quantitative evaluation of the viscoelastic contribution to the hysteresis. This can be done with experiments on amorphous gum rubber, lacking hysteresis due to filler or crystallization. If the crosslink junctions are covalently bonded carbon atoms, rather than polysulfidic linkages, and strains are limited to less than *ca.* 100%, the possibility of bond rupture can be negated. Mulling softening is illustrated in Figure 5.3 showing data for an SBR network; the stresses during retraction are significantly lower than the calculated values.

A simpler test employs a two-step deformation, with the stress from eqn (5.5) expressed as the sum of the stresses measured in single-step experiments

$$\sigma(t, t_1, \gamma_1, \gamma_2) = \sigma(t, \gamma_1 + \gamma_2) - \sigma(t, \gamma_2) + \sigma(t - t_1, \gamma_2) \tag{5.38}$$

where γ_1 is the shear strain applied at $t = 0$ and the total strain becomes $\gamma_1 + \gamma_2$ at $t = t_1$. As seen in data for peroxide-crosslinked NR (Figure 5.14) [80], the K-BKZ expression over-predicts the stress for $\gamma_2 < \gamma_1$ (strain reversal). This failure for reversing deformations of the (otherwise successful) Boltzmann superposition integral with time invariance implies that the calculation of the retraction stresses is in error; however, why this error becomes apparent only for reversing strains is unclear. Evidently there is an unaccounted-for contribution to the viscoelastic response manifested only during recovery.

This same anomalous hysteresis is also observed for polymer melts and is relevant to processing operations that involve strain reversals, such as calendering, extrusion, blow molding, etc. Constitutive models with the form of eqns (5.5) or (5.38) cannot quantitatively describe reversing deformations of uncrosslinked polymers [18,81–83]. This is illustrated in Figure 5.15 showing data for a concentrated solution of entangled polystyrene subjected to a double-step strain [84]. Similar to the results in Figure 5.3 and Figure 5.14 for rubbery networks, the calculated stresses are higher than the measured values, with the deviation growing with increase in the time interval prior to application of the second strain. Of course,

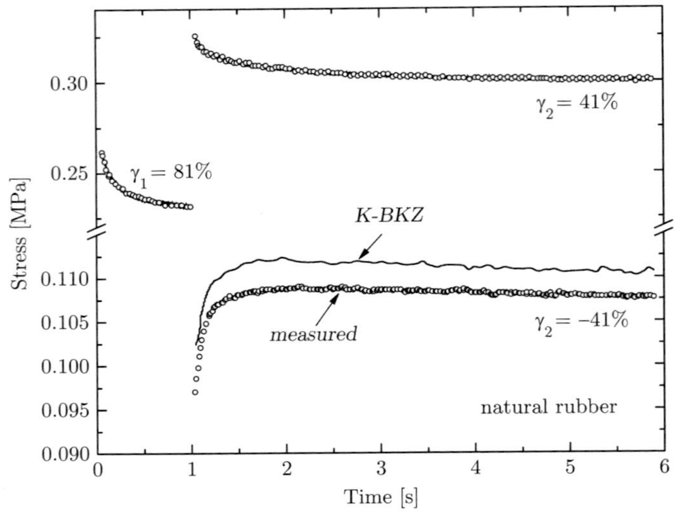

Figure 5.14 The stress measured for crosslinked NR subjected to double-step shear strains (circles) of the indicated magnitudes and sign. The second step was applied at $t_1 = 1$ s after the initial strain. The stresses calculated from eqn (5.38) for $t > t_1$ (solid line) are only evident for the reversing strain history ($\gamma_2 + \gamma_1 < \gamma_1$; lower curves), being masked by the experimental points for the stress-increasing deformation. Data from ref. [80]

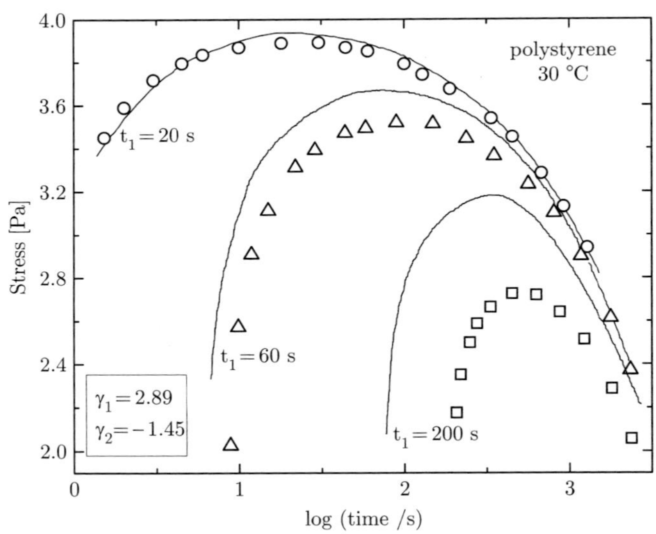

Figure 5.15 Shear stress for polystyrene ($M_w = 670$ kg/mol) solution for shear strain = 2.89 imposed at $t = 0$ and reduced to 1.44 after the indicated time intervals. Equation (5.38) (solid curves) increasingly overestimates the stresses with increasing time between successive steps. Data from ref. [84].

for sufficiently large t_1 the steps become independent since the initial strain fully relaxes and, in the limit $t_1 = 0$, the reversing-strain experiment reduces to simple stress relaxation at a strain equal to the sum of the two steps. The discrepancy in calculations involving a change in sign of the strain for entangled polymer melts is ascribed to chain-end retraction [2,9]. A chain is assumed to respond affinely to deformation, but since entanglements do not constrain motion along the chain contour, the equilibrium contour length is recovered when the chain retracts (according to the tube model, strain orients the chain segments without stretching them). This non-affine motion is very rapid, involving conformational transitions of independent chain segments. (It serves as the mechanism responsible for shear-thinning behavior of polymer melts [2].) If too fast to be included in the measurement ("viscoanelasticity" [18]), the reduction in stress is omitted from the memory kernel of the constitutive equation, with the error in the calculation only becoming evident for strain reversals.

There have been various attempts to improve on the K-BKZ description of reversing strains [85–88]. In the Doi–Edwards model this failing has been addressed by removing the independent alignment assumption [16,18,84]. Wagner and coworkers [81,89] introduced irreversibility into the calculation by requiring that the damping function retain its minimum value, attained at the highest strain. The underlying idea is that chain-end retraction causes a near-instantaneous loss of stress that is only sensed during a strain reversal. The effect is captured by assuming $g(\gamma)$ becomes constant and equal to the value at the reversal strain. Irreversible damping does not change the results for strain-increasing deformations but can improve the accuracy of predictions for reversing strain histories. This is seen in data for polyisobutylene in Figure 5.16 [10]. The overprediction of the stresses by eqn (5.38) is corrected by assuming that the value of $g(\gamma_1)$ is retained after the total strain becomes $\gamma_1 + \gamma_2$.

Since networks formed by random crosslinking have a significant number of non-load-bearing chain ends, Mullins softening in rubbers could have a similar anelastic

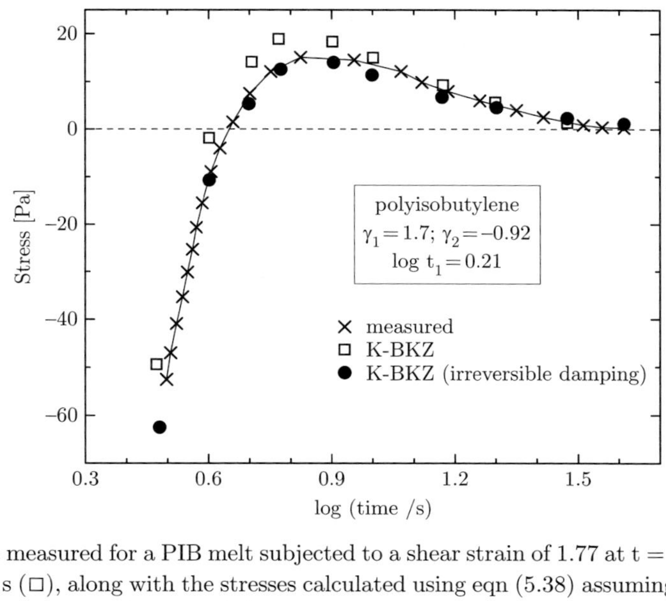

Figure 5.16 Stress measured for a PIB melt subjected to a shear strain of 1.77 at t = 0 that was reduced to 0.85 at $t = 1.62$ s (□), along with the stresses calculated using eqn (5.38) assuming reversible (∀) and irreversible (,) damping. Data from ref. [10].

origin. Harwood and Payne [78] originally ascribed the effect to non-affine displacement of the network junctions, in particular those near the end of chains. This idea can be tested by experiments on end-linked networks, which have a negligible chain-end concentration. However, results on end-linked PTHF [80] show similar discrepancies with K-BKZ calculations for reversing strains (see Figure 5.17); thus, the problem cannot be the chain ends *per se*. The accuracy of the calculation can be improved by invoking irreversible damping, but the agreement remains imperfect. Such experiments on end-linked, monodisperse chains also rule out a distribution of network strand lengths as the source of Mullins softening, although such network heterogeneity can make the response less affine.

5.3.2 Birefringence during stress-free recovery

During the recovery of a network or well-entangled polymer after the load is released, such as during recovery from creep flow, there is no external stress; nevertheless, measurable birefringence can be detected since there remains net orientation. This stress-free birefringence was first reported for natural rubber undergoing recovery from uniaxial stretching (Figure 5.18) [90]. As the applied stress was increased incrementally, proportionality was maintained between Δn and $\hat{\sigma}$, even though the mechanical response became highly non-linear. Upon removal of the load, a small birefringence persists, subsequently decaying slowly to zero as the chain segments assume an isotropic configuration. These results emphasize that birefringence arises due to segment orientation, which is not necessarily proportional to the stress.

Creep-recovery experiments on rubber have shown even more anomalous behavior. After removal of the load, the residual birefringence has a sign opposite to that under load, signifying a *reversal* in the orientation. This anomaly was seen in 1,4-polybutadiene networks (Figure 5.19) [27], in linear polyisobutylene (Figure 5.20) [91], as well as in glassy PVC [92].

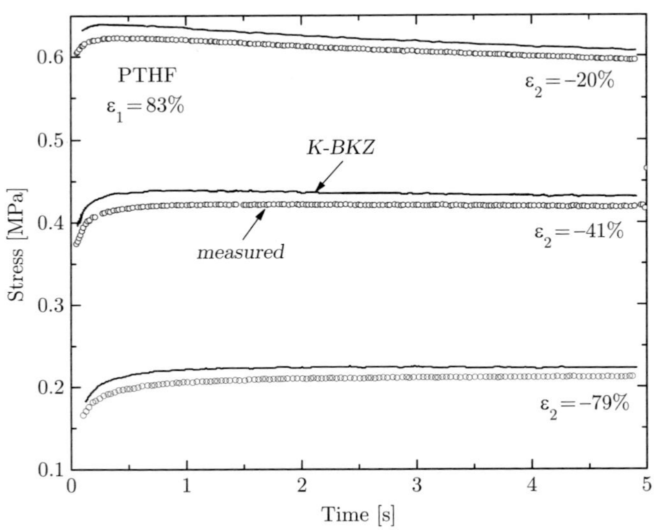

Figure 5.17 Experimental (circles) and calculated (solid lines) shear stresses measured in end-linked PTHF after double-step shear strains of the indicated magnitudes and sign. The second step was applied at $t_1 = 1$ s after the initial strain. Data from ref. [80].

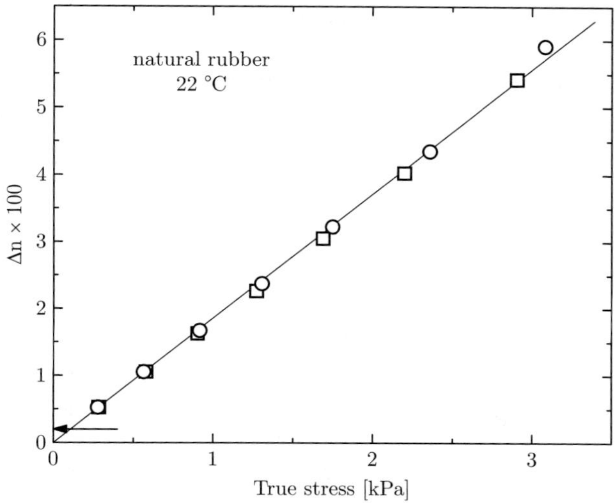

Figure 5.18 Birefringence from uncrosslinked NR under uniaxial tension. Measurements were made upon increment of the stress and after some subsequent creep. The arrow indicates the magnitude of Δn measured about 2 min after removal of all stress. Data from ref. [90].

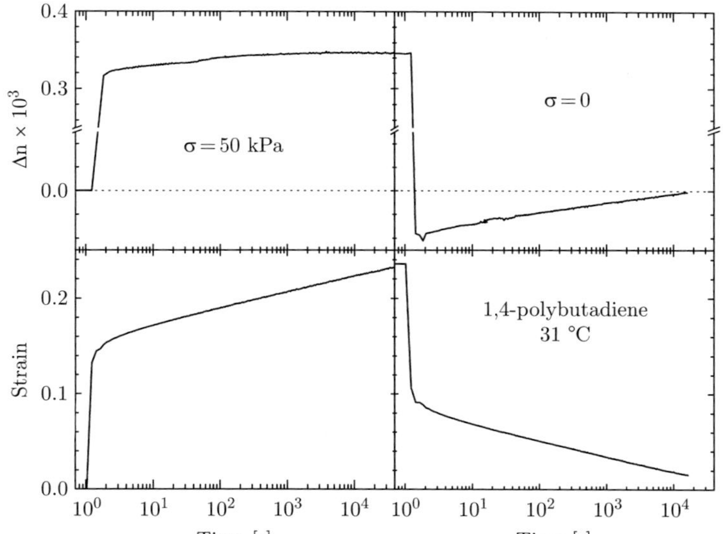

Figure 5.19 Shear strain and associated birefringence of crosslinked PBD during creep and recovery. The birefringence becomes measurably invariant at 10^4 s, although the compliance never reaches steady state. The sign of Δn changes upon removal of the load. Data from ref. [27].

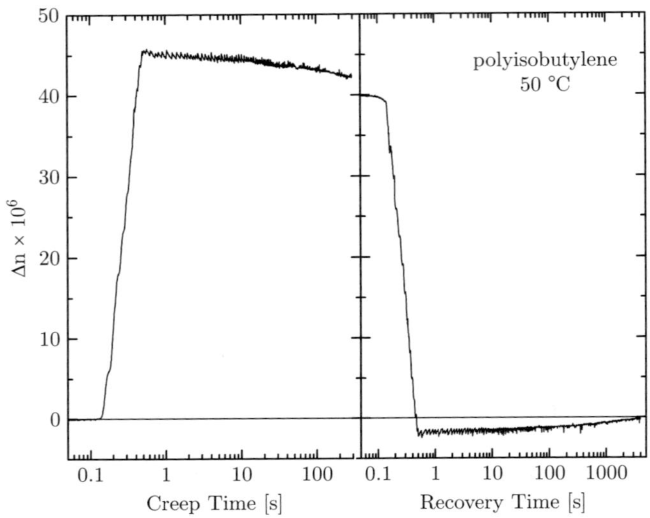

Figure 5.20 Birefringence for PIB ($M_v = 9 \times 10^5$ g/mol) under compression. Data below *ca.* 1 s reflect the instrumental rise time. The decrease in Δn during creep is caused by the reduction in stress due to the changing cross-sectional area; the stress-optical coefficient actually increases slightly with strain. The birefringence has an opposite sign during the stress-free recovery. Data from ref. [91].

The change in sign of Δn implies that the chain segments, net orientation is opposite both to their orientation under stress and to the macroscopic orientation during recovery.

There is no diminution in the magnitude of the recovery Δn when the molecular weight of the polymer is varied by more than an order of magnitude [91]. This argues against the above hypothesis for the anomalous relaxation of polymers melts subjected to strain reversals—rapid response of chain ends upon removal of stress—since the chain end concentration varies inversely with M_n. It has been suggested [37,93] that when the stress is removed prior to attainment of mechanical equilibrium, the recovery is governed by a competition between unrelaxed chains and those having equilibrated to the imposed strain. A similar idea was invoked to describe the recoil following partial stress relaxation of polymer melts [94]. Thus, the change in sign of the birefringence upon a strain reversal may reflect a delicate balance within the distribution of chain orientations during deformation, prior to attainment of steady state. It would be interesting to measure the birefringence during recovery from steady-state creep flow. Nevertheless, a change in sign of the net birefringence is not easily reconciled in this manner, and it represents another aspect of the behavior of polymers subjected to reversing deformations whose origin remains to be understood.

5.4 Empirical rules for non-linear flow

The zero-shear-rate limiting viscosity describes the shearing of unperturbed, fully entangled chains, and η_0 can be measured either in a dynamic experiment or during steady flow. At higher rates of flow, the viscosity of high molecular polymers decreases, a result of orientation and disentanglement of the chains. This non-Newtonian behavior is also observed in dynamic measurements, even though the amplitude of the oscillation is small and thus the strains are negligible. Shear-thinning behavior is governed by the shear rate, and as first discovered by Cox and Merz [95] in experiments on polystyrene, for equal values of the shear rate and frequency, the steady-state shear and complex dynamic viscosities are essentially the same

$$\eta^*(\omega) = \eta(\dot{\gamma}) \ (\omega = \dot{\gamma}) \tag{5.39}$$

This relation is strictly empirical, although efforts have been made to derive a theoretical basis [19,96,97], or at least determine the form of the relaxation function necessary for the K-BKZ equation (eqn (5.4)) to yield the Cox–Merz rule [98,99].

For linear polymers eqn (5.39) holds fairly well, especially for monodisperse, highly flexible chains [100,101]. A series of tests at 11 laboratories found that for neat polymers at modest shear rates the agreement was within the scatter of the experimental data [102]. At higher strain rates that induce chain stretching, dynamic viscosities are usually somewhat larger than the steady-state values [98,103–107]. Some results are shown in Figure 5.21 for various polyethylenes with broad molecular weight distributions [103]; the difference between $\eta(\dot{\gamma})$ and $\eta^*(\omega)$ is about 10%. In Figure 5.22 are viscosity data for neat polypropylene and a dynamically vulcanized blend of PP with EPDM [108]. Dynamic vulcanization refers to thermoplastic elastomers in which dispersed rubber particles are sulfur-crosslinked during mixing with the saturated polymer component [109]. Unlike the neat PP, the blend shows significant departures from eqn (5.39). Presumably, flow orients and deforms the dispersed phase, reducing the viscosity from the value obtained in a small-strain dynamic measurement. Similarly, in filled systems dynamic viscosities tend to exceed steady-state values (Figure 5.23) [102,110].

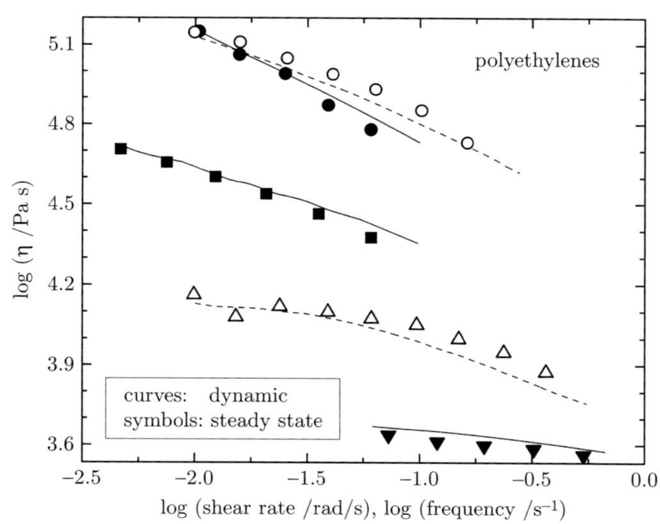

Figure 5.21 Comparison of $\eta^*(\omega)$ (symbols) and $\eta(\dot{\gamma})$ (lines) for high-density polyethylenes (HDPE, composed of primarily linear chains; solid symbols;) and low-density polyethylenes (LDPE, which has branching on about 2% of its carbon atoms; open symbols). Different symbols correspond to different M_w, in the range from 100–140 kg/mol with substantial polydispersities (≥ 4). Temperature for HDPE is 190°C and for LDPE = 175°C. Data from ref. [103].

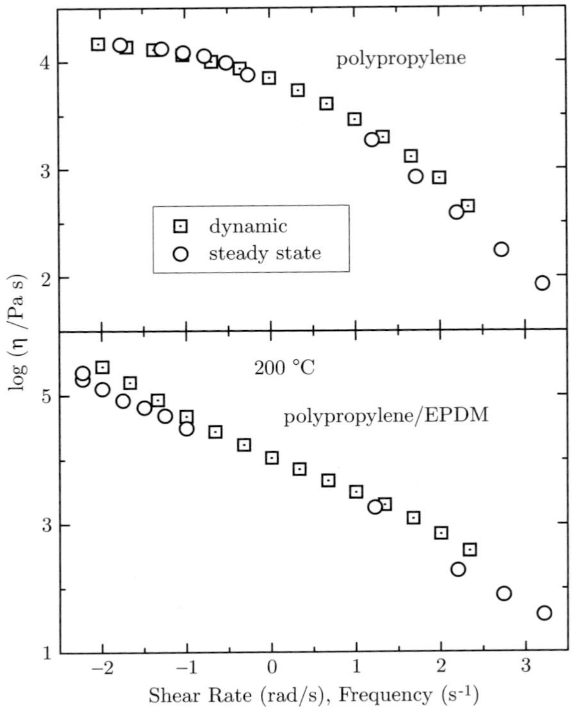

Figure 5.22 Comparison of $\eta^*(\omega)$ and $\eta(\dot{\gamma})$ at $T = 220$°C for polypropylene and a dynamically vulcanized elastomer of polypropylene and EPDM. Data from ref. [108].

Two empirical relationships of Laun [111] relate the dynamic moduli to the first normal stress coefficient measured during steady state flow

$$\psi_1^0(\dot{\gamma}) = 2(G'/\omega^2)\ [1 + (G'/G'')^2]^{0.7} \quad (\dot{\gamma} = \omega) \tag{5.40}$$

and to the steady-state recoverable strain

$$\gamma_r(\dot{\gamma}) = (G'/G'')\ [1 + (G'/G'')^2]^{1.5} \quad (\dot{\gamma} = \omega) \tag{5.41}$$

The former equation is more commonly used, with typical results shown in Figure 5.24 for polyethylene and polypropylene [111] and in Figure 5.25 for two polystyrenes with different molecular-weight distributions [112]. For the latter, the deviation of eqn (5.40) from the steady state values of ψ_1^0 is substantial. Often, better agreement with experiments can be achieved by adjusting the exponent from its original value of 0.7; however, when employed as a fitting curve, eqn (5.40) loses its predictive capability. The terminal dynamic moduli are

$$\lim_{\omega \to 0} \frac{G'(\omega)}{\omega^2} = \eta_0^2 J_s^0\ ;\ \lim_{\omega \to 0} \frac{G''(\omega)}{\omega} = \eta_0 \tag{5.42}$$

whereby eqn (5.40) reduces at low shear rates to the relation of Coleman and Markovitz [113]

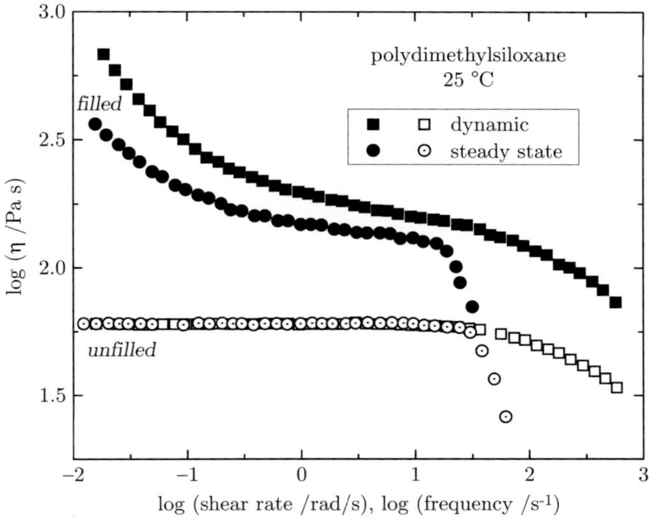

Figure 5.23 Comparison of $\eta^*(\omega)$ (squares) and $\eta(\dot{\gamma})$ (circles) for neat PDMS (open symbols) and hydroxyl-terminated PDMS containing 12.9% by volume calcium carbonate (solid symbols) ($M_w = 1.3\times10^5$ g/mol). Only the former conforms to the Cox–Merz rule. The steep decrease in the steady-state viscosity at shear rates > 20 rad/s is due to flow instabilities as the test specimen begins to fracture. Data from ref. [110].

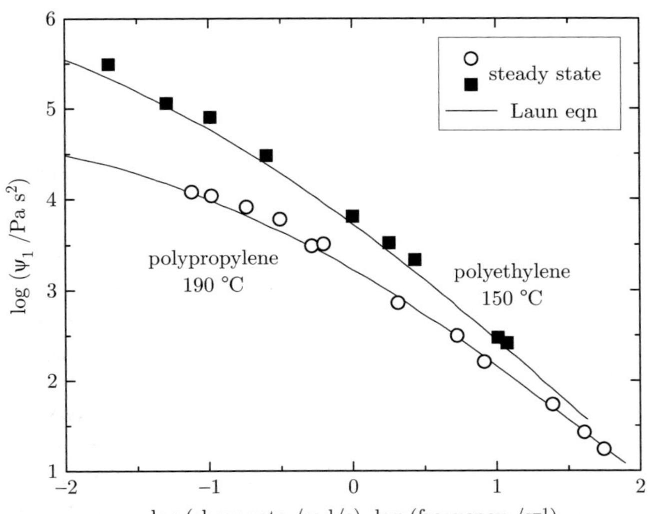

Figure 5.24 Experimental (symbols) and calculated (eqn (5.40); solid) first normal stress coefficient for polypropylene (circles) and HDPE (squares). Data from ref. [111].

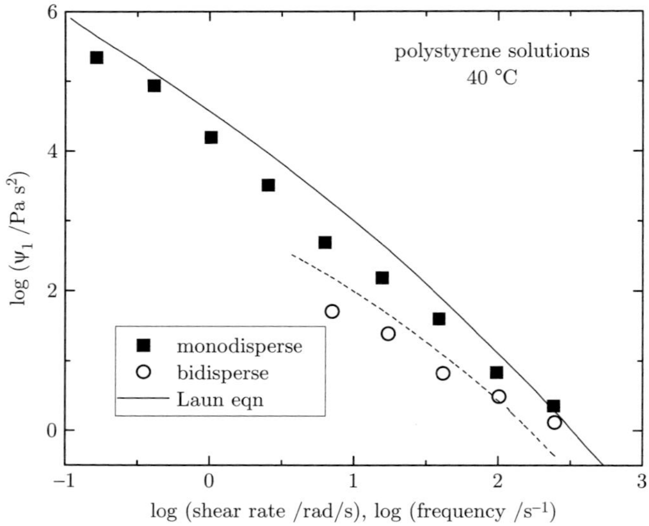

Figure 5.25 Experimental (symbols) and calculated (eqn (5.40); solid and dashed lines) first normal stress coefficient for monodisperse ($M_w = 2\times10^6$ g/mol; $P = 1.09$) and bidisperse ($M_w = 7\times10^5$ g/mol blended with 8% $M_w = 2\times10^6$ g/mol) polystyrene 20% solutions. Data from ref. [112].

$$\psi_1^0 \equiv \lim_{\dot{\gamma}\to\infty} \frac{N_1(\dot{\gamma})}{\dot{\gamma}^2} = 2J_s^0\eta_0^2 \tag{5.43}$$

The strain recovered by rubber subsequent to forming operations that squeeze the material (such as extrusion and calendaring) is known as die swell, $\hat{D}$, defined as the ratio of the final and initial dimensions transverse to the flow. Die swell is governed by the normal force,[14] although only for steady-state conditions is there a fundamental connection between $\hat{D}$ and strain rate. Two equations relate die swell to the first normal force [114]

$$N_1 = 2\sigma(2\hat{D}^6 - 2)^{1/2} \tag{5.44}$$

and [115]

$$N_1 = 2\sigma(2\hat{D}^4 - \hat{D} - 2) \tag{5.45}$$

These can be used with the Lodge–Meissner equation (eqn (5.8)) to connect the die swell to the shear strain. Or, in combination with Laun's rule, eqns (5.44) or (5.45) provide estimates of $\hat{D}$ from dynamic mechanical measurements. Some results are shown in Figure 5.26 for a filled EPDM rubber [116]. There is rough agreement with experimentally measured $\hat{D}$, which is satisfactory given the approximate nature of the derivation of eqn (5.44) or (5.45) [114,115].

An empirical equation was suggested by Gleissle [117] to estimate the steady-state viscosity and its dependence on shear rate from the stress growth during startup at a slow (Newtonian)

[14] Although die swell in polymers becomes marked at strain rates associated with non-Newtonian viscosities, even Newtonian fluids lacking elasticity exhibit die swell, with $\hat{D} \sim 1.2$ under isothermal conditions. This is caused by changes in the velocity profile.

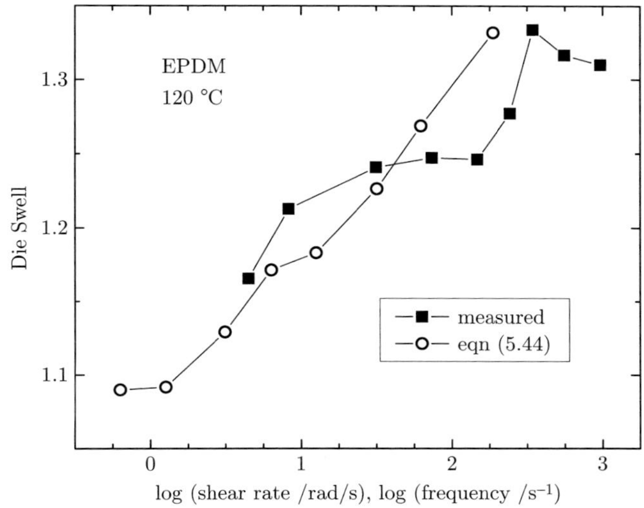

Figure 5.26 Experimentally measured $\hat{D}$ (solid square) for carbon-black-reinforced EPDM compound sheared in a capillary rheometer, along with the values (open circles) calculated from dynamic measurements using eqns. (5.40) and (5.44). Data from ref. [116].

shear rate. The transient viscosity, $\eta^+(t)$, during the approach to steady state is related to $\eta(\dot{\gamma})$ as

$$\lim_{\dot{\gamma}\to 0} \eta^+(t,\dot{\gamma}) = \eta(\dot{\gamma}) \ (t^{-1} = \dot{\gamma}) \tag{5.46}$$

Although the agreement with experimental data is not quantitative, this equation is quite useful because it provides estimates of viscosities over a range of shear rates from a single experiment. Figure 5.27 compares $\eta^+(t)$ to both $\eta(\dot{\gamma})$ and $\eta^*(\omega)$ for linear and branched PIB [107]. Results for a PDMS are displayed in Figure 5.28 [118]. Others assessments of the Gleissle relation have been reported for PIB [117] and polyethylene [111,119]. Gleissle [117] proposed an expression for the normal stress, analogous to eqn (5.46); however, in that case the relationship between reciprocal time and shear rate requires a numerical factor, which must be determined empirically for a given substance. This requires knowledge of $\eta(\dot{\gamma})$, obviating the purpose of the approximation, to circumvent such measurements.

5.5 Payne effect

Utilization of rubber often relies on the incorporation of small-particle fillers, which improve processability by reducing die swell and generally increase both the stiffness and strength of the elastomer. Since most fillers are less expensive than the polymer, they also reduce costs. Rubber mixed with carbon black is the most widely used polymeric composite. The use of carbon black as a pigment dates to 4000 BC and as a reinforcing agent for rubber since the early twentieth century. Methods to incorporate nanometer-sized fillers, the distribution and dispersion of the particles, and their effect on properties are aspects of the technology that have been widely reviewed [120–126]. Interest has been rekindled recently as part of the

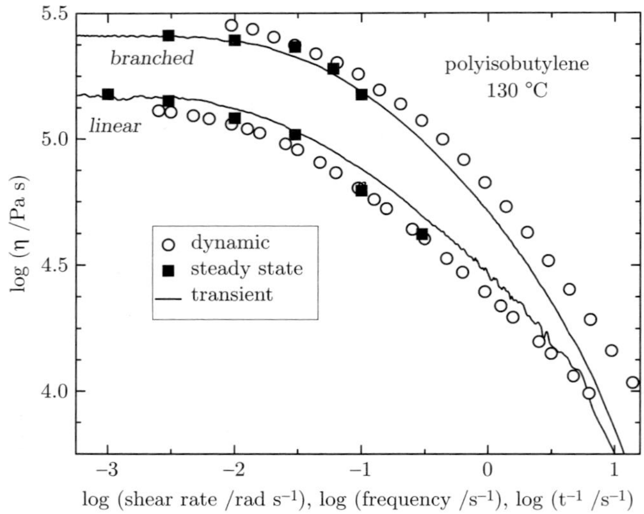

Figure 5.27 Viscosities determined from dynamic (open circles), steady-state (solid squares), and the transient (eqn (5.46); line) shear stress measurements on a linear ($M_w = 5.0 \times 10^5$ g/mol) and a highly branched ($M_w = 1.08 \times 10^6$ g/mol) PIB. Data from ref. [107].

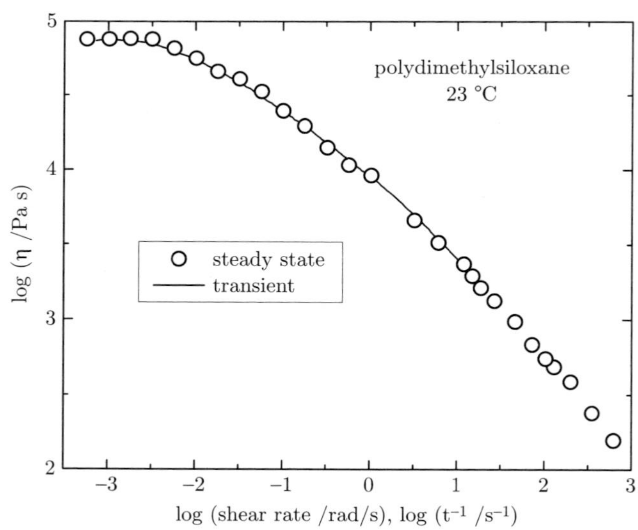

Figure 5.28 Viscosities determined from steady state (symbols) and the transient (eqn (5.46); line) shear stress measurements on a branched PDMS ($M_w = 4.7 \times 10^5$ g/mol). Data from ref. [118].

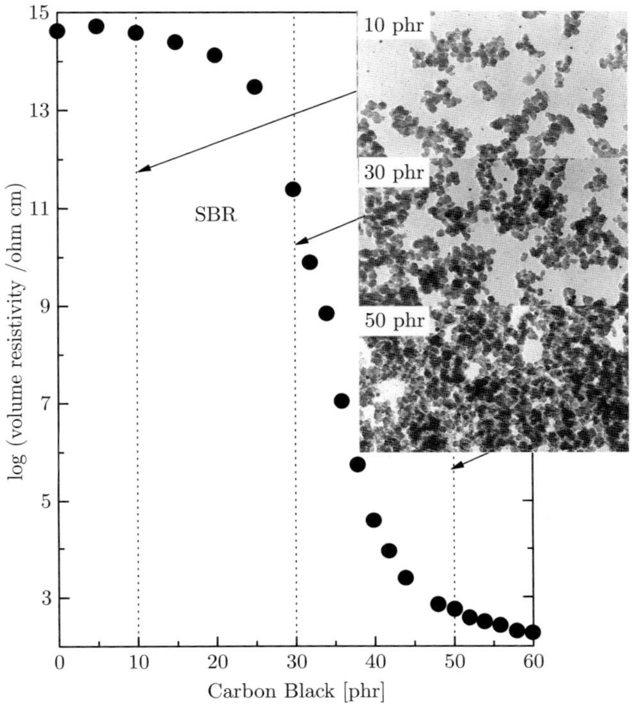

Figure 5.29 Transmission electron micrograph images of SBR containing the indicated quantity of N330 carbon black. The corresponding electrical resistivity behavior is indicated in the graph. Data from ref. [147].

burgeoning attention to nanoscience and technology (Chapter 10), with carbon nanotubes [127,128], silica nanoparticles [129,130], nanoclay [131–133], graphene [134,135], and diamond nanoparticles [136] investigated as potential fillers for polymers.

The effect of reinforcing fillers[15] on the modulus is very strain dependent. At all strains, stress amplification in the vicinity of the inextensible particles [137–139] increases the modulus. This hydrodynamic effect is enhanced by the occlusion of rubber at the particle interface [140,141]. At higher strains various mechanisms—specific to the material—become operative, including detachment of chains from the filler [142], orientation of the dispersed particles [143], and microscopic crack deviation around the filler phase [144]. All contribute to reinforcement and increase the non-linearity of the mechanical behavior of filled rubber.

Although flocculation of dispersed particles is common to virtually all fillers, agglomeration of carbon black has received the most attention. Carbon black exists as aggregates, formed during the manufacturing process and ranging in size from about 10 to 100 nm. When the filler concentration exceeds the percolation threshold, interaction among the aggregates leads to formation of a continuous network structure. This filler network causes an enormous increase

[15] Fillers that do not "reinforce" the polymer are referred to as extenders, since their chief purpose is to increase the quantity of material and thus lower its cost. Extenders can increase the modulus by amplifying local stresses; however, they are usually relatively large and interact negligibly with the polymer chains. The result is that the modulus enhancement is modest, and failure properties such as elongation and fatigue life usually become worse.

in electrical conductivity [145–147] (Figure 5.29), the appearance of yield stresses (thixotropy), and an inverse variation with strain amplitude of the dynamic modulus [148–150]. The latter is due to the breakup of the flocculated structure as the mechanical stress overcomes the van der Waals forces responsible for aggregate bonding [142–144,151]. This breakup of the filler network dissipates energy [152] and thus can be a significant contributor to, for example, the fuel consumption of motor vehicle tires [153].

In a typical dynamic experiment, harmonic shear deformation is imposed at an excitation frequency ω

$$\gamma^*(t) = \gamma_0(t)\exp(i\omega t) \tag{5.47}$$

After steady state is attained (requiring a duration exceeding the relevant relaxation time, with a few cycles usually sufficient), the linear response is also a single-harmonic function at the same frequency, albeit with a phase shift due to any loss

$$\sigma^*(t) = \sigma_0(t)\exp(i[\omega t + \delta]) \tag{5.48}$$

(This complex representation is a convenience; the actual motion is sinusoidal.) In terms of the dynamic moduli,

$$G^* = \sigma^*/\gamma^* \tag{5.49}$$

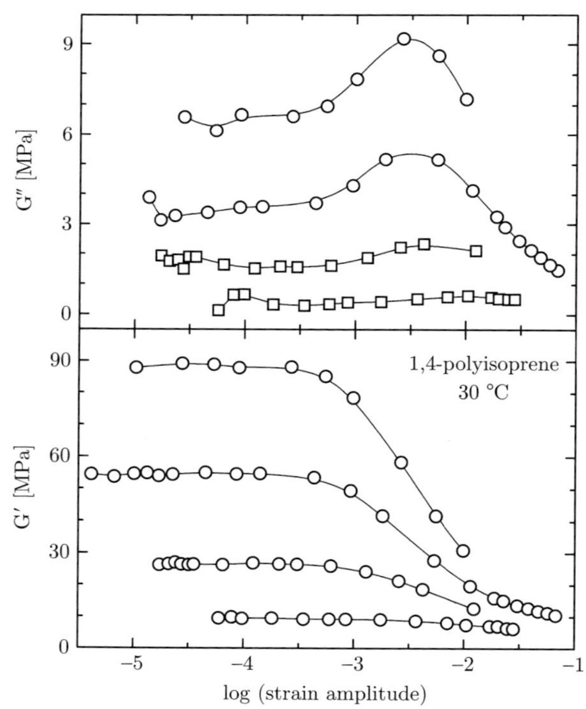

Figure 5.30 Dynamic storage (lower) and loss modulus (upper) variation with strain amplitude for PI at 0.1 Hz and 30°C. At low strains the modulus is constant, with higher amplitudes disrupting the filler network. Data from ref. [149].

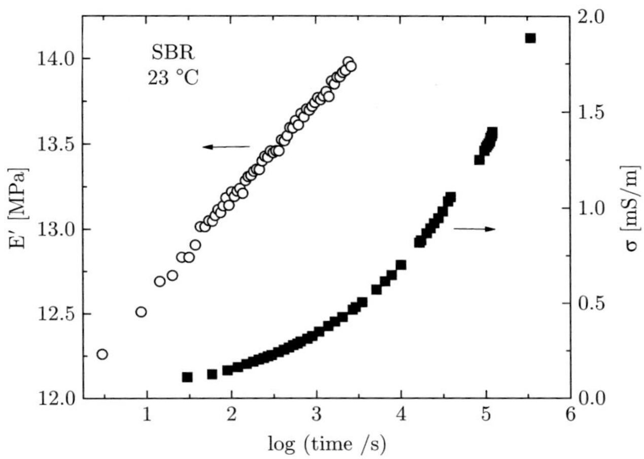

Figure 5.31 Low strain storage modulus (open circles) measured at 40 Hz for an SBR network with 82.5 phr N339 carbon black and 62.5 phr oil, following imposition of a static strain of 1.9%; also shown is the corresponding DC conductivity (filled squares). The values of E' (= 15.0 MPa) and σ_{DC} (= 4.2×10^{-3} S/m) prior to the static strain are substantially larger than attained after recovery (which extended to four days for the electrical measurement).

$$G' = (\sigma_0/\gamma_0)\cos(\delta) \tag{5.50}$$

and

$$G' = (\sigma_0/\gamma_0)\sin(\delta) \tag{5.51}$$

If the response depends either on the amplitude, γ_0, or frequency, it is no longer described by eqn (5.48). However, an arbitrary periodic function can be expressed as a sum of harmonics, so that departures from linearity give rise to intensity at frequencies other than the fundamental ω. If the non-linearity is approximated as a power series in the strain or strain rate, this superposition has the form [154]

$$\sigma(t) = \sigma_0\cos(\omega t) + \sigma_3\cos(3\omega t) + \sigma_5\cos(5\omega t) + \ldots \tag{5.52}$$

That is, the non-linearities appear as odd harmonics in the measured response.[16]

Although disruption of the agglomerated particles introduces strong non-linearities into the dynamic mechanical behavior, the response of filled rubber at any given strain remains linear; thus, the sinusoidal response lacks higher-order harmonics [157–159]. A steady-state degree of agglomeration is maintained, as determined by the maximum strain. At low strains the modulus is constant, as the filler network remains intact; higher strain amplitudes reduce the stiffness, with the appearance of a maximum in the loss modulus (Figure 5.30) [149]). This

[16] These higher harmonics resulting from material non-linearity or experimental distortions are distinct from the application of multiple input frequencies in order to simultaneously measure the linear response at several frequencies ("Fourier transform rheometry") [155]. The latter is an efficient experimental method, especially useful for transient structures; however, signal-to-noise issues are a consideration given the need to remain in the linear regime [156].

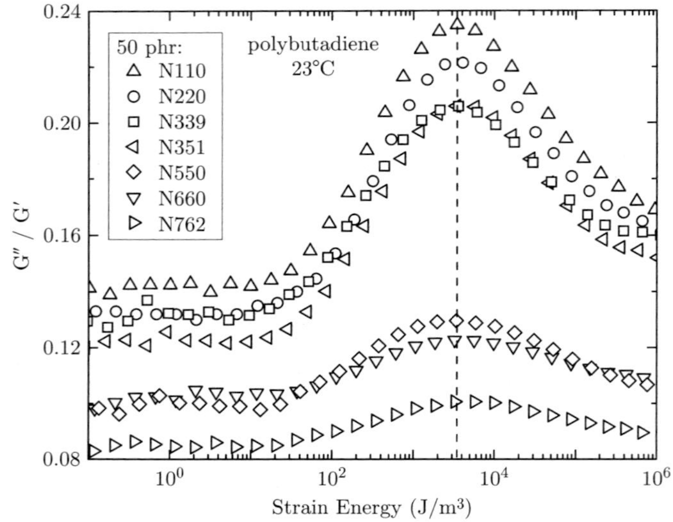

Figure 5.32 Loss tangent *versus* strain energy for polybutadiene with the indicated carbon-black content. The aggregate size of the carbon black varies from 54 nm (N110) to 188 nm (N762), the first two digits being a rough measure of the primary particle size (e.g., ~55 nm for N550, which has an aggregate size of 139 nm). The vertical dashed line denotes the characteristic strain energy, 3.5 kJ/m^3, which is independent of filler size. Dynamic data were obtained at 0.5 Hz at shear strain amplitudes from 10^{-4} to 0.3 Data from ref. [167].

characteristic behavior is known as the Payne effect [160].[17] The same phenomenon is seen in colloidal suspensions and other thixotropic fluids [163,164]. The Payne effect is not accompanied by any non-linearity during sinusoidal deformation at fixed amplitude, because the network does not recover on the timescale of the dynamic perturbation. This very slow recovery of the filler network is illustrated in Figure 5.31, showing the gradual recovery of the storage modulus and DC electrical conductivity upon reduction in the strain amplitude.

The Payne effect becomes evident at strain amplitudes in the range 0.1–1%. Interestingly, the disruption of the particle contacts transpires at a characteristic strain energy that is independent of the size or concentration of the particles [163,165–167]. This is illustrated in Figure 5.32 showing the superpositioning of the loss tangent curves for SBR with a fixed concentration of varying size of carbon-black particles [167]. The Payne effect is not specific to filled polymers, requiring only a flocculated particulate structure within a fluid medium; thus, oil containing silica particles shows the Payne effect, with strain energy again as the control parameter for the break-up of the filler structure, Figure 5.33 [165].

[17] The dependence on amplitude of the dynamic response unique to filled rubber was noted earlier by Gehman et al. [161] and by Fletcher and Gent [162].

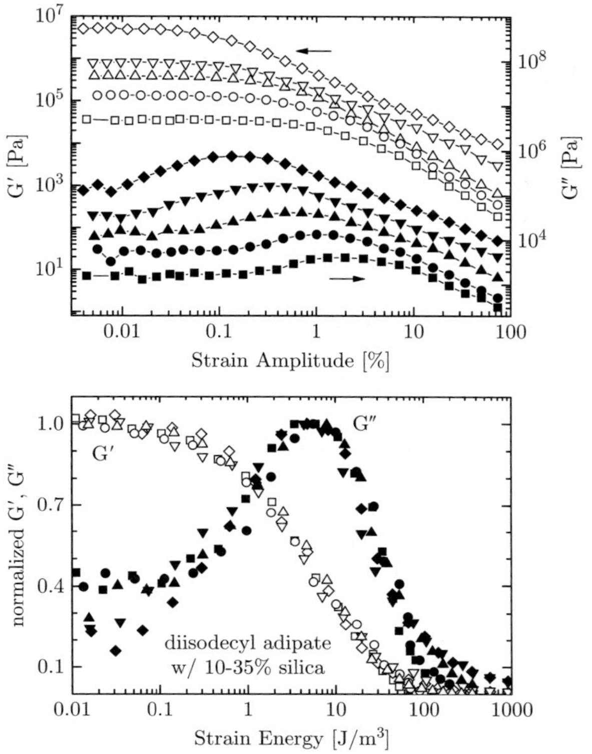

Figure 5.33 (upper) Dynamic storage (open symbols) and loss (filled symbols) moduli at 30°C and 0.5 Hz *versus* strain amplitude for the diisodecyl adipate containing 0.10 (squares), 0.14 (circles), 0.20 (triangles), 0.30 (down triangles), and 0.35 (diamonds) volume fraction silica particles. Data from ref. [166]. (lower) Same data with the storage modulus normalized by the low strain plateau value and the loss modulus by the maximum in G". Data from ref. [165].

References

1. R.B. Bird, R.C. Armstrong, and O. Hassager, *Dynamics of Polymeric Liquids*, Vol. 1, 2nd Edition, Wiley, New York (1987).
2. M. Doi and S.F. Edwards, *The Theory of Polymer Dynamics*, Clarendon Press, Oxford (1986).
3. H. Watanabe, *Prog. Polym. Sci.* **24**, 1253 (1999).
4. T.P. Lodge, N.A. Rotstein, and S. Prager, S. *Adv. Chem. Phys.* **79**, 1 (1990).
5. I.V. Yannas, *J. Polymer Sci. Macro. Rev.* **9**, 163 (1974).
6. W.V. Chang, R. Bloch, and N.W. Tschoegl, *Rheol. Acta* **15**, 367 (1976).
7. B. Bernstein, E.A. Kearsley, and L.J. Zapas, *Trans. Soc. Rheol.* **7**, 391 (1963).
8. H. Leaderman, *Trans. Soc. Rheol.* **6**, 361 (1962).
9. K.N.G. Fuller, in *Natural Rubber Science and Technology*, (ed. A.D. Roberts) Oxford University Press, New York (1988).
10. R.I. Tanner, *J. Rheol.* **32**, 673 (1988).

11. A.S. Lodge and J. Meissner, *Rheol. Acta* **11**, 351 (1972).
12. R.S. Rivlin, *Proc. 2nd Tech. Conf. (London)*, Heifer, Cambridge (1948).
13. J.S. Vrentas, D.C. Venerus, and C.M. Vrentas, *J. Polym. Sci. Polym. Phys. Ed.* **29**, 537 (1991).
14. B. D. Coleman and L. J. Zapas, *J. Rheol.* **33**, 501 (1989).
15. G.B. McKenna and L.J. Zapas, *J. Rheol.* **23**, 151 (1979).
16. M. Doi, *J. Polym. Sci. Polym. Phys. Ed.* **18**, 1891 (1980); **18**, 2055 (1980).
17. R.G. Larson, *J. Rheol,* **29**, 823 (1985).
18. R.G. Larson and V.A. Valesano, *J. Rheol.* **30**, 1093 (1986).
19. S.T. Milner, *J. Rheol.* **40**, 303 (1996).
20. L.R.G. Treloar, *Rep. Prog. Phys.* **36**, 755 (1973).
21. R.S. Rivlin, *Rubber Chem. Technol.* **65**, G51 (1992).
22. C.M. Roland, *J. Rheol.* **33**, 659 (1989).
23. A. Ciferri and P.J. Flory, *J. Appl. Phys.* **30**, 1498 (1959).
24. C.M. Roland, *Rubber Chem. Technol.* **62**, 880 (1989).
25. Y. Miyamoto, K. Fukao, H. Yamao, and K. Sekimoto, *Phys. Rev. Lett.* **88**, 225504 (2002).
26. W. Kuhn and F. Grün, *Kolloidzschr.* **101**, 248 (1942).
27. P.H. Mott and C.M. Roland, *Macromolecules* **29**, 8492 (1996).
28. M. Van Gurp, *Coll. Polym. Sci.* **273**, 607 (1995).
29. R.D.B. Fraser, *J. Chem. Phys.* **21**, 1511 (1953).
30. V. Balasubramanian, K. Bush, S. Smoukov, D.C. Venerus, and J.D. Schieber, *Macromolecules* **38**, 6210 (2005).
31. R. Muller and D. Froelich, *Polymer* **26**, 1477 (1985).
32. R. Muller and J.J. Pesce, *Polymer* **35**, 734 (1994).
33. C. Luap, C. Muller, T. Schweizer, and D.C. Venerus, *Rheol. Acta* **45**, 83 (2005).
34. J. van Meerveld, *J. Non-Newtonian Fluid Mech.* **123**, 259 (2004).
35. D. Hands, *Rubber Chem. Technol.* **53**, 80 (1980).
36. D.C. Venerus, J.D. Schieber, H. Iddir, J.D. Guzman, and A. W. Broerman, *Phys. Rev. Lett.* **82**, 366 (1999).
37. J.D. Ferry, *Viscoelastic Properties of Polymers*, 3rd edition, Wiley, New York (1980).
38. I.S. Choi and C.M. Roland, *Rubber Chem. Technol.* **70**, 202 (1997).
39. H. Okamoto, T. Inoue, and K. Osaki, *J. Polym. Sci. Polym. Phys. Ed.* **33**, 1409 (1995).
40. B.E. Read, *Polym. Eng. Sci.* **23**, 835 (1983).
41. T. Inoue, H. Okamoto, and K. Osaki, *Macromolecules* **24**, 5670 (1991).
42. T. Inoue, H. Hayashihara, H. Okamoto, and K. Osaki, *J Polym. Sci., Polym. Phys. Ed.* **30**, 409 (1992).
43. H. Okamoto, T. Inoue, and K. Osaki, *J. Polym. Sci., Polym. Phys.* **33**, 417 (1985).
44. K. Osaki, H. Okamoto, T. Inoue, and E.J. Hwang, *Macromolecules* **28**, 3625 (1995).
45. P.H. Mott, A.S. Argon, and U.W. Suter, *Philos. Mag. A* **67**, 931 (1993).
46. D.J. Plazek, *J. Rheol.* **36**, 1671 (1992).
47. C.M. Roland and P.H. Mott, *Macromolecules* **32**, 4728 (1999).
48. P.H. Mott and C.M. Roland, *Macromolecules* **31** 7095 (1998).

49. B. Deloche and E.T. Samulski, *Macromolecules* **14**, 575 (1981).
50. M.M. Jacobi, R. Stadler, and W. Gronski, *Macromolecules* **19**, 2884 (1986).
51. C.M. Ylitalo, J.A. Zawada, G.G. Fuller, V. Abetz, and R. Stadler, *Polymer* **33**, 2949 (1992).
52. I.S. Zemel and C.M. Roland, *Polymer* **33**, 4522 (1992).
53. R.C. Hedden, E. McCaskey, C. Cohen, and T. M. Duncan, *Macromolecules* **34**, 3285 (2001).
54. K. Kawabata, T. Fukuda, Y. Tsujii, and T. Miyamoto, *Macromolecules* **26**, 3980 (1993).
55. C.M. Ylitalo, J.A. Kornfield, G.G. Fuller, and D.S. Pearson, *Macromolecules* **24**, 749 (1991).
56. J.A. Kornfield, G.G. Fuller, and D.S. Pearson, *Macromolecules* **22**, 1334 (1989).
57. J.F. Tassin, A. Baschwitz, J.Y. Moise, and L. Monnerie, *Macromolecules* **23**,1879 (1990).
58. U. Seidel, R. Stadler, and G.G. Fuller, *Macromolecules* **28**, 3739 (1995).
59. J.A. Zawada, G.G. Fuller, R.H. Colby, L.J. Fetters, and J. Roovers, *Macromolecules* **27**, 6851 (1994).
60. B.H. Arendt, R. Krishnamoorti, J.A. Kornfield, and S.D. Smith, *Macromolecules* **30**, 1127 (1997).
61. P. Sotta, B. Deloche, J. Herz, A. Lapp, D. Durand, and J.C. Rabadeux, *Macromolecules* **20**, 2769 (1987).
62. A. Batra, R.C. Hedden, P. Schofield, A. Barnes, C. Cohen, and T.M. Duncan, *Macromolecules* **36**, 9458 (2003).
63. A.R.C. Baljon, G.S. Grest, and T.A. Witten *Macromolecules* **28**, 1835 (1995).
64. S. Barsky, and G.W. Slater, *Macromolecules* **32**, 6348 (1992).
65. J.-P. Queslel, B. Erman and L. Monnerie, *Polymer* **29**, 1818 (1988).
66. L. Mullins, *Rubber Chem. Technol.* **42**, 339 (1969).
67. K.D. Suh, H. Oikawa, and K. Murakami, *Rubber Chem. Technol.* **59**, 541 (1986).
68. C.J. Ohlemacher and G.R. Hamed, *Rubber Chem. Technol.* **81**, 650 (2008).
69. C.M. Roland and W.J. Sobieski, *Rubber Chem. Technol.* **62**, 683 (1989).
70. J.A.C. Harwood and A.R. Payne, *J. Appl. Polym. Sci.* **10**, 1203 (1966).
71. F. Clement, L. Bokobza, and L. Monnerie, *Rubber Chem. Technol.* 74, 847 (2001).
72. G.R. Hamed, *Rubber Chem. Technol.* **73**, 524 (2000).
73. F. Bueche, *Rubber Chem. Technol.* **35**, 259 (1962).
74. J.A.C. Harwood and A.R. Payne, *J. Appl. Polym. Sci.* **11**, 1825 (1967).
75. P.J. Flory, *J. Chem. Phys.* **15**, 397 (1947).
76. S. Toki, I. Sics, S. Ran, L. Liu, B.S. Hsiao, S. Murakami, M. Tosaka, S. Kohjiya, S. Poompradub, Y. Ikeda, and A.H. Tsou, *Rubber Chem. Technol.* **77**, 317 (204).
77. J.A.C. Harwood, L. Mullins, and A.R. Payne, *J. Appl. Polym. Sci.* **9**, 3011 (1965).
78. J.A.C. Harwood and A.R. Payne, *J. Appl. Polym. Sci.* **10**, 315 (1966).
79. J.A.C. Harwood and A. Schallamach, *J. Appl. Polym. Sci.* **11**, 1835 (1967).
80. P.G. Santangelo and C.M. Roland, *Rubber Chem. Technol.* **65**, 965 (1992).
81. M.H. Wagner and P. Ehrecke, *J. Non-Newtonian Fluid Mech.* **76**, 183 (1998).
82. D.C. Venerus, C.M. Vrentas, and J.S. Vrentas, *J. Rheo1.* **34**, 657 (1990).
83. C. Sui and G.B. McKenna, *J. Rheol.* 51, 341 (2007).
84. K. Osaki, S. Kimura, and M. Kurata, *J. Rheol.* **25**, 549 (1981).
85. M. Takahaski, T. Masuda, and S. Onogi, *J. Rheol.* **22**, 285 (1978).

86. L.J. Zapas, in *Deformation and Fraction of High Polymers*, (eds H.H. Kaush, J.A. Hassell, and R.L. Jaffe), Plenum, New York (1973).
87. H.C. Ottinger and A.N. Beris, *J. Chem. Phys.* **110**, 6593 (1999).
88. G. Marrucci, *J. Non-Newtonian Fluid Mech.* **21**, 329 (1986).
89. M.H. Wagner and S.E. Stephenson, *J. Rheol.* **23**, 489 (1979).
90. D.W. Saunders, *Trans. Faraday Soc.* **52**, 1414 (1956).
91. P.H. Mott, A. Rizos, and C.M. Roland, *Macromolecules 34*, 4476 (2001).
92. R.D. Andrews and Y. Kazama, *J. Appl. Phys.* **39**, 4891 (1968).
93. C.R. Taylor and J.D. Ferry, *J. Rheol.* **23**, 533 (1979).
94. L.K. Djiauw and A.N. Gent, *J. Polym. Sci.* **C48**, 159 (1974).
95. W.P. Cox and E.H. Merz, *J. Polym. Sci.* **28**, 619 (1958).
96. G. Ianniruberto and G. Marrucci, *J. Non-Newtonian Fluid Mech.* **65**, 241 (1996).
97. G. Marrucci, *J. Non-Newtonian Fluid Mech.* **62**, 279, (1996).
98. H.C. Booij, P. Leblans, J. Palmen, and G. Tiemersma-Thoone, *J. Polym. Sci. Polym. Phys. Ed.* **21**, 1703 (1983).
99. M. Renardy, *J. Non-Newtonian. Fluid Mech.* **68**, 133 (1997).
100. R.M. Schulken, R.H. Cox, and L.A. Minnick, *J. Appl. Polym. Sci.* **25**, 1341 (1980).
101. D. Ferri and P. Lomellini, *J. Rheol.* **43**, 1355 (1999).
102. M. Gahleitner, *Proc. XII Intl. Congr. Rheol.* (eds A. Ait-Kadi, J.M. Dealy, D.F. James and M.C. Williams) Aug. 23, page 70 (1996).
103. R.B. Bird, R.C. Armstrong, and O. Hassager, *Dynamics of Polymeric Liquids*, Vol. 1, 2nd edition, Wiley, New York (1987).
104. S. Venkatraman, M. Okano, and A. Nixon, *Polym. Eng. Sci.* **30**, 308 (1990).
105. D. Ferri and P. Lomellini, *J. Rheol.* **43**, 1355 (1999).
106. C.C. Hua, *J. Chem. Phys.* **112**, 8176 (2000).
107. C.R. Robertson, C.M. Roland, and J.E. Puskas, *J. Rheol.* **46**, 307 (2002).
108. P.K. Han and J.L. White, *Rubber Chem. Technol.* **68**, 728 (1995).
109. A.Y. Coran, *Rubber Chem. Technol.* **68**, 351 (1995).
110. X. Xu, X. Tao, C. Gao, and Q. Zheng, *J. Appl. Polym. Sci.* **107**, 1590 (2008).
111. H.M. Laun, *J. Rheol.* **30**, 459 (1986).
112. Y.H. Wen, H.C. Lin, C.H. Li, and C.C. Hua, *Polymer* **45**, 8551 (2004).
113. B.D. Coleman and H. Markovitz, *J. Appl. Phys.* **35**, 1 (1964).
114. R.I. Tanner, *J. Polym. Sci. Polym. Phys. Ed.* **8**, 2067 (1970).
115. E.B. Bagley and H.J. Duffey, *Trans. Soc. Rheol.* **14**, 545 (1970).
116. H.W. M llner, G. Ernst, and J. Eberhardsteiner, *J. Appl. Polym. Sci.* **110**, 76 (2008).
117. Gleissle, W., in *Rheology*, (eds G. Astarita, G. Marucci, and L. Nicolais) Plenum, New York, **2**, 457 (1980).
118. N.E. Kissi, J.M. Piau, R Attane, and G. Turrel, *Rheol. Acta* **32**, 293 (1993).
119. P.M. Wood-Adams, *J. Rheol.* **45**, 203 (2001).
120. A.E. Medalia, *Rubber Chem. Technol.* **60**, 45 (1987).
121. G. Hamed, *Rubber Chem. Technol.* **73**, 524 (2000).

122. J.-B. Donnet and E. Custodero, in "Science and Technology of Rubber," 3rd Edition, (eds J.E. Mark, B. Erman, and F.R. Eirich) Academic Press, Burlington, MA (2005).
123. *Carbon Black-Polymer Composites*, (ed. E.K. Sichel), Marcell Dekker, New York (1982).
124. G. Heinrich, M. Kluppel, and T.A. Vilgis, *Curr. Opin. Solid State Mater. Sci.* **6**, 195 (2002).
125. W.H. Waddell and L.R. Evans, *Rubber Chem. Technol.* **69**, 377 (1996).
126. G.R. Hamed, *Rubber Chem. Technol.* **80**, 533 (2007).
127. L. Bokobza, *Polymer* **48**, 4907 (2007).
128. J.N. Coleman, U. Khan, W.J. Blau and Y.K Gunko *Carbon*, **44**, 1624 (2006).
129. A. Bansal, H.C. Yang, C.Z. Li, K.W. Cho, B.C. Benicewicz, S.K. Kumar, and L.S. Schadler, *Nature Mater.* **4**, 693 (2005).
130. R.B. Bogoslovov, C.M. Roland, A.R. Ellis, A.M. Randall, and C.G. Robertson, *Macromolecules* **41**, 1289 (2008).
131. Y. T. Vu, J. E. Mark, L. H. Pham, and M. Engelhardt, *J. Appl. Polym. Sci.* **82**, 1391 (2001).
132. S. Sadhu and A.K. Bhowmick, *Rubber Chem. Technol.* **76**, 860 (2003).
133. J. Carretero-Gonzalez, H. Retsos, R. Verdejo, S. Toki, B.S. Hsiao, E.P. Giannelis, and M.A. Lopez-Manchado, *Macromolecules* **41**, 6763 (2008).
134. S. Stankovich, D.A. Dikin, G.H.B. Dommett, K.M. Kohlhaas, E.J. Zimney, E.A. Stach, R.D. Piner, S.T. Nguyen, and R.S. Ruoff, *Nature* **442**, 282 (2006).
135. K. Wakabayashi, C. Pierre, D.A. Dikin, R.S. Ruoff, T. Ramanathan, L.C. Brinson, and J.M. Torkelson, *Macromolecules* **41**, 1905 (2008).
136. K.D. Behler, A. Stravato, V. Mochalin, G. Korneva, G. Yushin, and Y. Gogotsi, *ACS Nano* **3**, 363 (2009).
137. H.M. Smallwood, *J. Appl. Phys.* **15**, 758 (1944).
138. E. Guth, *Rub. Chem. Technol.* **18**, 596 (1945).
139. G.K. Batchelor and J.T. Green, *J. Fluid Mech.* **56**, 401 (1972).
140. G. Kraus, *J. Poly. Sci.* **B8**, 601 (1970).
141. A. Medalia, *Rubber Chem. Technol.* **47**, 411 (1974).
142. F. Bueche, *Physical Properties of Polymers*, Wiley, New York (1962).
143. J.-B. Donnet and A. Voet, *Carbon Black – Physics, Chemistry, and Elastomer Reinforcement*, Marcel Dekker, New York (1976).
144. A. Medalia, *Rubber Chem. Technol.* **60**, 45 (1987).
145. A.I. Medalia, *Rubber Chem. Technol.* **59**, 432 (1986).
146. C.P. O'Farrell CP, M. Gerspacher, and L. Nikiel, *Kautschuk Gummi Kunststoffe* **53**, 701 (2000).
147. L. Nikiel, W. Wampler, J. Neilsen, and N. Hershberger, *Rubber Plastics News*, page 12, March 23 (2009).
148. A.I. Medalia, *Rubber Chem. Technol.* **51**, 411 (1978).
149. C.M. Roland, *J. Rheol.* **34**, 25 (1990).
150. Z.Y. Zhu, T. Thompson, S.Q. Wang, E.D. von Meerwall, and A. Halasa, *Macromolecules* **38**, 8816 (2005).
151. D. Goritz, H. Raab, J. Frohlich, and P.G. Maier, *Rubber Chem. Technol.* **72**, 929 (1999).
152. J. D. Ulmer, W. L. Hergenrother, and D. F. Lawson, *Rubber Chem. Technol.* **71**, 637 (1998).
153. D.J. Schuring and S. Futamura, *Rubber Chem. Technol.* **63**, 315 (1990).

154. M. Wilhelm, D. Maring, and H.W. Spiess, *Rheol. Acta* **37**, 399 (1998).
155. E.E. Holly, S.K. Venkataraman, F. Chambon, and H.H. Winter, *J. Non-Newtonian Fluid Mech.* **27**, 17 (1988).
156. A. Ya. Malkin, *Rheol. Acta* **43**, 1435 (2004).
157. G. Heinrich and M. Klüppel, *Adv. Polym. Sci.* **160**, 1 (2002).
158. L. Chazeau, J.D. Brown, L.C. Yanyo, and S.S. Sternstein, *Polym. Compos.* **21**, 202 (2000).
159. C. G. Robertson and X. Wang, *Europhys. Lett.* **76**, 278 (2006).
160. A.R. Payne, *J. Appl. Polym. Sci.* **6**, 57 (1962).
161. R.D. Gehman, D.E. Woodford, and R.B. Stambaugh, *Ind. Eng. Chem.* **33**,1032 (1941).
162. W.P. Fletcher and A.N. Gent, *Trans. Inst. Rubber Ind.* **29**, 266 (1953).
163. E.R. Fitzgerald and J.F. Ferry, *Rubber Chem. Technol.* **55**, 1569 (1982).
164. M. Caggioni, P.T. Spicer, D.L. Blair, S.E. Lindberg, and D.A. Weitz, *J. Rheol.* **51**, 851 (2007).
165. C.G. Robertson and X. Wang, *Phys. Rev. Lett.* **95**, 075703 (2005).
166. X. Wang and C.G. Robertson, *Phys. Rev. E* **72**, 031406 (2005).
167. C.G. Robertson, R. Bogoslovov, and C.M. Roland, *Phys. Rev. E* **75**, 051403 (2007).

6

Reduced variables and characteristic relaxation times

6.1 Time–temperature superpositioning

An important aspect of the viscoelasticity of polymers is the effect of temperature and pressure. In 1949, samples of a polyisobutylene ($M_w = 1.56 \times 10^6$ g/mol) were measured in twenty-seven laboratories in several countries, employing a range of mechanical techniques to yield results covering 15 decades of the time/frequency scale [1]. This early work, in particular the stress-relaxation measurements of Andrews and Tobolsky [2] and the dynamic compliance experiments of Fitzgerald, *et al.* [3], demonstrated that polymers were thermorheologically simple. Time–temperature superpositioning can be applied to viscoelastic data obtained over a narrow range to calculate master curves extending over many decades of reduced time or frequency [4].

The time–temperature superposition principle is based on the idea that if all motions underlying a property are governed by the same friction coefficient, changes in temperature equate to changes in rate. A similar statement applies to the pressure dependence. In addition to effects operating solely on ζ, three further requirements for successful superpositioning are: (i) The material itself remains unchanged as temperature or pressure is varied; i.e. there is no crystallization, degradation, physical aging, etc. (ii) The strain rate does not exceed the rate at which stress waves propagate through the material, so that the stress remains spatially uniform. (iii) Heat build-up, more pronounced for low temperatures and high rates, causes no deviation from isothermal conditions.

An example of time–temperature superpositioning is Figure 1.1 , showing dynamic modulus data spanning 15 decades of frequency for high molecular weight 1,4-polyisoprene. In Figure 6.1 is a stress-relaxation master curve encompassing 13 decades of time for a polyurea; the curve was constructed from measurements at a series of temperatures, each covering 3.3 decades of real time. This illustrates the utility of the method for extrapolation to longer times and higher frequencies. However, the stress-relaxation behavior of the polyurea is monotonic, making the horizontal shifting procedure somewhat ambiguous. Errors can arise due to the direct influence of temperature on the modulus, which are unrelated to rate–temperature equivalency and must be accounted for by vertical shifting (for example, the ordinate values in Figure 6.1 were multiplied by the ratio of the reference temperature to the measurement temperature). This direct effect of temperature is twofold: The orientational component of the strain is entropic, making a contribution to the modulus proportional to absolute temperature. Additionally, the mass density decreases with temperature, with the reduced number of chain segments per unit volume contributing to lower modulus. The uncertainty in the modulus correction factor can be significant when there is minimal

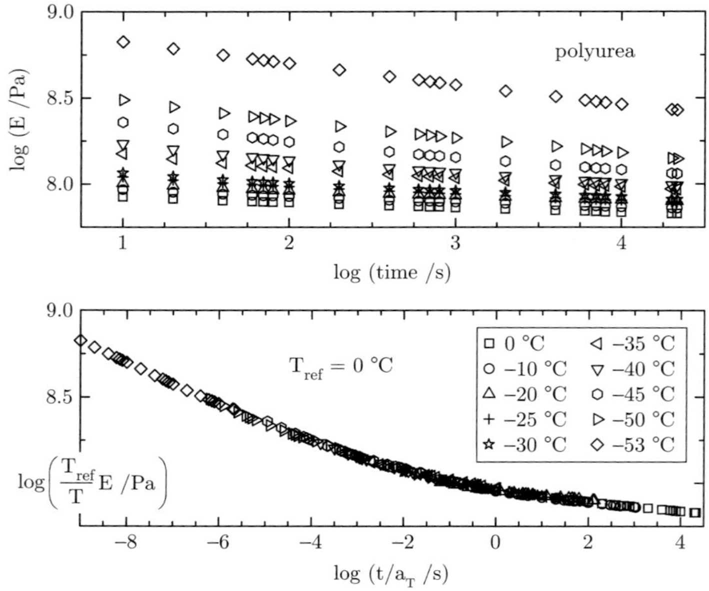

Figure 6.1 (upper) Tensile stress relaxation modulus of a polyurea ($T_g = -60$ °C) measured over 3.3 decades of time at the indicated temperatures. The abscissa is the actual time. (lower) Master curve obtained by shifting the measured data along the abscissa, which spans over 13 decades of reduced time. Data courtesy of W.G. Knauss.

"structure" in the curves. The problem can be avoided by obtaining time–temperature shift factors (a_T, the ratio of the measurement frequency, or time, to its corresponding value at the reference temperature) from analysis of the dynamic mechanical loss tangent, for which there is no modulus correction. And as seen below (Figure 6.13), the loss tangent for this same polyurea reveals a breakdown of time–temperature superpositioning, not evident in the featureless stress-relaxation data of Figure 6.1.

A master curve of the dynamic bulk modulus of a polyurethane[1] [5] is shown in Figure 6.2, with the B^* calculated from G^* and the sound speed. Such results illustrate that equations relating elastic quantities can be extended to the viscoelastic properties; the a_T are the same for dilational and shear deformations.

The original derivation of an equation for a_T was based on the idea that the rate of molecular motions is determined by the available free volume (unoccupied space). The Doolittle equation [6,7]

$$\log \eta(T) = \log A_D + B_D / f_V(T) \tag{6.1}$$

where A_D and B_D are constants, relates the viscosity of molecular liquids to their fractional free volume, f_V ($\equiv \frac{V}{V_0} - 1$ where V is the specific volume and V_0 the occupied volume normalized by the total mass). Experimental data deviate from eqn (6.1) [8–10]. By assuming that free

[1] Polyurethanes are formed from the reaction of an isocyanate with an alcohol, whereas reaction of isocyanate with an amine yields a polyurea; these reactants are di- or polyfunctional to yield a network.

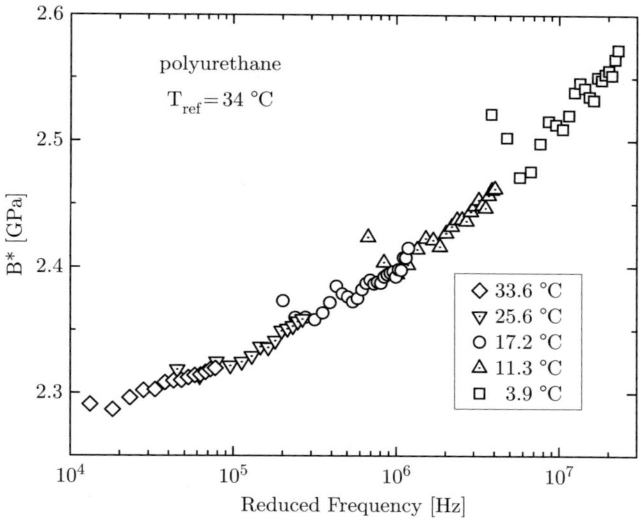

Figure 6.2 Complex bulk modulus for a polyurethane calculated from shear modulus and sound speed measurements at the indicated temperatures; the scatter is due to the experimental uncertainties in these underlying data. The master curve was constructed using time–temperature shift factors determined from superpositioning the dynamic shear modulus. The material exhibits thermorheological complexity at lower temperatures, closer to T_g. Data from ref. [5].

volume expands linearly with temperature, $\frac{df_V}{dT}\big|_P \propto T$, the Williams–Landel–Ferry (WLF) equation is obtained [11]

$$\log a_T \equiv \log\Big(\eta(T)/\eta(T_{ref})\Big) = \frac{-c_1(T - T_{ref})}{c_2 + T - T_{ref}} \tag{6.2}$$

where T_{ref} is an arbitrary reference temperature. An equivalent expression applies to relaxation times, since η and τ are directly proportional except near the glass transition (Chapter 2). Originally the constants c_1 and c_2 were related to the free volume and its change with temperature, with "universal values" of $c_1 = 8.86$ and $c_2 = 101.6$ K [11]; in practice these serve as empirical fitting parameters.

To account for the variation of the dynamics with pressure, the free volume must be assumed to compress differently than the total compressibility of the material. However, in analyzing experimental data this leads to the unphysical result that the free volume, $V - V_0$, is less compressible than the occupied volume [4,12]. Extensions of free-volume models yield modifications of eqn (6.2); however, the time–temperature superposition principle does not rely on any particular interpretation of its origin. For example, an alternative expression for the T dependence is the VFTH relation (eqn (2.4)), which can be derived from configurational entropy models [13,14], yet is mathematically equivalent to the WLF equation, with $B = c_1c_2$ and $c_2 = T_{ref} - T_\infty$ [15].[2]

[2] An incorrect relationship between the WLF and VFTH parameters was given in ref. [4] and this error is oft-repeated in subsequent literature.

Time–temperature superpositioning has been applied not only to linear viscoelastic measurements on polymers, but also to various other physical properties. To the extent the latter involve the same molecular motions, these properties will have the same a_T as the low-strain modulus or viscosity. This is illustrated in Figure 6.3 and Figure 6.4 [16], in which the shift

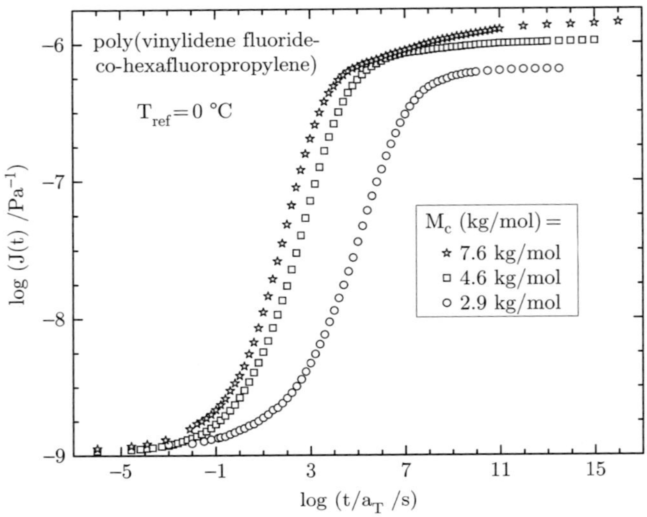

Figure 6.3 Creep compliance master curves for Viton® flouroelastomers having the indicated molecular weight between crosslinks. The measurements spanned 45 deg. and actual times from 0.1 to 10^5 s. Data from ref. [16].

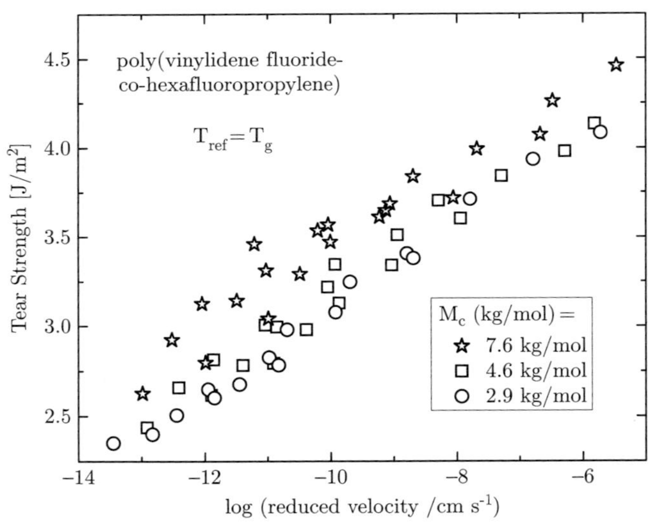

Figure 6.4 Tearing energy of Viton® flouroelastomers *versus* cross-head speed with the latter multiplied by the shift factor determined from superpositioning of linear creep data (Figure 6.3). The original measurements covered 2 decades of rate, which is extended to more than 7 decades by superpositioning. Data from ref. [16].

factors used to construct master curves of the creep compliance of different fluoroelastomers also superpose their respective tearing energies; thus, the temperature-dependence is the same for the response at low strains and at failure strains. Time–temperature-superpositioning of tearing energies has been shown for other elastomers, including PBD [17], SBR [17], and PDMS [18]. Figure 6.5 is a master curve of the failure strain in tension for a crosslinked SBR; the same values of a_T also superpose the tensile strength of the material [19]. The fact that the shift factors are the same for stress and strain at break leads to the idea of a failure envelope, in which the strength is plotted as a function of the ultimate strain. An example is given in Figure 6.6 for a polyethylene [20]. The results in Figures 6.4–6.6 imply that the viscoelastic response of the amorphous chains dominates the failure properties.

Another application of time–temperature superpositioning to non-linear properties is the frictional resistance to sliding, as shown in Figure 6.7 for a nitrile rubber [21]. The obtained master curve for the friction coefficient exhibits a maximum versus sliding velocity, associated with the energy-dissipative glass-transition zone (Chapter 2). These data raise an issue discussed below – are the molecular motions underlying the response at very high rates (10^6 s^{-1}) and failure strains of less than 10% the same as those active at slow rates (10^{-9} s^{-1}) and deformations exceeding 600%? The latter involve displacements of large sections of the chains, which do not respond at high frequencies associated with brittle behavior. As discussed below, the chain modes and local segmental modes respond differently to changes in thermodynamic conditions, which can cause a breakdown of the superposition principle. The problem is that superpositioning *per se* does not guarantee the resulting master curve represents the actual response for measurements made over the range of rates at the reference temperature.

Similar to the effect of temperature, hydrostatic pressure changes the timescale of the dynamics, leading to the possibility of time–pressure superpositioning. However, there is a paucity of published mechanical results at elevated pressure. Tschoegl and coworkers [22–24] have demonstrated the reduction of stress-relaxation data to a single curve for various polymers, including PBD, SBR, and a fluoroelastomer, at pressures up to ~ 0.5 GPa. In

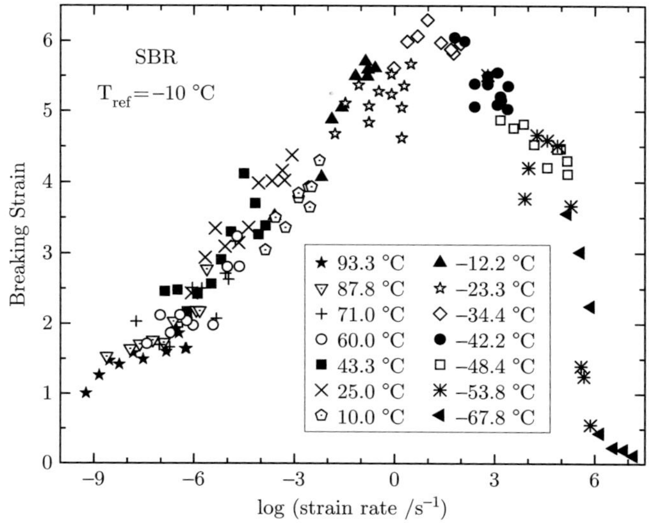

Figure 6.5 Failure strain in tension for an SBR plotted versus the reduced strain rate. The actual measurement temperatures are indicated. Data from ref. [19].

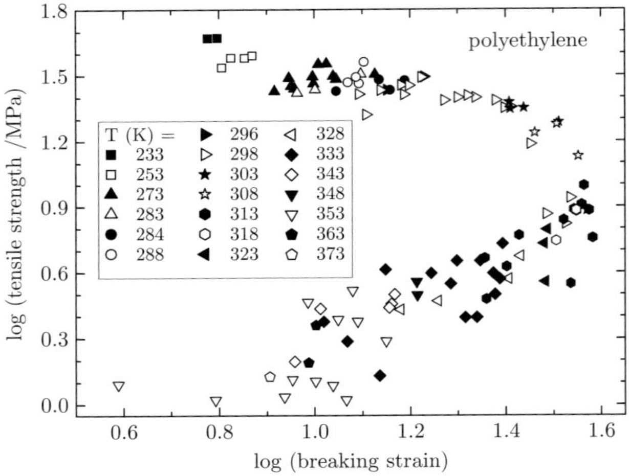

Figure 6.6 Failure envelope of semicrystalline polyethylene ($M_w = 73$ kg/mol) for the indicated temperatures and nominal strain rates from 4×10^{-3} to 4×10^{-2} s^{-1}. The test specimen was designed to avoid necking and achieve uniform deformation. Data from ref. [20].

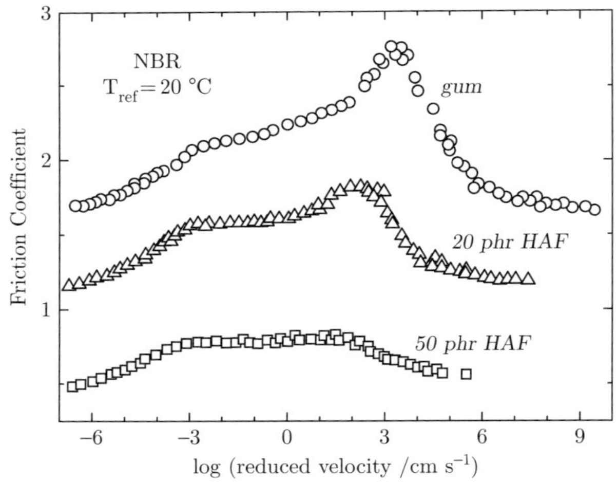

Figure 6.7 Master curves of the friction coefficient for nitrile rubbers against dry silicon carbide. The carbon black content is as indicated. The actual measurements covered 4 decades of sliding velocity, the latter not exceeding a few cm/s to minimize frictional heating. Data from ref. [21].

Figures 6.8 and 6.9 are respective master curves of the stress-relaxation modulus of an SBR and a filled NR at pressures as high as 350 MPa [25]. Even with carbon black present, the data superpose well. Generally, the only effect of reinforcing fillers on the response to pressure of rubber is an increase in the bulk modulus due to the reduction in the polymer content; the degree of agglomeration of the filler and its interaction with the polymer have a negligible

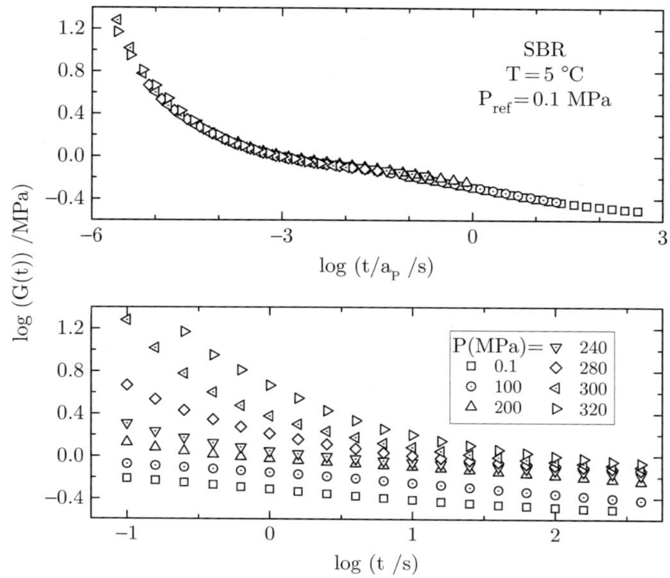

Figure 6.8 Stress-relaxation modulus of an SBR measured at the indicated pressures (lower panel) and superposed to form a master curve (upper panel). Data from ref. [25].

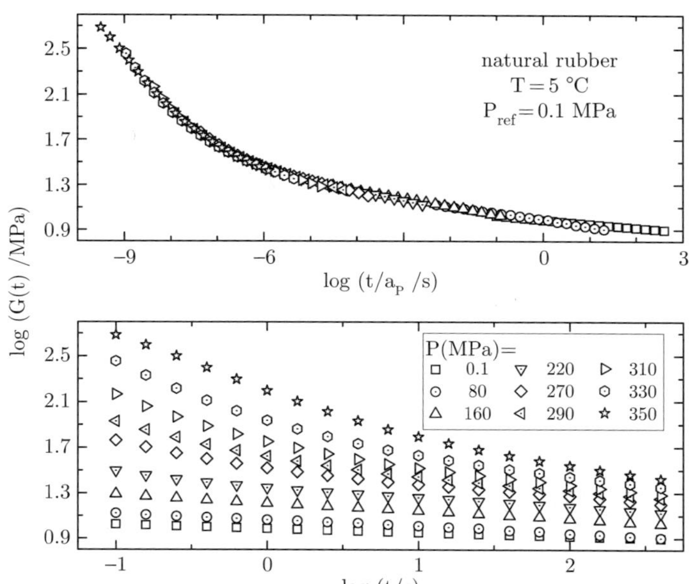

Figure 6.9 Stress-relaxation modulus of a filled NR vulcanizate measured at the indicated pressures (lower panel) and superposed to form a master curve (upper panel). Data from ref. [25].

effect [26,27]. If the pressure shift factors, a_P, are independent of strain rate, a master curve of viscosity versus rate can be constructed, as shown in Figure 6.10 for two polymers for $P \leq 80$ MPa [28]. Note that since the viscosity is the ratio of the stress and shear rate, in constructing the master curve of $\eta(\dot{\gamma})$, the ordinate values are divided by the same a_P used to multiply the abscissa values.

Dielectric relaxation measurements under pressure [29] are much more common than mechanical experiments. If a polymer has a dipole moment parallel to the chain, both the global chain and local segmental dynamics give rise to peaks in the dielectric loss spectrum, the shapes of which are invariant to changes in T or P if time–temperature or time–pressure superpositioning applies. Dielectric loss peaks are usually fit to the equation of Havriliak and Negami [30]

$$\varepsilon''(\omega) = \frac{\Delta\varepsilon \sin(\beta_{HN}\psi)}{[1 + 2(\omega\tau_{HN})^{\alpha_{HN}} \cos(\pi\alpha_{HN}/2) + (\omega\tau_{HN})^{2\alpha_{HN}}]^{\beta_{HN}/2}} \tag{6.3}$$

where

$$\psi = \arctan\frac{(\omega\tau_{HN})^{\alpha_{HN}} \sin(\pi\alpha_{HN}/2)}{1 + (\omega\tau_{HN})^{\alpha_{HN}} \cos(\pi\alpha_{HN}/2)} \tag{6.4}$$

τ_{HN} is a relaxation time[3] and α_{HN} and β_{HN} are shape factors describing the peak breadth. Their product is related approximately to the Kohlrausch exponent as $\alpha_{HN}\beta_{HN} = \beta_K^{1.23}$ [31]. In

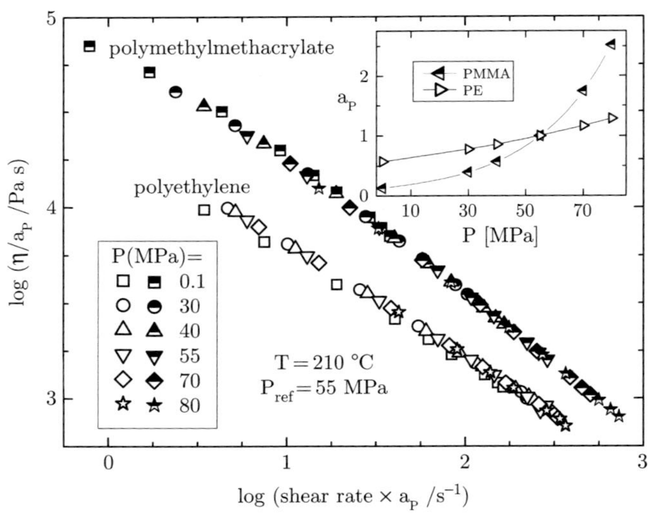

Figure 6.10 Master curves of the viscosity of PMMA (half-filled symbols) and a low density PE (open symbols). The actual data covered 1.7 decades of shear rate. The range of the reduced shear rate is greater for the PMMA because its viscosity is more pressure sensitive, as seen in the inset, showing the pressure-shift factors. Data from ref. [28].

[3] $\tau_{HN} = \tau_{\max}\left[\sin\left(\frac{\pi\alpha_{HN}\beta_{HN}}{2+2\beta_{HN}}\right)/\sin\left(\frac{\pi\alpha_{HN}}{2+2\beta_{HN}}\right)\right]^a$ where τ_{max} is the inverse of the peak frequency (in rad/s).

Figure 6.11 this product is shown for both the normal and segmental modes in polyoxybutylene [32]. The shape of the normal-mode peak is constant with changes in either T or P, demonstrating adherence to the superposition principle. The same is true for the segmental relaxation peak, but only over a narrower range at lower temperatures.

6.2 Thermorheological complexity

The constraints on applicability of the superposition principle are that the response to changes in temperature or pressure is governed by only one viscoelastic mode (or if more than one, all have the same T or P dependence), the material itself remains the same during the measurements, and that effects due to heating or stress-wave propagation are negligible. However, while successful superpositioning of data is consistent with these requirements, the ability to construct a master curve does not guarantee thermorheological (or barorheological) simplicity. In fact, as first discovered for polystyrene [33], t–T superpositioning breaks down in the softening zone (glass-transition region) of polymers. At higher frequencies, successively shorter length scales are involved in the underlying motions, until eventually no polymeric modes contribute to the response (the relaxation behavior becoming indistinguishable from that of a molecular liquid). The behavior of these local modes changes more with changes in thermodynamic variables than do the polymeric modes. Consequently, in the glass-transition zone, where both contribute to the viscoelastic properties, measured data do not superpose. This differing sensitivity to temperature is illustrated in Figure 6.12,

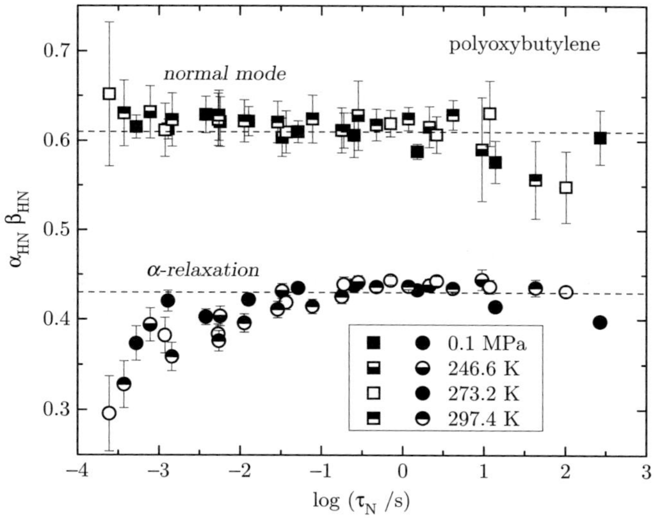

Figure 6.11 Havriliak–Negami shape factors (eqn (6.3)) plotted *versus* the relaxation time for the global chain modes (squares) and the local segmental modes (circles) of polyoxybutylene. The product $\alpha_{HN}\beta_{HN}$ equals the value of the exponent of a power law fit to the high-frequency flank of the dielectric loss peak. The invariance of the ordinate variable to changes in P or T indicates frequency–temperature and frequency–pressure superpositioning is valid. Data from ref. [32].

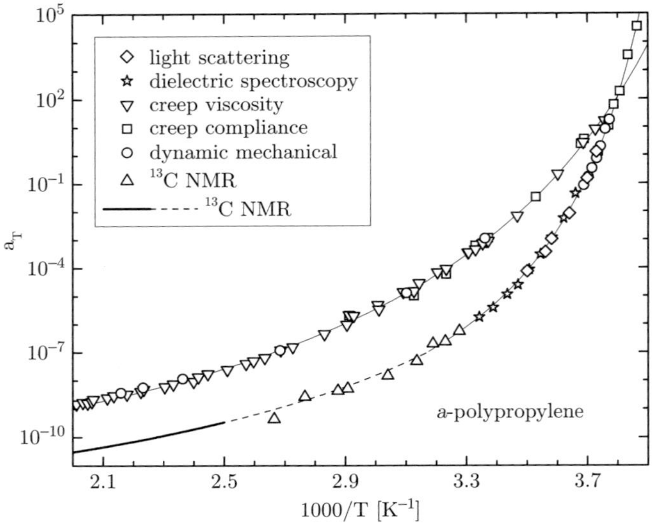

Figure 6.12 Time–temperature shift factors for atactic polypropylene global chain modes (upper curve) and local segmental modes (lower curve), obtained by the indicated methods. The vertical position is arbitrary and only indicates the relative temperature dependence of the two modes. The NMR data denoted by lines have higher uncertainty than those denoted by symbols. Data from ref. [34].

comparing the a_T for the chain modes in atactic polypropylene to those for the segmental modes [34]. The latter show a much stronger variation near T_g, which will cause break down of t–T superpositioning. At higher temperatures, these differences diminish and a_T for the two modes become parallel when plotted versus T.

This breakdown of t–T superpositioning is especially evident in the loss tangent, which exhibits a peak in the transition zone and is unaffected by the modulus shift factors referred to above. An extreme example is Figure 1.14; the loss tangent of polyisobutylene changes dramatically with temperature because of the unusual prominence of the sub-Rouse modes [35]. In Figure 6.13 [36] the changing shape of the loss tangent with temperature is shown for the same polyurea for which the relaxation modulus superposes to form an ostensibly satisfactory master curve (Figure 6.1). The apparent conformance to t–T superpositioning results from the monotonic form of the $E(t)$ curves. A similar decrease in the height of the loss tangent with increasing temperature is shown for PI in Figure 6.14 [37]; this breakdown of time–temperature superpositioning is not apparent in the master curve of the corresponding dynamic moduli (Figure 1.1). Deviations from t–T superpositioning in the glass-transition zone have also been reported for polyvinylacetate [38], polymethylmethacrylate [39], 1,4-polybutadiene [40], and polypropylene glycol [41].

Another manifestation of thermorheological complexity is the decrease of the steady-state recoverable compliance as temperature is reduced towards T_g. This effect is less common, since measurement of J_s^0 is difficult, particularly near the glass transition. Figure 6.15 [42] shows for

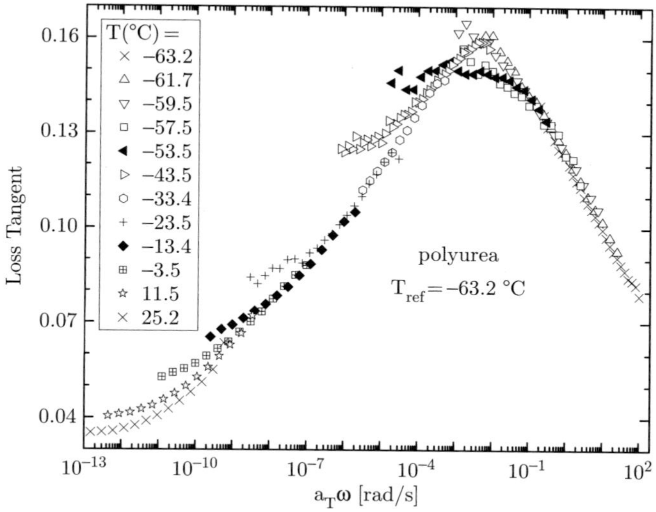

Figure 6.13 Mechanical loss tangent of PU measured at the indicated temperatures and plotted *versus* reduced frequency (data shifted to superpose the peak maxima). There is a lack of superpositioning due to thermorheological complexity in the glass-transition zone, as well as possibly additional complications from changes in hydrogen bonding or crystallinity with temperature. Data from ref. [36].

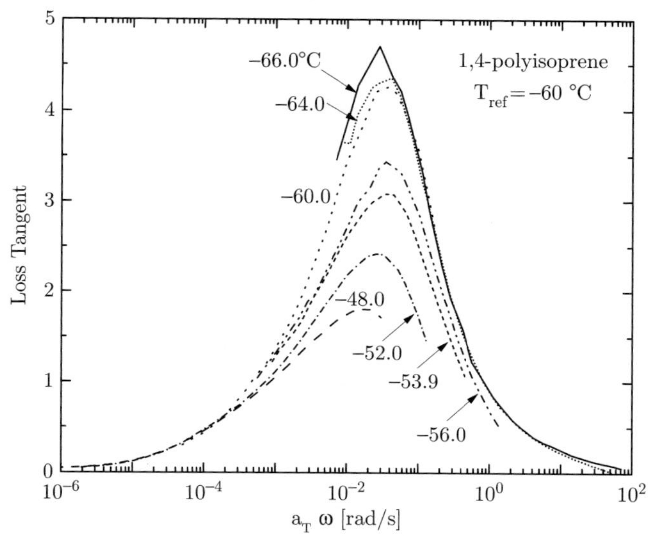

Figure 6.14 Mechanical loss tangent of PI measured at the indicated temperatures and plotted *versus* reduced frequency (data shifted to superpose the peak maxima). There is a lack of superpositioning, notwithstanding the ostensibly satisfactory master curve of the dynamic modulus for this same material (Figure 1.1). Data from ref. [37].

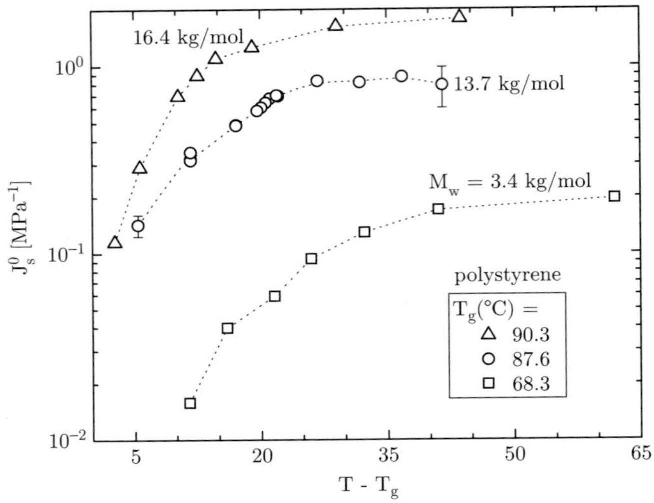

Figure 6.15 Steady-state recoverable compliance measured for PS having the indicated molecular weights and glass-transition temperatures. At high temperatures J_s^0 becomes constant. Data from ref. [42].

three polystyrenes J_s^0 decreasing by an order of magnitude from the respective plateau values.[4] Since $\eta \sim \tau_\eta / J_s^0$, this decrease causes a divergence in the T dependences of the terminal relaxation time and the viscosity, as illustrated in Figure 6.16 [43]. The variation with temperature of η, τ_η, and τ_α (the latter calculated as the inverse frequency of the maximum of the local segmental relaxation peak) can all be different. Note in Figure 6.16 that shift factors for the reciprocal of the diffusion constant, however, have the same temperature dependence as τ_η; that is, the enhancement of translational motions relative to molecular reorientations seen in small molecules (Chapter 2) is absent for unentangled polymer chains.

Thermorheological complexity of polymers in the softening zone does not invalidate the use of t–T superpositioning for longer times; i.e. the rheological properties of linear polymers should superpose, since the underlying molecular mechanisms have the same temperature dependence (excepting some polymers with long-chain branching [44–47]). However, when high-temperature measurements are used to obtain terminal (or chain) relaxation times and these are combined with low-temperature measurements of the segmental dynamics, the extrapolation of the combined shift factors can be in error by orders of magnitude. An example of this is the use of t–T superpositioning to determine the impact response of polymers from measurements at low strain rates and low temperatures. The problem is exacerbated if the polymer undergoes a transition to the glassy state during impact loading [48]. The curves in Figures 6.5–6.7 combine chain and segmental shift factors, and for this reason the data are at best qualitative at either end of the frequency range.

[4] This plateau in the compliance equals G_N^{-1} for entangled polymers. However, data showing the decrease in J_s^0 for temperatures close to T_g are available only for low molecular weight, unentangled polymers, since the duration of the measurement must exceed the terminal relaxation time. For unentangled polymers well above T_g, J_s^0 is the compliance of the longest Rouse mode, having a magnitude proportional to the chain length.

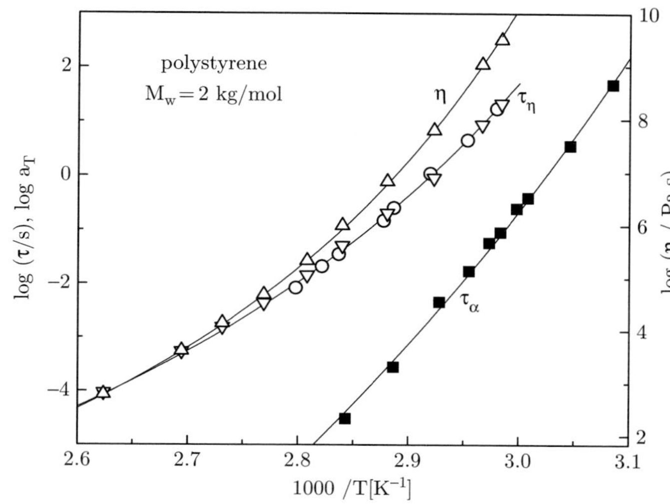

Figure 6.16 Global relaxation times calculated as $\tau_\eta = (15/\pi^2)\eta J_s^0$ (*cf.* eqns (3.2) and (3.4)) (inverted triangles), shift factors for the reciprocal of the diffusion constant (circles), local segmental relaxation times (solid squares), and the viscosity (triangles) of unentangled PS. The ordinates scales were adjusted to superpose τ_η and η at the highest temperature. Data from ref. [43].

6.3 Density scaling of the dynamics

An alternative to time–temperature superpositioning drawn from free-volume models of polymer dynamics is to assume that relaxation and related properties are governed by the interactions among segments. This seems especially appropriate in the "supercooled" regime near the glass transition, where molecular mobilities can be described as the result of (infrequent) correlated jumps over barriers that are large compared to the available thermal energy [14,49,50]. As discussed in Chapters 1 and 2, such cooperativity is inherent to vitrifying materials; motion of any molecule or polymer segment involves competing interactions with neighboring segments. Since the multidimensional potential-energy landscape is extremely complex, any description of the intermolecular interactions requires simplifying assumptions. Commonly, the intermolecular potential is approximated with a Lennard-Jones (LJ) form [51], which expresses the potential energy as the sum of independent two-body interactions

$$U(r) = 4\varepsilon_{LJ}[(r_{LJ}/r)^{m_{LJ}} - (r_{LJ}/r)^6] \tag{6.5}$$

where r is the separation distance and ε_{LJ}, a microscopic energy, and r_{LJ}, a molecular size, are constants (Figure 6.17). Equation (6.5) assumes the interaction is spherically symmetric, although the potential energy between molecules with dipole or higher moments depends on their relative orientation. For dispersion interactions (no Coulombic forces), the exponent of 6 in the attraction term is universal.[5] Repulsive forces are strong and short ranged (typically $m_{LJ} \geq 12$) in comparison to the attractions. The parameter m_{LJ} in the exponent accounts for

[5] The van der Waals attractive forces are Coulombic interactions due to the instantaneous dipoles arising from electron correlation between different atoms. Although dipole forces vary as r^{-2}, accounting for the charge separation

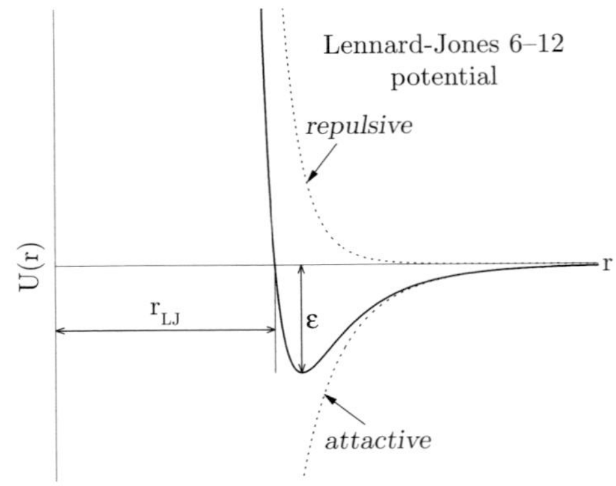

Figure 6.17 Lennard-Jones interaction potential (eqn (6.5)) with $m_{\mathrm{LJ}} = 12$.

the steepness of the repulsive interaction, which varies among materials [53,54]. The value of $m_{LJ} = 12$ was adopted historically because of the computational convenience of the term equaling the square of the attractive term. Since the force on any molecule is the vector sum of the contributions from all other molecules, and in condensed matter each molecule (or polymer segment) has many neighbors, the attractive forces tend to cancel locally [55]. This leads to the classic van der Waals picture of liquids, in which the arrangement of molecules (liquid structure) is governed by steric effects and local packing, with the attractions manifested primarily as a uniform background pressure [56–58]. Accordingly, at least in analyzing local properties, the details of the attractive interactions are often neglected, since they primarily exert a mean-field, density-dependent pressure, serving to cohere the material [55]. This simplification is also consistent with the fact that in liquids, the static structure factor at intermediate and large wave vectors is sensitive only to the repulsive part of the potential [59,60].

Dropping the attractive term in eqn (6.5) gives the inverse-power-law (IPL) repulsive potential [61–63]

$$U(r) = \left[4\varepsilon_{LJ} r_{LJ}^{m_{LJ}}\right] r^{-m_{LJ}} \tag{6.6}$$

where the factor in brackets is a constant. If the IPL accurately represents the local interaction energy between segments, then thermodynamic properties, such as energy, volume, and entropy, as well as dynamic properties governed by the local potential, can be expressed as unique functions of $r^{m_{LJ}}$ or $V^{m_{LJ}/3}$ [64,65].[6] In practice, not accounting accurately for the attractive forces limits consideration to local properties; for example, eqn (6.6) does not yield an accurate equation of state (EOS) [66,67]. Moreover, as discussed below, the effective slope

on each induced dipole, expanding in a Taylor series, and then truncating after the first term gives a dispersion force varying as r^{-7}, or $U(r) \sim r^{-6}$ for the van der Waals potential [52].

[6] Strictly speaking, the scaling implied by an IPL applies to quantities expressed in reduced units; e.g., $t^* \sim t(T^{1/2} V^{1/3})$, $\eta^* \sim \eta(T^{-1/2} V^{2/3})$, $D^* \sim D(T^{-1/2} V^{-1/3})$, etc., although the distinction is usually ignored in analyses of experimental data near T_g. Using reduced units yields very similar values of γ, but with slightly superior superpositioning.

of the repulsive part of the potential is not the same as that of the repulsive term, due to the other interactions.

Neglecting the question of whether the segmental dynamics, having cooperative length scales extending a few nanometers [68–72], can be regarded as sufficiently local to adopt the IPL approximation, the properties of an IPL lead to the expectation that the local segmental relaxation times are a function of the variable TV^{γ}

$$\tau_{\alpha} \sim f(TV^{\gamma}) \tag{6.7}$$

where $\gamma \sim m_{LJ}/3$. This scaling[7] of τ_{α} is illustrated in Figure 6.18 for polymethyltolylsiloxane [73], with γ a material-specific constant, having a value determined empirically from collapse of the segmental relaxation times onto a single curve versus TV^{γ}. Although eqn (6.7) implies that the relaxation involves thermally activated surmounting of potential barriers, whose height has a volume dependence represented by V^{γ}, Arrhenius behaviour (log τ_{α} proportional to TV^{γ}) is not obtained.

Similar results have been found for many molecular (Figure 6.19) and polymeric (Figure 6.20) liquids [32,74–78]. The τ_{α} in these figures span a broad range, with each datum representing a unique state point (condition of T, P, and V). Only the constant γ is required to construct each master curve. The magnitude of γ yielding superposition ranges from 1.9 for polyvinylethylene to 5.6 for polymethylphenysiloxane. Generally, the values of the scaling exponent are smaller for polymers than for molecular liquids, for which $3.7 \leq \gamma \leq 8.5$. A smaller γ implies a softer repulsive potential, consistent with the larger activation energy ratios (weaker volume effects) of polymers (Chapter 2). For hydrogen-bonded materials, such

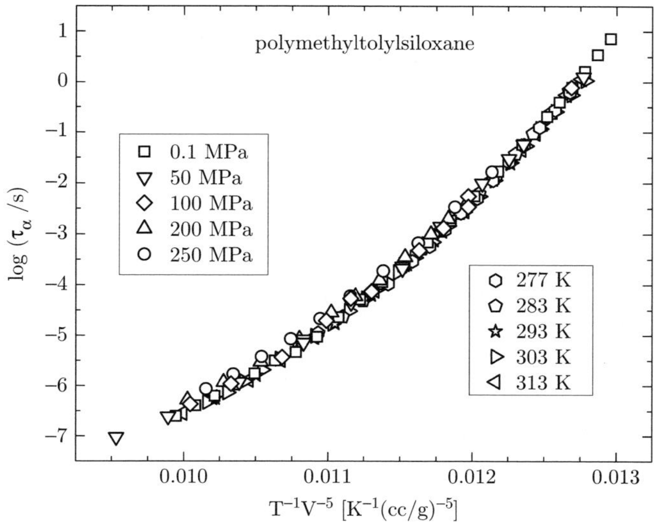

Figure 6.18 Scaling plot of the segmental relaxation times for PMTS from Figure 2.18. Each datum represents a different state point. Data from ref. [73].

[7] The shape of a "scaling function" is unchanged when its argument is multiplied by a constant factor; i.e., $f(cx) \propto f(x)$ where c is a constant. Power law functions are polynomials that have this property of "scale invariance."

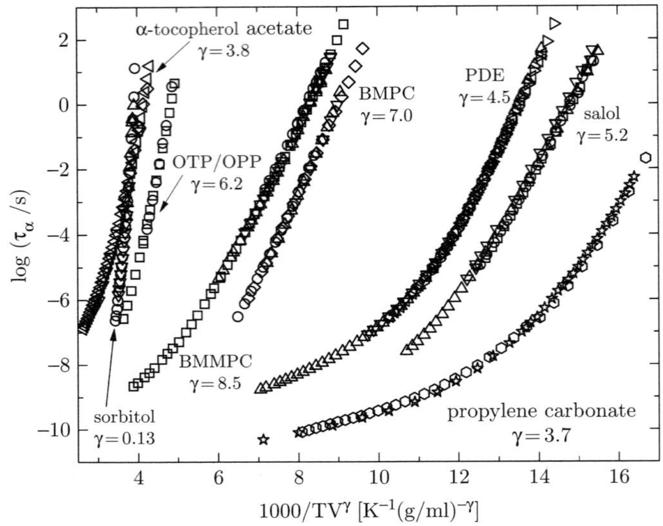

Figure 6.19 Dielectric α-relaxation times of molecular liquids as a function of the reciprocal of temperature times the specific volume, the latter raised to the indicated power. Data from refs. [74–76].

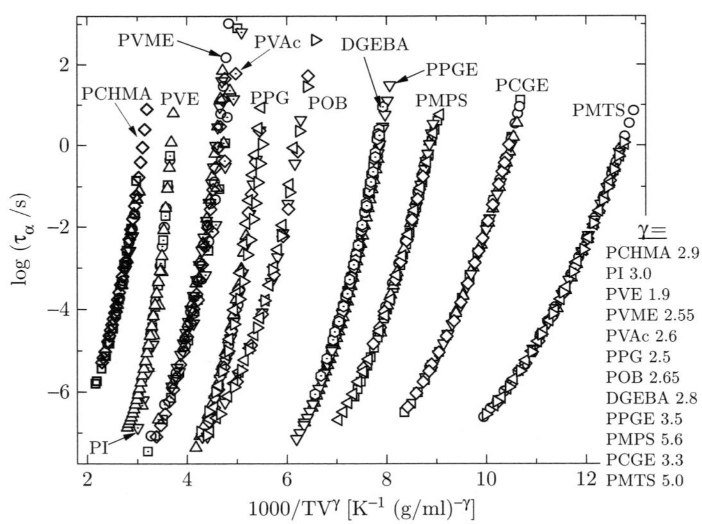

Figure 6.20 Local segmental relaxation times of polymers measured by dielectric spectroscopy as a function of the reciprocal of temperature times the specific volume, the latter raised to the indicated power. Data from refs. [32,75,77,78].

as the polyalcohol sorbitol in Figure 6.19, γ can be very small; however, their relaxation times scale only over limited ranges due to changes in the degree of H-bonding with T and P [79].

6.3.1 Repulsive potential and correlating liquids

The superpositioning of relaxation times as a function of TV^{γ} is an empirical fact, not contingent on any particular representation of the intermolecular potential. There are experimental results supporting the putative dependence of the magnitude of γ on the steepness of the potential [80,81] (Figure 6.21), but the connection is not rigorous. The accuracy of the IPL approximation can be assessed using molecular-dynamics simulations, which afford the ability to modify the interactions systematically. Using the general LJ potential (eqn (6.5)), diffusion constants were calculated for a binary particle mixture (components having different sizes to suppress crystallization) with different values used for the repulsive exponent, $m_{LJ}=8, 12, 24$, and 36 [82]. Figure 6.22 shows the reduced diffusion constants obtained from the simulations; the data for each m_{LJ} superpose versus TV^{γ}. Relaxation times calculated for the same systems also scale using the same γ [83]. The values of the scaling exponents maximizing overlap among different isobaric data sets, tabulated in Table 6.1, show that γ increases with increasing m_{LJ}; however, its actual value is systematically larger than $m_{LJ}/3$.

This discrepancy between γ and $m_{LJ}/3$ is due to the fact that a purely repulsive interaction is realized only in the asymptotic region of small interparticle distances, which is normally inaccessible. In the region of r close to the minimum, the effective potential is steeper than in the $r \rightarrow 0$ limit, due to the contribution of the attractive term in eqn (6.5) [82,84,85]. This is illustrated in Figure 6.23 for $m_{LJ} = 24$. Fitting an IPL to the potential in the range of r between 0.95 and 1.01 gives a value of 27.5 for the effective slope, in good agreement with the

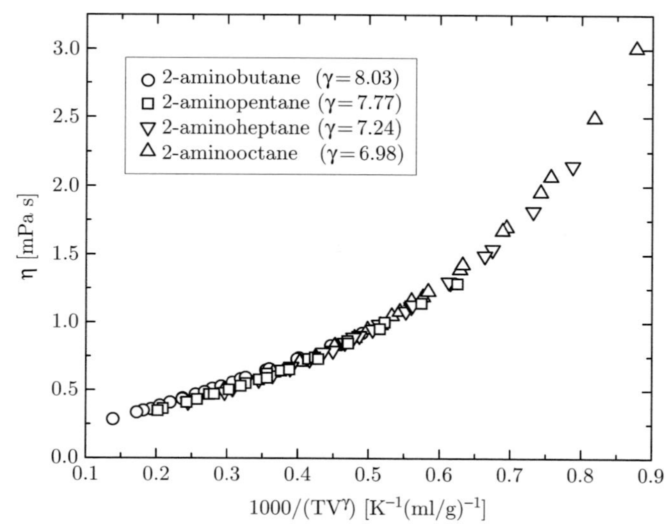

Figure 6.21 Viscosities for sec-alkylamines plotted *versus* TV^{γ}; the pressures range from 0.1 to 100 MPa and the temperatures from 293 to 353 K. The scaling exponent decreases with increasing length of the alkyl group, consistent with a less-steep repulsive curve. Interestingly, the scaling function (f in eqn (6.7)) is the same for the four compounds; that is, all η collapse to a single curve. Data from ref. [81].

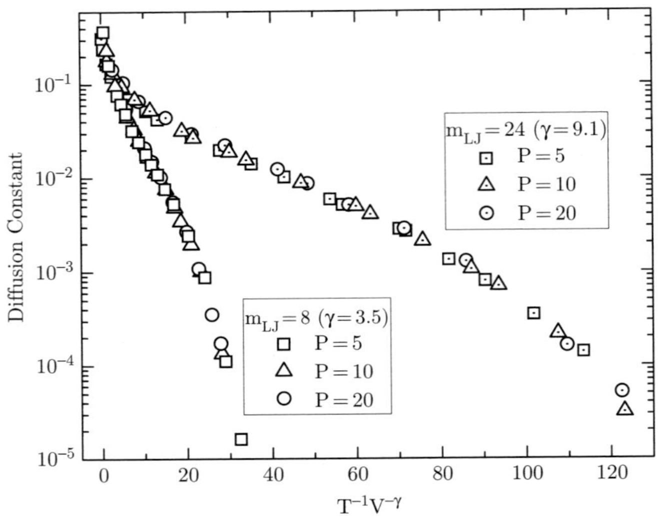

Figure 6.22 Reduced diffusion coefficients as a function of the inverse product of temperature and specific volume, the latter raised to the indicated value of γ, for two LJ liquids for which $m_{LJ} = 8$ and 24. The pressures are given in LJ units with the data spanning a range of temperatures. Data from ref. [82].

Table 6.1 MD simulation results for LJ liquids [82,88].

m_{LJ} (eqn (6.5))	3γ (eqn (6.7))	m_{LJ} (eqn (6.6))*	m_{LJ} (eqn (6.10))^
8	10.5 ± 0.2	10.9	11.0 ± 0.006
12	15.0 ± 0.2	14.9	15.3 ± 0.01
24	27.3 ± 0.3	27.2	27.2 ± 0.03
36	40.2 ± 0.6	39.9	39.4 ± 0.03

* fitting IPL to typical region of closest approach between particles ($0.95 \leq r \leq 1.01$)
^ slope of $\Delta p_v(t)$ versus ΔU

scaling value, $3\gamma = 27.3 \pm 0.3$. This range of r corresponds to typical distances of closest approach between particles probed within the simulation conditions; that is, values of r between the first non-zero value of the radial distribution function[8] and the position of the half-height of the first peak in $g(r)$, (Figure 6.23). Similar results are found for MD simulations employing other values of m_{LJ} (Table 6.1).

Thus, with inclusion of a constant or weakly varying attractive term, the repulsive core can still be described by a power law, but it is steeper than the bare $r^{-m_{LJ}}$ of the IPL potential. Only the asymptote at small r reflects the value of the LJ repulsive exponent. For polymers, the use of an IPL entails an additional assumption, neglect of contributions from intrachain motions. These chain-stretching and bending modes can be described by a

[8] The radial distribution function (the distribution function for an isotropic system) is the probability of finding two particles a distance r apart, divided by the corresponding probability for a random distribution; thus, $g(r)$ measures the deviation from a random structure.

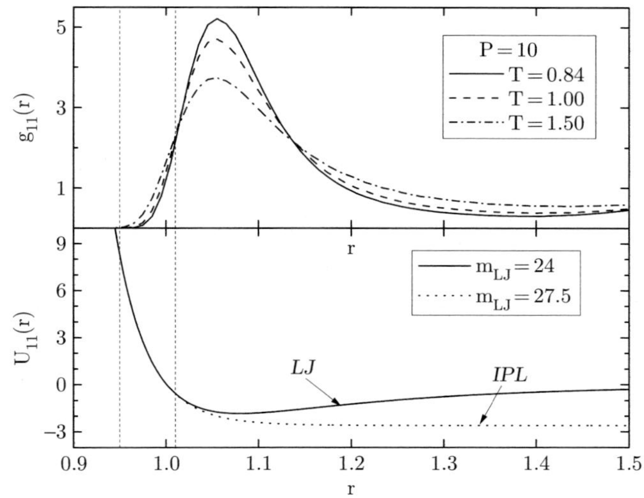

Figure 6.23 (lower panel) Radial pair distribution functions at one pressure and three temperatures in LJ units. (bottom panel) Pair potential for LJ (eqn (6.5) and IPL (eqn (6.6) with added constant) fitted in the range of particle separations indicated by the vertical dashed lines. Data from ref. [82].

harmonic potential, varying as the square of the instantaneous bond length. Using the LJ potential with the conventional value of $m_{LJ} = 12$, Tsolou *et al.* [86] obtained $\gamma = 2.8$ for the segmental relaxation times of simulated 1,4-polybutadiene. This is *less* than $m_{LJ}/3$, indicating that the intramolecular degrees of freedom make the repulsive interaction softer, consistent with the smaller experimental values of γ for polymers in comparison to molecular liquids.

If an IPL is strictly valid, there is an exact relation between the potential energy and the virial pressure (the configurational parts of the energy and pressure). The virial pressure, p_v, is the non-ideal gas part of the pressure of a fluid

$$p = \frac{NT}{v} + p_v \tag{6.8}$$

and

$$p_v = -\frac{1}{3} r \frac{dU}{dr} \tag{6.9}$$

For an IPL a simple relationship exists between the equilibrium fluctuations of the potential energy and the virial pressure

$$\frac{\Delta p_v}{\Delta U} = \frac{m_{LJ}}{3} \tag{6.10}$$

where $\Delta p_v = \frac{p_v - \langle p_v \rangle}{\sqrt{\langle \Delta p_v^2 \rangle}}$ and $\Delta U = \frac{U - \langle U \rangle}{\sqrt{\langle \Delta U^2 \rangle}}$. This relationship has been demonstrated in MD simulations of various materials [85,87,88] and is illustrated in Figure 6.24 for an LJ fluid with $m_{LJ} = 36$. The slope of the linear regression of ΔU and Δp_v yields one-third the value of m_{LJ} (Figure 6.25). Results for LJ liquids with various values of m_{LJ} are given in Table 6.1.

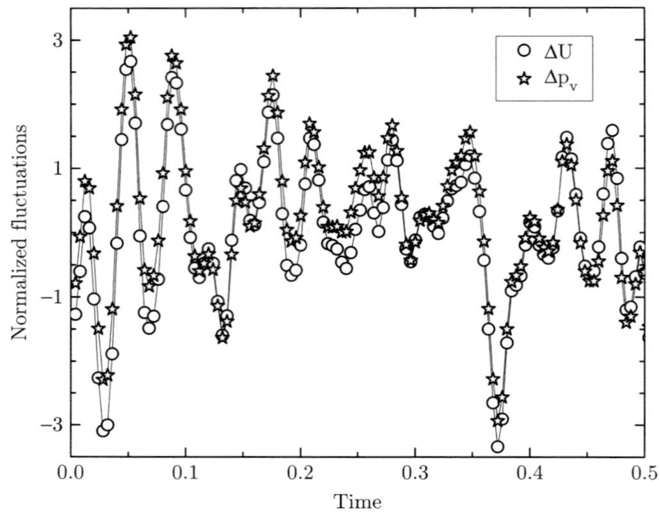

Figure 6.24 Instantaneous fluctuations normalized by the average values of the potential energy (circles) and the virial (stars) for an LJ liquid with $m_{LJ} = 36$ at a fixed state point (constant temperature and pressure). Data from ref. [88].

6.3.2 Scaling of chain dynamics

For polymers, the experimental results described above pertain only to segmental motions; i.e. bond rotations and coordinated motion of a few backbone atoms. Since the chain dynamics, as discussed above, have different T and P dependences than the segmental modes, they cannot have the same dependence on the scaled variable TV^γ, nor do they necessarily superpose versus TV^γ. The viscosity of three oligomeric compounds are plotted in Figure 6.26 [89], indicating compliance with the scaling relationship (eqn (6.7)). This suggests that the scaling extends to longer length scales, notwithstanding the local approximation inherent in the presumed connection to an IPL potential. However, if dielectric spectra of the segmental and normal modes are obtained at different state points for which τ_α is constant, the normal-mode peaks do not superpose (Figure 6.27 [90]). This means that the scaling behavior of the latter differs somewhat from the TV^γ behavior for segmental relaxation.

In Figure 6.28 are shown the volume dependences of the local and global relaxation times for polyoxybutylene [32]. Note that the V dependences of the τ_α are stronger than for the normal-mode relaxation times. When these data are plotted versus TV^γ (Figure 6.29), master curves are obtained for both modes with the same value of γ [32]. Similarly for 1,4-polyisoprene and polypropylene glycol (Figure 6.30) [91], the same exponent superposes the local segmental and the global chain relaxation times. Thus, while τ_α and τ_α have different scaling behaviors in principle (Figure 6.27), the differences are small over the range of measurements encompassing both modes. Thus, the same γ superposes both relaxation times, although their different T and V dependences are still manifested in different functional forms (i.e. the f in eqn (6.7)) [92]; τ_α does vary more with changes in TV^γ, with the difference more pronounced at longer τ_α.

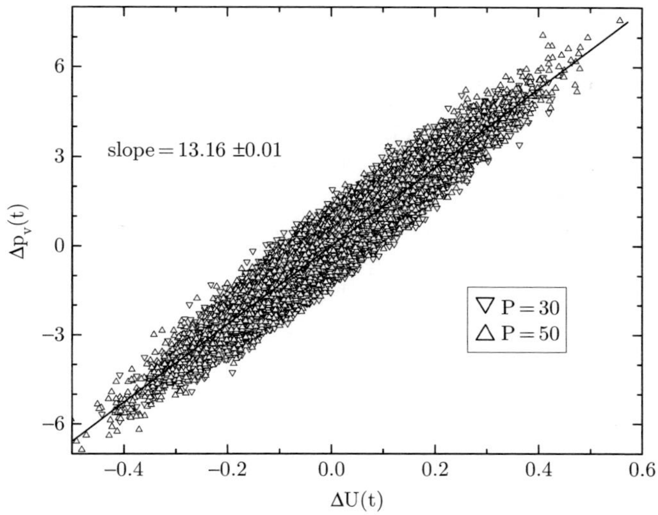

Figure 6.25 Fluctuations of the virial *versus* fluctuations of the potential energy for an LJ liquid with $m_{LJ} = 36$. The two pressures are in reduced LJ units, with temperatures over a range corresponding to densities between 1.20 and 1.75. The solid line is the linear fit yielding the indicated slope. Data from ref. [88].

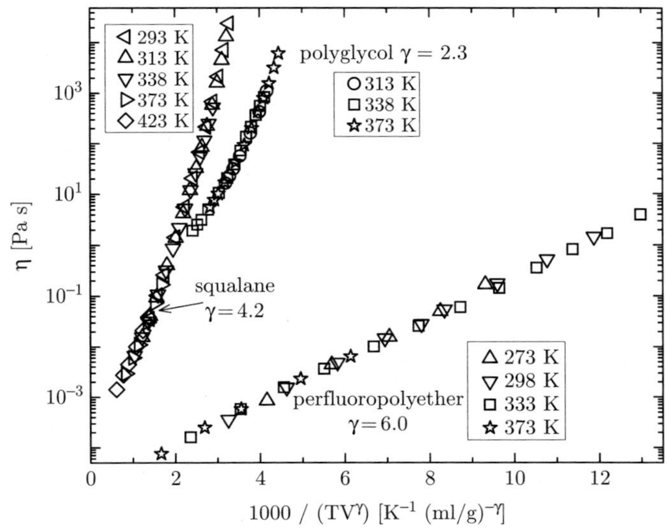

Figure 6.26 Superpositioned viscosities of a random copolymer of polyethylene and polypropylene glycol, squalane (2,6,10,15,19,23-hexamethyltetracosane), and a perfluorinated random copolymer of polyethylene oxide and polymethylene oxide (Fomblin® Z25). Pressure range of the data was from ambient to as high as 1.2 GPa. Data from ref. [89].

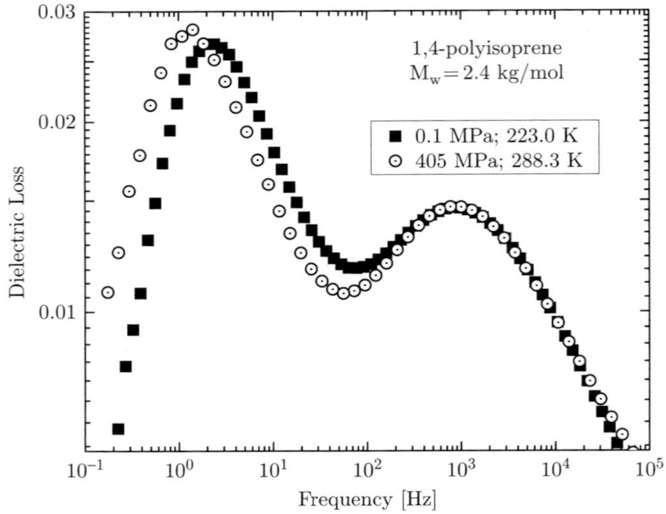

Figure 6.27 Dielectric spectra showing the normal and segmental modes in PI at two conditions for which τ_α is constant. These conditions yield the same τ_α but there is a small change in frequency of the normal-mode peak. The ordinate was shifted to superpose the peak maxima. Data from ref. [90].

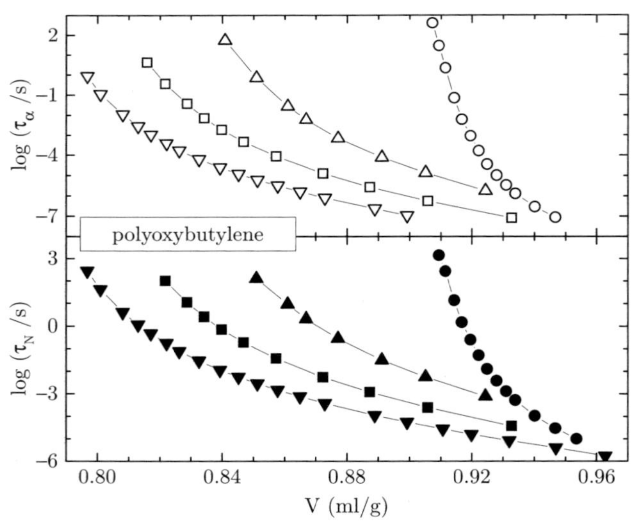

Figure 6.28 Local segmental (upper) and normal mode (lower) relaxation times for POB as a function of the specific volume. Dielectric data at constant pressure (circles: 0.1 MPa) and constant temperature (triangles: 246.6 K; squares: 273.2 K; inverted triangles: 297.4 K). Data from ref. [32].

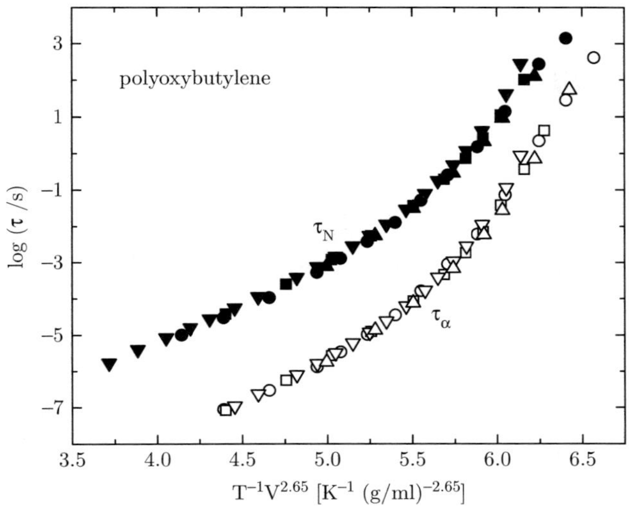

Figure 6.29 Data from Figure 6.28 replotted *versus* the scaling variable $TV^{2.65}$. Data from ref. [32].

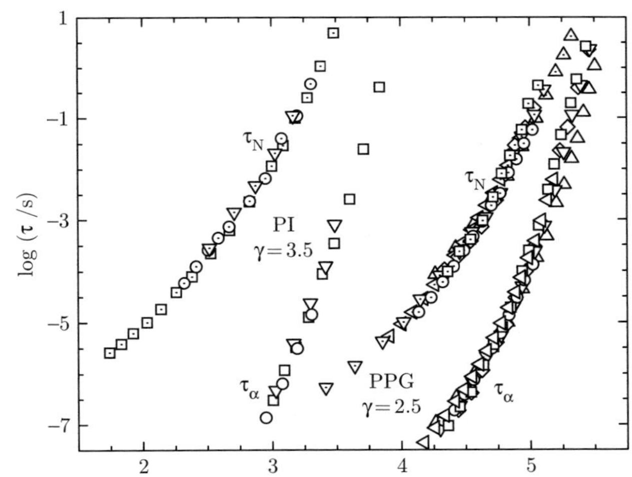

Figure 6.30 Scaling plot of the local segmental relaxation times (hollow symbols) and normal mode relaxation times (dotted symbols) for 1,4-polyisoprene ($M_w = 11.1$ kg/mol) at pressures up to 0.350 GPa and $283 \le T\,(\mathrm{K}) \le 298$ and for polypropylene glycol ($M_w = 4$ kg/mol) at pressures up to 1.2 GPa and $258 \le T\,(\mathrm{K}) \le 313$. Data from ref. [41].

6.3.3 Determination of the scaling exponent from PVT data

There is a simple relationship between the scaling exponent and the activation energy ratio (Chapter 2) [74]. From the definition of the activation energies, the respective derivatives of eqn (6.7) give

$$E_V = R\frac{d\Im}{dT^{-1}}(-T^2)\,V^\gamma \tag{6.11}$$

and

$$E_a = R\frac{d\Im}{dT^{-1}}(-T^2)\,V^\gamma(1 + T\gamma\,V^{\gamma-1}\left.\frac{dV}{dT}\right|_P) \tag{6.12}$$

The ratio is then[9]

$$E_V/E_a = (1 + \gamma\alpha_P T)^{-1} \tag{6.13}$$

The scaling exponent can also be obtained from PVT data [94]. The temperature derivative of the volume at constant value of the relaxation time is

$$\left.\frac{dV}{dT}\right|_{\tau_\alpha} = \left.\frac{dV}{dT}\right|_P + \left.\frac{dV}{dP}\right|_T \left.\frac{dP}{dT}\right|_{\tau_\alpha} \tag{6.14}$$

Dividing through by the derivative of the volume with respect to temperature at constant pressure, eqn (6.14) becomes

$$\frac{\alpha_\tau}{\alpha_P} = 1 + \left(\left.\frac{dV}{dP}\right|_T \Big/ \left.\frac{dV}{dT}\right|_P\right)\left.\frac{dP}{dT}\right|_\tau \tag{6.15}$$

By the rule of implicit partial derivatives,

$$\left.\frac{dT}{dP}\right|_V = -\left(\left.\frac{dV}{dP}\right|_T \Big/ \left.\frac{dV}{dT}\right|_P\right) \tag{6.16}$$

so that

$$\frac{\alpha_\tau}{\alpha_P} = 1 - \left.\frac{dT}{dP}\right|_V \left.\frac{dP}{dT}\right|_\tau \tag{6.17}$$

From the definition of the activation energy [95,96]

$$E_V = E_a - RT^2\left.\frac{dP}{dT}\right|_V \left.\frac{d\ln\tau}{dP}\right|_T \tag{6.18}$$

Using Euler's chain rule

$$\left.\frac{d\ln\tau}{dP}\right|_T = -\left.\frac{d\ln\tau}{dT}\right|_P \left.\frac{dT}{dP}\right|_\tau \tag{6.19}$$

and the definition of E_a gives

[9] For polymers the product $\alpha_P T$ at T_g has a near universal value $= 0.18 \pm 0.02$ (the "Boyer–Spencer rule") [73,93], whereby eqn (6.13) at T_g simplifies to $E_V/E_a = (1 + 0.18\,T_g)^{-1}$.

$$\frac{E_V}{E_a} = 1 - \left.\frac{dP}{dT}\right|_V \left.\frac{dT}{dP}\right|_\tau \tag{6.20}$$

Combining eqns (6.17) and (6.20) yields

$$E_V/E_a = (1 - \frac{\alpha_P}{\alpha_\tau})^{-1} \tag{6.21}$$

which inserted into eqn (6.13) gives

$$\gamma = \frac{-1}{T\alpha_\tau(T)} \tag{6.22}$$

Since at T_g the relaxation time is constant (independent of pressure), from the definition of the isochronal thermal expansion coefficient

$$\alpha_\tau = V_g^{-1} \left.\frac{dV_g}{dT_g}\right|_\tau \tag{6.23}$$

and from eqn (6.22) the scaling exponent is obtained in terms of T_g and the volume at the glass transition

$$\gamma = \left(T_g V_g^{-1} \frac{dV_g}{dT_g} \right)^{-1} \tag{6.24}$$

We illustrate this approach with data for poly(phenylene oxide) [97]. In Figure 6.31 are PVT data, from which $\alpha_\tau = -2.36\times10^{-4}$ K^{-1} at $T_g = 459.8$ K and $P = 0.1$ MPa. Eqn (6.24) then gives $\gamma = 9.2 \pm 0.1$. Relaxation times measured for this material at ambient pressure in

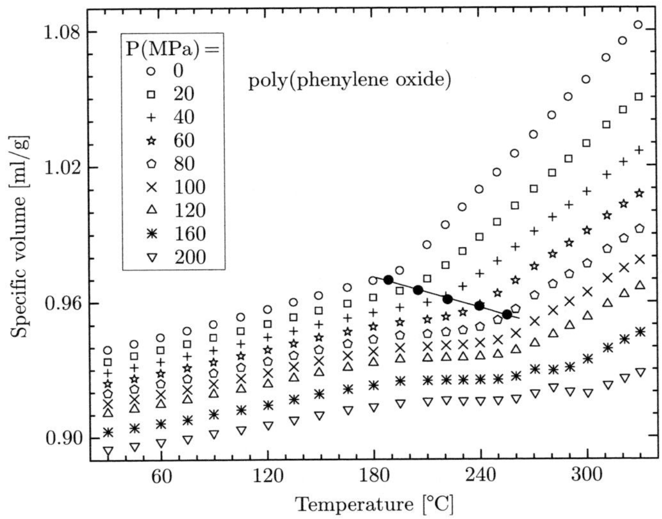

Figure 6.31 Specific volume *versus* temperature for PPO at the indicated pressures. The solid circles represent $T_g(P)$ as determined from the intersection of the liquid and glassy $V(T)$ data (the weak minimum near T_g is an experimental artifact). Solid circles represent (V_g, T_g), the fit of which yields $\alpha_\tau = 2.36\times10^{-4}$ K^{-1}. Data from ref. [97].

combination with the EOS (obtained from Figure 6.31) enable the value of TV^γ associated with each state point (T, 0.1 MPa) to be calculated; τ_α is then known for any P and T corresponding to an ambient pressure value of TV^γ. This method is quite powerful because it allows determination of relaxation times for any thermodynamic condition (T, P, or V) provided that the same value of τ_α is available at ambient pressure.

6.3.4 Relation of the scaling exponent to the Grüneisen parameter and the entropy

The method described above, using PVT data rather than empirical scaling of relaxation times to determine γ, relies on no assumptions other than the experimental fact that the dynamic variable (e.g. τ_α) is uniquely defined by TV^γ. However, it does require knowledge of the EOS and the behavior at the pressure-dependent glass transition. An alternative method of estimating γ avoids this by exploiting the idea that density scaling is a consequence of the intermolecular potential being approximated by an IPL in the relevant region of intermolecular separations. The Grüneisen parameter, γ_G, a measure of the anharmonicity of lattice vibrations, is formally defined in terms of the volume-dependence of the phonon frequency, ω [98]

$$\gamma_G = -\frac{d\ln\omega}{d\ln V} = \frac{r d\omega}{3\omega dr} \tag{6.25}$$

The force constant for the vibrations, $K \equiv \frac{d^2U}{dr^2}$, in the harmonic approximation is proportional to ω^2. Differentiating eqn (6.6) to obtain K and substituting into eqn (6.25) gives [99]

$$\gamma_G = \frac{1}{2}\gamma + \frac{1}{3} \tag{6.26}$$

which, similar to the discussion in Section 6.3.1, relates γ to the steepness of the repulsive portion of the intermolecular potential as represented by an IPL.

Values for the Grüneisen parameter can be calculated from its thermodynamic definition [100]

$$\gamma_G = \frac{V\alpha_P}{C_V\kappa_T} \tag{6.27}$$

in which C_V is the isochoric heat capacity and γ_G is assumed to be the same for all vibrational modes. Table 6.2 compares the scaling exponent γ calculated from superpositioning of experimental $\tau_\alpha(T, V)$ data to the value calculated using eqns (6.26) and (6.27), for two polymers and three molecular liquids.

Substituting in the definitions of the quantities in eqn (6.27) and using one of Maxwell's thermodynamic relations,[10] leads to

$$\gamma_G = V\left.\frac{dS}{dV}\right|_T \left(T\left.\frac{dS}{dT}\right|_V\right)^{-1} \tag{6.28}$$

From the properties of an IPL, the entropy (as well as other thermodynamic quantities) should be a function of the same product variable [101]

[10] $\left.\frac{dP}{dT}\right|_V = \left.\frac{dS}{dV}\right|_T$

Table 6.2 Scaling exponent values.

	γ			
	τ_α	$-\frac{d\ln\omega}{d\ln V}$	γ_G	S–S_{ref}
	[eqn (6.7)]	[eqns (6.26) and (6.27)]	[eqn (6.30)]	[eqn (6.29)]
polyvinylacetate	2.5	0.7	0.7	0.9
polymethylmethacrylate	1.25	0.7	0.7	—
o-terphenyl	4	1.7	1.2	1.2
salol	5.2	3.1	1.9	1.7
propylene carbonate	3.7	2.1	1.4	2

$$S \sim f(TV^\gamma) \tag{6.29}$$

as the relaxation time. Evaluating the derivatives in eqn (6.28) gives

$$\gamma_G = \frac{T\gamma V^\gamma f' + k_B}{TV^\gamma f' + 3k_B/2} \approx \gamma \tag{6.30}$$

in which f' represents the derivative of the unknown functional dependence of S on TV^γ. Results for eqn (6.30) are included in Table 6.2. A similar result, $\gamma_G \sim \gamma$, can be derived from an entropy model of the glass transition [101]. The limitations of this approach are twofold: (i) The mode Grüneisen parameter varies with T, whereas the scaling exponent is a constant. (ii) It is not obvious that a parameter such as γ_G characterizing vibrational motions should relate to relaxational dynamics, notwithstanding any mutual dependence on the intermolecular potential.

A more obvious route is to attempt superposition of the entropy versus TV^γ to obtain the scaling exponent more directly than using eqns. (6.28) and (6.30). For any condition of T and P, S can be calculated from thermodynamic data using

$$S(T,P) - S_{ref}(T_{ref}, P=0) = \int_{T_{ref}}^{T} \frac{C_P}{T}\, dT - \int_0^P \left.\frac{\partial V}{\partial T}\right|_P dP \tag{6.31}$$

where S_{ref} is the entropy at the reference temperature T_{ref} and zero pressure. The first term on the right-hand side can be determined from the heat capacity at atmospheric pressure, and the second term evaluated using the EOS. The necessary data are only available for a few materials. In Figure 6.32 results are shown for PVAc [102]. The entropy changes substantially over the range of temperatures and pressures (figure inset), but collapses to a single curve when plotted versus $TV^{0.9}$. This value of the exponent, 0.9, is close to the Grüneisen parameter for PVAc, $\gamma_G = 0.7$, but about a factor of three smaller than the scaling exponent for τ_α (Table 6.2). At least some of this discrepancy is due to contributions in eqns (6.27)–(6.31), such as vibrations and secondary relaxation processes, that are unrelated to relaxation. An approximate correction would be to subtract the crystal contribution from the heat capacity, etc., assuming this contribution were known. The results of entropy scaling for small molecule liquids (Figure 6.33) are similar to those for PVAc [102,103], including the disparity among the various exponents (Table 6.2).

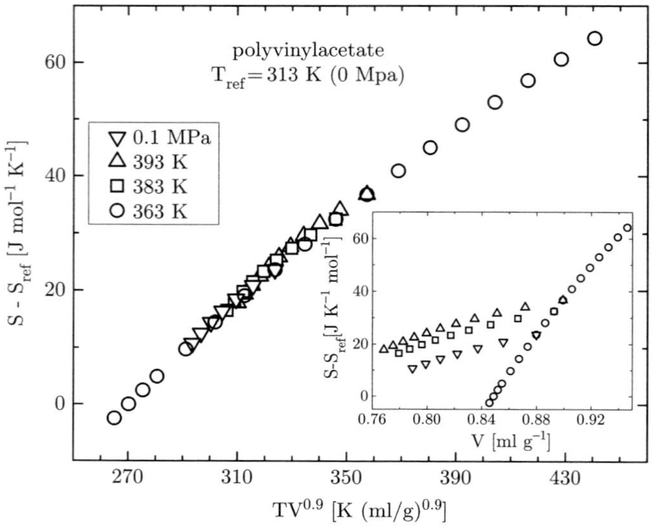

Figure 6.32 Entropy above that at 313K and atmospheric pressure for PVAc *versus* $TV^{0.9}$, with the variation with specific volume shown in the inset. Data from refs. [102].

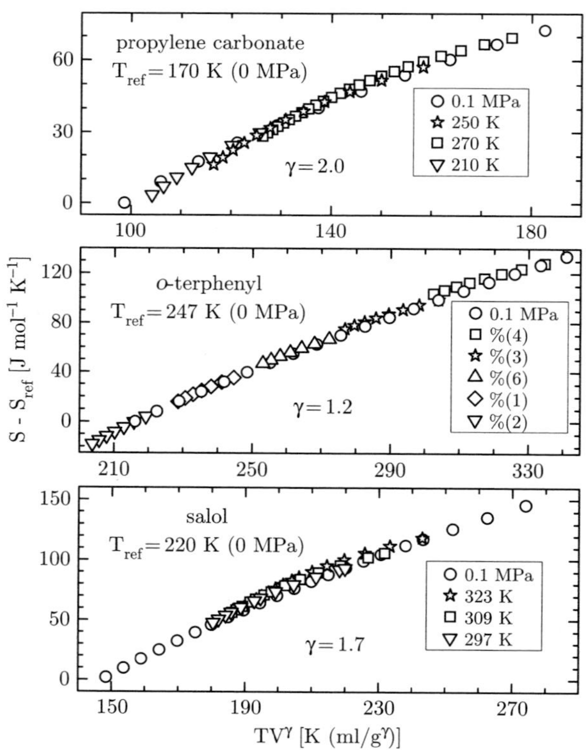

Figure 6.33 Scaling plot of the entropy above a reference value for (top) propylene carbonate, (middle) *o*-terphenyl, and (bottom) salol (phenyl salicylate) at constant pressure and at various temperatures. S–S_{ref} was calculated from eqn (6.31) using the atmospheric pressure heat capacity and the EOS. Data from refs. [102, 103].

6.4 Characteristic relaxation times

Different ranges of temperature and pressure are associated with different regimes of dynamical behavior – the independent motions at high T and low P become increasingly intermolecularly correlated, leading eventually to vitrification. Over this range liquids and polymers exhibit transitions in their dynamic behavior: the onset of a non-Arrhenius temperature dependence, followed by a "dynamic crossover," and ultimately the glass transition. An intriguing feature is that the relaxation time at these transitions is a material constant; that is, the crossover to different dynamic properties transpires at an essentially fixed value of τ_α, independent of T, P, and V. Since the dynamic behavior changes, temperature or pressure extrapolations from one regime of τ_α to another can lead to large errors in the predicted properties. For example, as discussed below, a single VFTH equation is usually insufficient to describe $\tau_\alpha(T)$ over the entire range of the liquid state. This is an advantage of the scaling property (eqn (6.7)), which is valid throughout this range. In this section we enumerate the different dynamical regimes and show that the relaxation time associated with each dynamic transition is a material constant.

The most important change occurs at the glass transition, at which translational and rotational fluidity is lost. The particular values of T_g and $\tau_\alpha(T_g)$ depend on the experimental protocol (e.g., onset or inflection point in the response to heating at an arbitrary rate), and different dynamic variables, such as the heat capacity or dielectric relaxation time, may respond differently. Nevertheless, τ_α at the pressure-dependent glass-transition temperature is constant for a given method of determining T_g [29], with its particular value governed by the experimental rate or timescale.

6.4.1 Arrhenius behavior

At high temperatures molecular interactions are sufficiently weak that relaxation becomes exponentially fast. For small molecules this leads to a relaxation function having the Debye form; however, for polymers the contribution of intrachain correlations complicates the behavior, so that even isolated chains exhibit decay that is only approximately exponential [104–106]. Exponential relaxation implies motion involving a single, constant activation barrier and thus an Arrhenius temperature-dependence of the relaxation time (eqn (1.12)). As a liquid or polymer is cooled, the loss of thermal energy and concomitant molecular crowding due to the volume contraction constrain the dynamics, increasing the degree of intermolecular cooperativity [14]. The viscous liquid is comprised of fast- and slow-moving molecules (or polymer segments), which interchange their roles at times on the order of the structural relaxation time. This makes the many-body dynamics inherently cooperative, and the consequent dynamic heterogeneity (i.e. spatial fluctuations of the relaxation time for durations comparable to its average value [107,108]) leads to non-Debye and non-Arrhenius behaviors that arise at a characteristic temperature, T_A.

For polymers, Arrhenius behavior is difficult to attain, since their long terminal relaxation times require measurements at very high temperatures, with possible chemical degradation. In Figure 6.34 are Arrhenius plots of the zero-shear viscosity at high temperatures for several PDMS [109,110]. This polymer has a very low T_g and is relatively stable, facilitating experiments above T_A. The activation energy for the viscosity is about 15 kJ/mol and varies only weakly with molecular weight. This is about an order of magnitude larger than van der Waals bond energies. The activation energy of PDMS is about the same as for polyethylene oxide, for

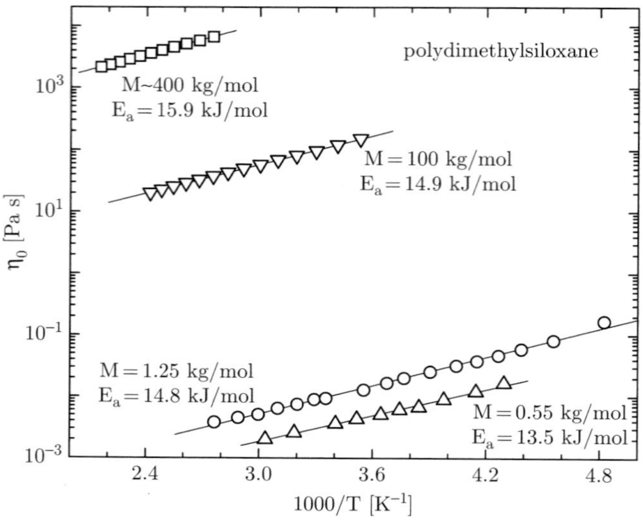

Figure 6.34 The zero-shear viscosities at high temperatures for linear PMDS of various molecular weights; the activation energies obtained from the slopes of the fitted lines are indicated. Data from refs. [109,110].

which $E_a = 16$ kJ/mol [111], but significantly smaller than the value of 27 kJ/mol for polyethylene [112]. Presumably the facile rotation about the oxygen linkage in PDMS and PEO causes their lower activation energies. Since at high temperature the T dependence of the global chain modes and local segmental modes become the same (as seen in the parallelism of the two curves in Figure 6.12), the E_a for the viscosity is expected to be the activation energy for the segmental process at high temperature. Near T_g the activation energy for global motions cannot be measured, and for local segmental relaxation E_a becomes much larger; e.g. *ca.* 240 kJ/mol for PDMS [113].[11]

In Figure 6.34 the viscosity at which the logarithm of η_0 and reciprocal temperature become proportional depends on molecular weight; that is, $\eta_0(T_A)$ is not a constant even for the same polymer differing only in chain length. The relaxation time or viscosity at T_A is not universal for molecular liquids either. In Table 6.3 relaxation times at the onset of Arrhenius behavior are tabulated for various materials; the values span about one decade centered around 100 ps. Note that $\tau_\alpha(T_A)$ is about 2 orders of magnitude longer than the inverse Debye frequency (Chapter 1), since the relevant length scale for liquid relaxation is the intermolecular distance, rather than the interatomic distances relevant to the vibrational motion associated with ω_D.

To determine whether $\eta_0(T_A)$ or $\tau_\alpha(T_A)$ is a material constant requires data at elevated pressure. Since measurements above T_A are difficult at elevated pressures, there are no data available for a single material as a function of pressure. However, the scaling relation, eqn (6.7), together with the EOS can be used to calculate $\tau_\alpha(T_A)$ for various combinations of T and P. Illustrative results are shown in Figure 6.35 for salol at three pressures and at constant volume [114]. The data achieve Arrhenius behavior at the highest temperatures. The

[11] E_a for η is slightly smaller than for τ due to the temperature dependence of the plateau shear modulus (see eqn (2.3), the Maxwell relation). The activation energy of G_N is on the order of 1–2 kJ/mol [109].

Table 6.3 Relaxation times at the onset of Arrhenius behavior.

Material	T_A (K)	$\log(\tau(T_A)/s)$
salol	348	–10.03 ±0.03
butyl benzene	268	–10.2
propyl benzene	240	–10.1
propylene carbonate	290	–9.67 ±0.03
a-polypropylene	476	–10.3
DGEBA	352	–10.3 ±0.2
poly[(phenyl glycidyl ether)-co-formaldehyde]	390	–11.0 ±0.3
trinaphthylbenzene	588	–10.7
benzophenone	309	–9.8 ±0.1
glycerol	413	–11.3 ±0.3
o-terphenyl	443	–12*

* using $\eta = 10^{-3}$ Pa s and assuming $G_\infty = 1$ GPa

onset of an Arrhenius T dependence is most accurately assessed using the Stickel derivative function [115,116] introduced in Chapter 2, the slope of which is zero for Arrhenius behavior. This function is plotted in Figure 6.35, from which $\log(\tau_\alpha)$ is determined to equal -10.03 ± 0.03 at T_A, independent of P or V. Similar analyses for propylene carbonate and *ortho*-terphenyl confirm this result – for a given material, the characteristic relaxation time at the onset of Arrhenius behavior appears to be invariant to thermodynamic conditions.

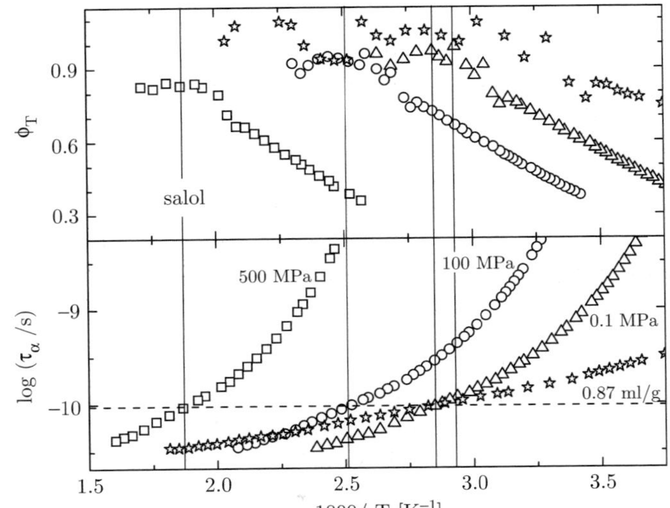

Figure 6.35 (lower panel) Relaxation times of salol measured at atmospheric pressure and calculated for higher pressure or constant volume using the scaling relation. (upper panel) Corresponding derivative plots (eqn (2.6)), with the intersection of the linear segments yielding the value of T_A indicated by the vertical lines and $\tau_\alpha(T_A)$ by the horizontal dashed line. Data from ref. [114].

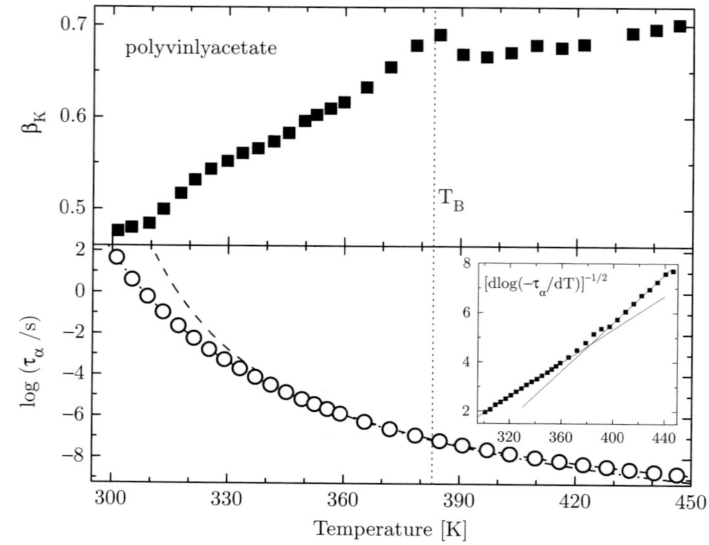

Figure 6.36 (upper) Kohlrausch exponent (eqn (1.7)) for PVAc, showing the change in temperature-sensitivity at the dynamic crossover at $T_B = 383$ K, the latter indicated by the vertical dotted line. (lower) Segmental relaxation times of PVAc, along with the respective fits to the VFTH equation above (dashed line) and below (dot-dash line) T_B. The inset is the derivative plot, showing the change in the behavior of $\tau_\alpha(T)$ at T_B. Data from ref. [123].

6.4.2 Dynamic crossover

As the relaxation time of a liquid becomes longer with further cooling or densification, additional characteristic behaviors arise at a temperature, T_B, which is 10 to 50% above the glass-transition temperature. This change in dynamics has various signatures: (i) There is a deviation from the Stokes–Einstein relation (eqn (2.19)), predicting inverse proportionality between the viscosity and translational diffusion [117,118]. This was first discovered by Plazek and Magill [119] from comparison of transport-dominated crystal-growth rates and $\eta(T)$ data for trinaphthylbenzene. (ii) A similar breakdown occurs in the Maxwell relation (eqn (2.3)) between the viscosity and orientational relaxation times [120,121]. (iii) The structural relaxation function, or its counterpart, the dispersion in the frequency domain, shows a different temperature-dependence from that for $T_A > T > T_B$, with continued strong deviation from Debye behavior [122–124]. (iv) The structural relaxation splits into a fast, secondary relaxation, identified as the Johari–Goldstein process (chapters 1 and 2), and the slower structural or primary α-relaxation. (v) The nature of the non-Arrhenius temperature dependence of τ_α changes (hence the term "dynamic crossover"), whereby a single VFTH equation (eqn (2.4)) cannot describe $\tau_\alpha(T)$ [125].

The dynamic crossover can be detected in plots of τ_α versus the reciprocal of the β_K stretch exponent [126], or from the departure of $\tau_\alpha(T)$ from one set of VFTH parameters prevailing at higher temperatures to a second set of parameters at lower temperatures [125]. These behaviors are illustrated in Figure 6.36 for PVAc, with a change in dynamics at ambient pressure occurring at $T_B = 383$ K for which $\tau_\alpha = 8\times10^{-8}$ s [123]. This is in the range of values found for other materials, which span from $\tau_\alpha(T_B) = 5\times10^{-11}$ s for polymethacrylate [127] to 2×10^{-4} s for propanol [128].

When pressure is the independent variable, the equation corresponding to eqn (2.6) of Chapter 2 is [129]

$$\phi_P = \left[\frac{d\log(\tau_\alpha)}{dP}\right]^{-\frac{1}{2}} \tag{6.32}$$

By changing pressure $\tau_\alpha(T_B)$ can be evaluated for different thermodynamic conditions. This is shown in Figure 6.37 for the molecular liquid phenylphthaleindimethylether (PDE); the relaxation time at $T_B(P)$ is invariant to pressure. The changes of T and P in Figure 6.37 correspond to density variations of 3%. For comparison, the density change induced by cooling from T_B to T_g at atmospheric pressure is only half this amount, yet τ_α would change by more than 5 orders of magnitude. Results for various liquids are collected in Table 6.4. The constancy of τ_α at the crossover is a general result, valid also for $\eta(T_B)$. The change in dynamics does not arise at a critical temperature or volume, but rather is governed by the timescale of the relaxation.

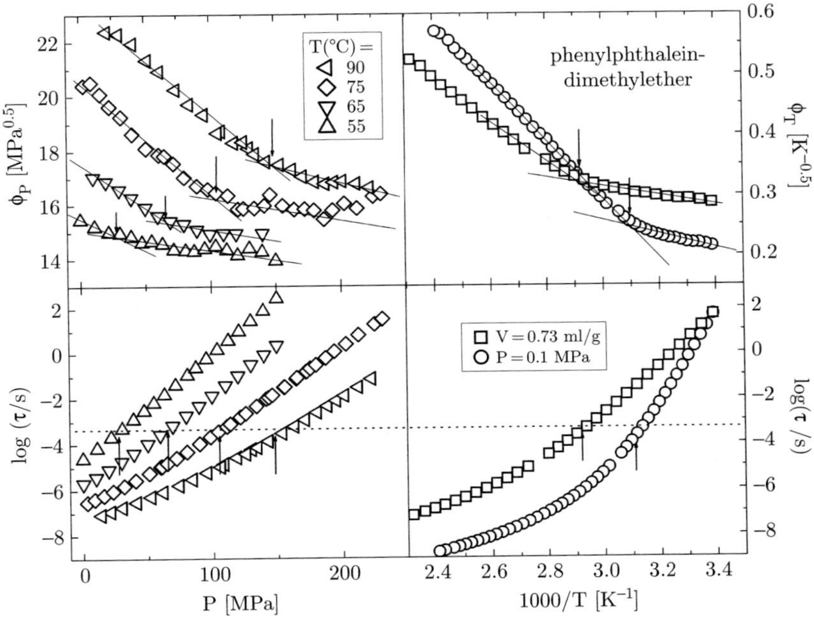

Figure 6.37 Relaxation data for PDE with the dynamic crossover indicated by vertical arrows: (lower left panel) isothermal variation of τ_α with pressure; (upper left panel) derivative function (eqn (6.32)) of $\tau_\alpha(P)$ showing the break at the crossover; (lower right panel) isobaric (circles) and isochoric (squares) variation of τ_α with temperature; (upper right panel) derivative function (eqn (2.6)) of $\tau_\alpha(T)$ showing the break at the crossover. For all conditions $\tau_\alpha(T_B) = -4(\pm1)\times10^{-3}$ s. Data from ref. [114].

Table 6.4 Relaxation times or viscosities at the dynamic crossover [114]

Material	P range [GPa]	T range [K]	$\log(\tau(T_B)$ [s])	$\log(\eta(T_B)$ [Pa s])
PDE	≤ 0.23	295–363	–3.4 ± 0.1	—
polychlorinated biphenyl*	≤ 0.24	325–344	–4.4 ± 0.2	—
polychlorinated biphenyl^	≤ 0.25	263–283	–3.8 ± 0.2	—
cresolphthalein-dimethylether	≤ 0.16	313–455	–5.9 ± 0.2	—
propylene carbonate	≤ 1.7	159–270	–7.4 ± 0.2	—
salol	≤ 0.95	303–363	–6.3 ± 0.7*	1.3 ±0.1
o-terphenyl	≤ 0.8	383–423	—	1.5 ±0.4
polyphenyl thioether	≤ 0.85	293–373	—	2.0 ±0.1

* 62% chlorinated
^ 42% chlorinated

References

1. R.S. Marvin in *Proc. 2nd Inter. Congress on Rheology*, (ed. V.G.W. Harrison), Buttonworths Sci. Pub. London (1953).
2. R.D. Andrews and A. Tobolsky, *J. Polym. Sci.* **1**, 221 (1951).
3. E.R. Fitzgerald, L.D. Grandine, and J.D. Ferry, *J. Appl. Phys.* **24**, 650 (1953); J.D. Ferry, L.D. Grandine, and E.R. Fitzgerald, *J. Appl. Phys.* **24**, 911 (1953).
4. J.D. Ferry, *Viscoelastic Properties of Polymers*, Wiley, New York, Chapter 10 (1980).
5. P.H. Mott, C.M. Roland, and R.D. Corsaro, *J. Acoust. Soc. Am.* **111**, 1782 (2002).
6. A.K. Doolittle and D.B. Doolittle, *J. Appl. Phys.* **28**, 901 (1957).
7. M.H. Cohen and D. Turnbull, *J. Chem. Phys.* **31**, 1164 (1959).
8. S. Corezzi, P.A. Rolla, M. Paluch, J. Ziolo, and D. Fioretto, *Phys. Rev. E* **60**, 4444 (1999).
9. M. Paluch, *J. Chem. Phys.* **115**, 10029 (2001).
10. K.U. Schug, H.E. King, and R. Böhmer, *J. Chem. Phys.* **109**, 1472 (1998).
11. M.L. Williams, R.G. Landel, and J.D. Ferry, *J. Amer. Chem. Soc.* **77**, 3701 (1955).
12. J.T. Bendler, J.J. Fontanella, M.F. Shlesinger, and M.C. Wintersgill, *Electrochim. Acta* **49**, 5249 (2004).
13. G. Adam and J.H. Gibbs, *J. Chem. Phys.* **43**, 139 (1965).
14. M. Goldstein, *J. Chem. Phys.* **51**, 3728 (1969).
15. A.J. Kovacs, *Adv. Polym. Sci.* **3**, 394 (1964).
16. D.J. Plazek, I.C. Choy, F.N. Kelley, E. von Meerwaal, and L.J. Su, *Rubber Chem. Technol.* **56**, 866 (1983).
17. D.J. Plazek and M. Rosner, *Rubber Chem. Technol.* **71**, 679 (1998).
18. L.C. Yanyo and F.N. Kelley, *Rubber Chem. Technol.* **61**, 100 (1988).
19. T.L. Smith, *J. Polym. Sci.* **32**, 99 (1958).
20. K. Nitta and T. Ishiburo, *J. Polym. Sci. Polym. Phys. Ed.* **40**, 2018 (2002).
21. K.A. Grosch, *Rubber Chem. Technol.* **69**, 495 (1996).

22. R.W. Fillers and N.W. Tschoegl, *J. Rheol.* **21**, 51 (1977).
23. W.K. Moonan and N.W. Tschoegl, *Macromolecules* **16**, 55 (1983).
24. W.K. Moonan and N.W. Tschoegl, *J. Polym. Sci. Polym. Phys. Ed.* **23**, 623 (1985).
25. A. Kralj, T. Prodan, and I. Emir, *J. Rheol.* **45**, 929(2001).
26. C.G. Robertson, R. Bogoslovov, and C.M. Roland, *Phys. Rev. E* **75**, 051403 (2007).
27. R.B. Bogoslovov, C.M. Roland, A.R. Ellis, A.M. Randall, and C.G. Robertson, *Macromolecules* **41**, 1289 (2008).
28. R. Cardinaels, P. Van Puyvelde, and P. Moldenaers, *Rheol. Acta* **46**, 495 (2007).
29. C.M. Roland, S. Hensel-Bielowka, M. Paluch, and R. Casalini, *Rep. Prog. Phys.* **68** 1405 (2005).
30. S. Havriliak and S. Negami, *Polymer* **8**, 161 (1967).
31. F. Alvarez, A. Alegria, and J. Colmenero, *Phys. Rev. B* **44**, 7306 (1991).
32. R. Casalini and C.M. Roland, *Macromolecules* **38**, 1779 (2005).
33. D.J. Plazek, *J. Phys. Chem.* **69**, 3480 (1965).
34. C.M. Roland, K.L. Ngai, P.G. Santangelo, X.H. Qiu, M.D. Ediger, and D.J. Plazek, *Macromolecules* **34**, 6159 (2001).
35. D.J. Plazek, I.-C. Chay, K.L. Ngai, and C.M. Roland, *Macromolecules* **28**, 6432 (1995).
36. J.A. Pathak, J.N. Twigg, K.E. Nugent, D.L. Ho, E.K. Lin, P.H. Mott, C.G. Robertson, M.K. Vukmir, T.H. Epps, and C.M. Roland, *Macromolecules* **41**, 7543 (2008).
37. P.G. Santangelo and C.M. Roland, *Macromolecules* **31**, 3715 (1998).
38. D.J. Plazek, *Polymer J.* **12**, 43 (1980).
39. D.J. Plazek, M.J. Rosner, and D.L. Plazek, *J. Polym. Sci. Polym. Phys.* **26**, 473 (1988).
40. C.G. Robertson and C.M. Rademacher, *Macromolecules* **37**, 10009 (2004).
41. C.M. Roland, M. Paluch, and R. Casalini, *J. Polym. Sci. Pt. B-Polym. Phys.* **42**, 4313 (2004).
42. P.G. Santangelo and C.M. Roland, *J. Rheol.* **45**, 583 (2001).
43. C.M. Roland, K.L. Ngai, and D.J. Plazek, *Macromolecules* **37**, 7051 (2004).
44. G.C. Berry and T.G. Fox, *Adv. Polym. Sci.* **5**, 261 (1968).
45. W.W. Graessley, *Acct. Chem. Res.* **10**, (1977).
46. C.A. Bero and C.M. Roland, *Macromolecules* **29**, 1562 (1996).
47. P.G. Santangelo and C.M. Roland, *J. Non-Cryst. Solids* **235**, 709 (1998).
48. R.B. Bogoslovov, C.M. Roland, and R.M. Gamache, *Appl. Phys. Lett.* **90**, 221910 (2007).
49. F.H. Stillinger and T.A. Weber, *Science* **225**, 983 (1984).
50. M. Sampoli, P. Benassi, R. Eramo, L. Angelani, and G. Ruocco, *J. Phys. Condens. Matter* **15**, S1227 (2003).
51. J.E. Jones, *Proc. Roy. Soc. Lond. A* **106**, 441 (1924); ibid. **106**, 462 (1924).
52. J.F. Dobson, K. McLennan, A. Rubio, J. Wang, T. Gould, H.M. Le, and B.P. Dinte, *Aust. J. Chem.* **54**, 513 (2001).
53. E.A. Moelwyn-Hughes, *Physical Chemistry*, 2nd edition, Pergamon Press, New York (1961).
54. V.Y. Bardik and V.M. Sysoev, *Low-Temp. Phys.* **24**, 601 (1998).
55. B. Widom, *Physica A* **263**, 500 (1999).
56. B. Widom, *Science* **157**, 375 (1967).
57. D. Chandler, J.D. Weeks, and H.C. Andersen, *Science* **220**, 4599 (1983).

58. H. Stillinger, P.G. Debenedetti, and T.M. Truskett, *J. Phys. Chem. B* **105**, 11809 (2001).
59. J.D. Weeks, D. Chandler, and H.C. Andersen, *J. Chem. Phys.* **54**, 5237 (1971).
60. A. Tölle, H. Schober, J. Wuttke, O.G. Randl, and F. Fujara, *Phys. Rev. Lett.* **80**, 2374 (1998).
61. H.C. Longuet-Higgins and B. Widom, *Mol. Phys.* **8**, 549 (1964).
62. R.J. Speedy, *J. Phys. Condens. Matter* **15**, S1243 (2003).
63. M.S. Shell, P.G. Debenedetti, E. La Nave, and F. Sciortino, *J. Chem. Phys.* **118**, 8821 (2003).
64. W.G. Hoover and M. Ross, *Contemp. Phys.* **12**, 339 (1971).
65. Y. Hiwatari, H. Matsuda, T. Ogawa, N. Ogita, and A. Ueda, *Prog. Theor. Phys.* **52**, 1105 (1974).
66. C.M. Roland, J.L. Feldman, and R. Casalini, *J. Non-Cryst. Solids* **352**, 4895 (2006).
67. A. Grzybowski, M. Paluch, and K. Grzybowska, *J. Phys. Chem. B* **113**, 7419 (2009).
68. C.J. Ellison, M.K. Mundra, and J.M. Torkelson, *Macromolecules* **38**, 1767 (2005).
69. M.J. Schroeder and C.M. Roland, *Macromolecules* **35**, 2676 (2002).
70. S.A. Reinsberg, X.H. Qiu, M. Wilhelm, H.W. Spiess, and M.D. Ediger, *J. Chem. Phys.* **114**, 7299 (2001).
71. E. Hempel, G. Hempel, A. Hensel, C. Schick, and E. Donth, *J.Phys. Chem. B* **104**, 2460 (2000).
72. U. Tracht, M. Wilhelm, A. Heuer, and H.W. Spiess, *J. Magn. Reson.***140**, 460 (1999).
73. R. Casalini and C.M. Roland, *Coll. Polym. Sci.* **283** 107 (2004).
74. R. Casalini and C.M. Roland, *Phys. Rev. E* **69**, 062501 (2004).
75. C.M. Roland and R. Casalini, *J. Chem. Phys.* **121**, 11503 (2004).
76. K. Kaminski, S. Maslanka, J. Ziolo, M. Paluch, K.J. McGrath, and C.M. Roland, *Phys. Rev. E* **75**, 011903 (2007).
77. R. Casalini and C.M. Roland, *Phys. Rev. E* **72**, 031503 (2005).
78. C.M. Roland and R. Casalini, *Macromolecules* **40**, 3631 (2007).
79. C.M. Roland, R. Casalini, R. Bergman, and J. Mattsson, *Phys. Rev. B* **77**, 012201 (2008).
80. A.S. Pensado, A.A.H. Padua, M.J.P. Comunas, and J. Fernandez, *J. Phys. Chem. B* **112**, 5563 (2008).
81. M. Yoshimura, C. Boned, G. Galliero, J.-P. Bazile, A. Baylaucq, and H. Ushiki, *Chem. Phys.* **369**, 126 (2010).
82. D. Coslovich and C. M. Roland, *J. Phys. Chem. B* **112**, 1329 (2008).
83. D. Coslovich and C. M. Roland, *J. Chem. Phys.* **131**, 151103 (2009).
84. D. Ben-Amotz and G. Stell, *J. Phys. Chem. B* **119**, 10777 (2003).
85. N.P. Bailey, U. R. Pedersen, N. Gnan, T. B. Schrøder, and J. C. Dyre, *J. Chem. Phys.* **129**, 184507 (2008).
86. G. Tsolou, V.A. Harmandaris, and V.G. Mavrantzas, *J. Chem. Phys.* **124**, 084906 (2006).
87. U.R. Pedersen, N.P. Bailey, T.B. Schrøder, and J.C. Dyre, *Phys. Rev. Lett.* **100**, 015701 (2008).
88. D. Coslovich and C. M. Roland, *J. Chem. Phys.* **130**, 014508 (2009).
89. C.M. Roland, S. Bair, and R. Casalini, *J. Chem. Phys.* **125**, 124508 (2006).
90. S. Pawlus, A.P. Sokolov, M. Paluch, and M. Mierzwa, *Macromolecules* **43**, 5845 (2010).
91. C.M. Roland, R. Casalini, and M. Paluch, *J. Polym. Sci. Polym. Phys.* **42**, 4313 (2004).
92. K.L. Ngai, R. Casalini, and C.M. Roland, *Macromolecules* **38**, 4363 (2005).
93. R.F. Boyer and R.S. Spencer, *J. Appl. Phys.* **15**, 398 (1944).

94. R. Casalini and C.M. Roland, *J. Chem. Phys.* **119**, 4062 (2003).
95. G. Williams, *Trans. Faraday Soc.* **61**, 1564 (1965).
96. M. Naoki, H. Endou, and K. Matsumoto, *J. Phys. Chem.* **91**, 4169 (1987).
97. C.M. Roland and R. Casalini, *Macromolecules* **38**, 8729 (2005).
98. E. Grüneisen, *Ann. Phys. (Leipzig)* **39**, 257 (1912).
99. K.A. Moelwyn-Hughes, K.A. *J. Phys. Coll. Chem.* **55**, 1246 (1951).
100. G. Hartwig, *Polymer Properties at Room and Cryogenic Temperatures*, Plenum Press, New York (1994), chapter 4.
101. R. Casalini, U. Mohanty, and C.M. Roland, *J. Chem. Phys.* **125**, 014505 (2006).
102. R. Casalini and C.M. Roland, *Philos. Mag.* **87**, 459 (2007).
103. C.M. Roland and R. Casalini, *J. Phys. Condens. Matter* **19**, 205118 (2007).
104. C.K. Hall and E. Helfand, *J. Chem. Phys.* **77**, 3275 (1982).
105. T. Haliloglu, I. Bahar, B. Erman, and W.L. Mattice, *Macromolecules* **29**, 8942 (1996).
106. M. Guenza and K.F. Freed, *J. Chem. Phys.* **105**, 3823 (1996).
107. K. Schmidt-Rohr and H.W. Spiess, *Phys. Rev. Lett.* **66**, 3020 (1991).
108. R. Richert, *J. Phys. Condens. Matter* **14**, R703 (1998).
109. C.M. Roland and P.G. Santangelo, *J. Non-Cryst. Sol.* **307–310**, 835 (2002).
110. Y. Ding, V.N. Novikov, A.P. Sokolov, R. Casalini, and C.M. Roland, *Macromolecules* **37**, 9273 (2004).
111. T. P. Liao and H. Morawetz, *Macromolecules* **13**, 1228 (1980).
112. G.B. McKenna, D.J. Plazek, and K.L. Ngai, *Polymer* **26**, 1651 (1985).
113. C.M. Roland and K.L. Ngai, *Macromolecules* **29**, 5747 (1996).
114. C.M. Roland, *Soft Matter* **4**, 2316 (2008).
115. F. Stickel, E.W. Fischer, and R. Richert, *J. Chem. Phys.* **102**, 6251 (1995).
116. C. Hansen, F. Stickel, R. Richert, and E. W. Fischer, *J. Chem. Phys.* **108**, 6408 (1998).
117. M.D. Ediger, C.A. Angell, and S.R. Nagel, *J. Phys. Chem.* **100**, 13200 (1996).
118. K.L. Ngai, J.H. Magill, and D.J. Plazek, *J. Chem. Phys.* **112**, 1887 (2000).
119. D.J. Plazek and J.H. Magill, *J. Chem. Phys.* **45**, 3038 (1996).
120. I. Chang and H. Sillescu, *J. Phys. Chem. B* **101**, 8794 (1997).
121. L. Comez, D. Fioretto, L. Palmieri, L. Verdini, P.A. Rolla, J. Gapinski, T. Pakula, A. Patkowski, W. Steffen, and E.W. Fischer, *Phys. Rev. E* **60**, 3086 (1999).
122. K.L. Ngai, *J. Phys. Chem. B* **103**, 5895 (1999).
123. K.L. Ngai and C.M. Roland, *Polymer* **43**, 567 (2002).
124. R. Casalini, K.L. Ngai, and C.M. Roland, *Phys. Rev. B* **68**, 014201 (2003).
125. F. Stickel, E. W. Fischer, and R. Richert, *J. Chem. Phys.* **104**, 2043 (1996).
126. K. Trachenko, C.M. Roland, and R. Casalini, *J. Phys. Chem. B* **112**, 5111 (2008).
127. J. Rault, *J. Non-Cryst. Sol.* **271**, 177 (2000).
128. C. Hansen, F. Stickel, T. Berer, R. Richert, and E.W. Fischer, *J. Chem. Phys.* **107**, 1086 (1997).
129. R. Casalini, M. Paluch, and C.M. Roland, *J. Chem. Phys.* **118,** 5701 (2003).

7 Blends

Although most polymer blends are immiscible, thermodynamically miscible blends, having an equilibrium morphology homogeneous down to the segmental level, exhibit the more interesting dynamics. Among miscible blends, in only a very few are both components rubbers ($T_g <$ ambient), although sometimes miscibility can be obtained by using low molecular weights or copolymer blends in which the components share common repeat units [1–6]. The focus of this chapter is miscible polymer blends, since they exhibit phenomena having no counterpart in neat materials. Selected aspects of phase-separated blends are discussed at the chapter's end.

7.1 Dynamic properties of miscible polymer blends

7.1.1 Dynamic heterogeneity

The local segmental relaxation function for neat polymers is described over at least a few decades by the Kohlrausch function (eqn (1.7)), which has a characteristic skewing towards higher frequencies. For a miscible blend this dispersion becomes more symmetric or even splits into two peaks. The less-mobile segments contribute toward lower frequencies, and vice versa for the more mobile component. The mobility is determined by a segment's intrinsic relaxation behavior, reflected mainly in its T_g, and by the composition of its local environment, the latter varying in a blend due to concentration fluctuations. (Unique for polymers is local self-enrichment due to chain connectivity.) The relaxation function for a blend is the superposition of the response from all local environments, but the lower frequency contributions tend to be broader (for reasons discussed in Section 7.2.5 below), so that the customary asymmetric KWW broadening toward higher frequency is diminished. The fact that the dynamics for different local environments are emphasized at different frequencies implies that the components themselves exhibit different relaxation behavior. When the disparity in the mobilities is very large, usually because T_g of the neat materials differ a lot, distinct relaxation peaks are present in the spectra, a phenomenon known as "dynamic heterogeneity" [7].

This dynamic heterogeneity is seen in dielectric loss spectra for PIP mixed with 25% PVE (Figure 7.1 [8]). There are two maxima, with the response of the higher T_g PVE at lower frequencies than the peak for the PI. The corresponding segmental relaxation times are shown in Figure 7.2. Similar behavior is shown for blends of PVME and poly(2-chlorostyrene) (P2CS) [9] in Figure 7.3. In both these cases the component glass-transition temperatures differ substantially ($\Delta T_g = 61$ K for PVE/PI and 149 K for P2CS/PVME). Distinct dispersions are not always observed in dielectric or mechanical spectra of miscible blends; instead, the segmental peak is broadened but has only one maximum. Even in these cases, the component dynamics can be resolved by taking advantage of a difference in the component

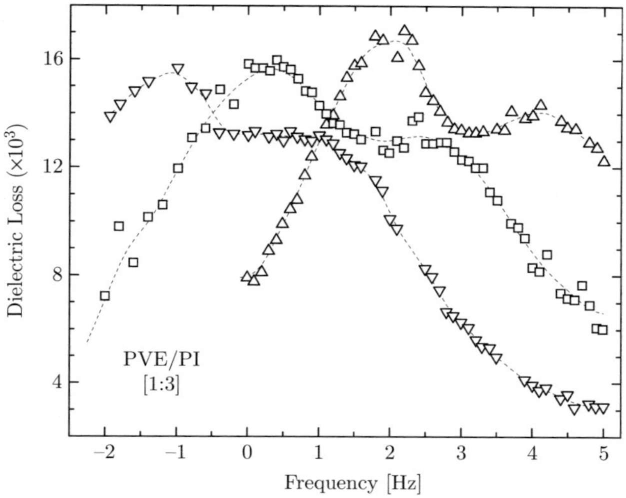

Figure 7.1 Local segmental relaxation spectra measured dielectrically for 25% PVE in PI at 223K (inverted triangles), 232K (squares), and 243K (triangles). The respective contributions of the components are evident as distinct maxima in the dielectric loss spectra, the PVE at lower frequencies and the PI at higher. The dashed lines are to guide the eye. Data from ref. [8].

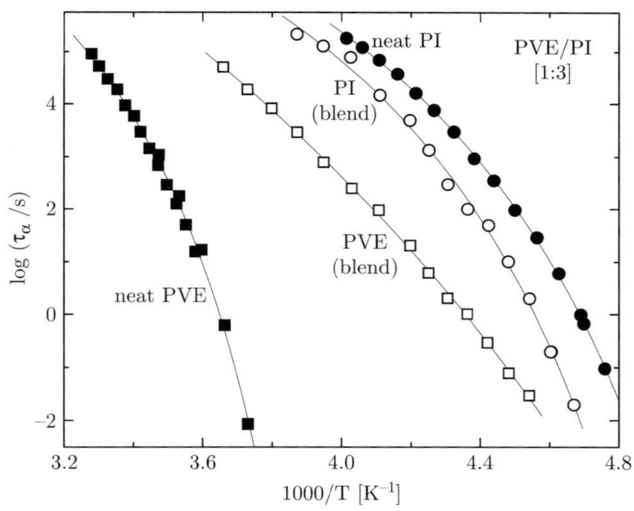

Figure 7.2 Local segmental relaxation times for the PVE (open squares) and PI (open circles) deduced from dielectric and mechanical spectroscopy for same blend as in Figure 7.1. τ_α for the neat components (filled symbols) are shown for comparison. Data from ref. [8].

susceptibilities [8,10] or, using NMR, different chemical shifts [7,11]. Since the relative contributions depend on the nature of the measurement, different experiments can yield different spectra. For example, in blends with non-polar poly-α-methylstyrene (PαMS), PCHMA dominates the dielectric response and there is only a single peak due to motion of the

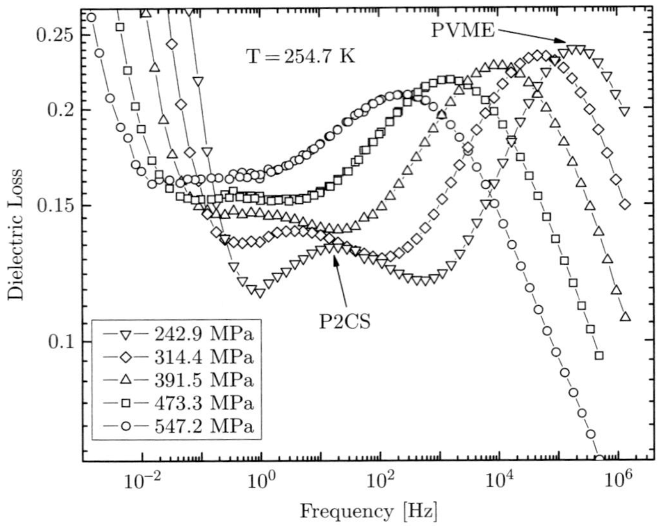

Figure 7.3 Local segmental relaxation spectra measured dielectrically for an equal weight blend of polyvinylmethylether and poly-2-chlorostyrene at various pressures. The higher-frequency peak is due to the PVME; the peak from the P2CS is masked at higher *P* by the dc-conductivity. Data from ref. [9].

PCHMA segments. However, this peak is broader than for neat PCHMA (Figure 7.4), since the segments in the blend exist in a range of local environments having different concentrations of PαMS [12].

Evident in the spectra in Figure 7.1 and Figure 7.3 is a breakdown of time-temperature (or time-pressure) superpositioning in miscible blends, due to the different T and P dependences of the components. This thermorheological complexity is a characteristic of miscible polymer blends. Even when one component is present in trace amounts, so that concentration fluctuations have no effect, the relaxation spectra can still change with temperature because of intrinsic differences in the T dependences of the respective relaxation times of the components (Figure 7.5 [13]).

7.1.2 Unusual concentration dependences

The dynamics of a blend component is affected by the mobility of its local environment. If motion of surrounding segments is slower, the relaxation peak for a given component tends to broaden at lower frequencies and the mean τ_α becomes longer. The latter effect is described by various "mixing rules" for the concentration dependence of T_g for mixtures. Common is the Fox equation [14]

$$T_g(\phi_A) = \left(\frac{\varphi_A}{T_{g,A}} + \frac{1-\varphi_A}{T_{g,B}}\right)^{-1} \tag{7.1}$$

where the subscripts refer to respective components A and B. This equation has no adjustable parameter and some limited applicability (Figure 7.6 [15,16]), although eqn (7.1) is based on an untenable free-volume interpretation of the glass transition. Other expressions for the composition-dependence of T_g have been derived, reflecting various conceptions of the cause of

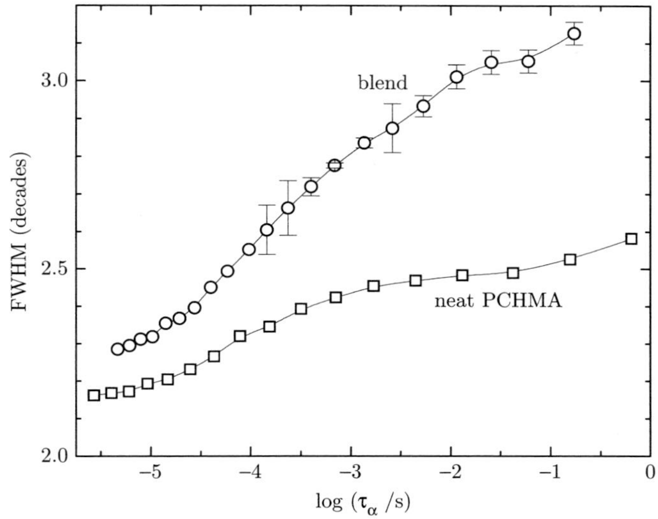

Figure 7.4 The breadth of the segmental relaxation peak (full width at half-maximum) for polycyclohexylmethacrylate neat (squares) and mixed with an equal weight of PαMS (circles). The dielectric response is dominated by the more polar PCHMA; nevertheless, the peak for the blend is broader and changes more with temperature. Data from ref. [12].

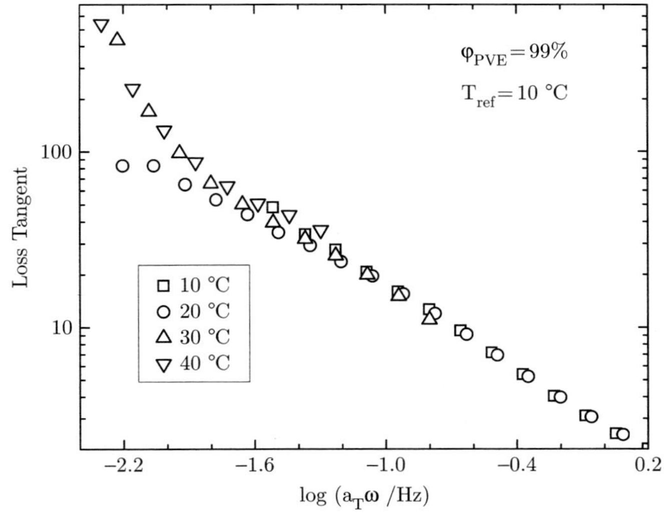

Figure 7.5 Master curves of the mechanical loss tangent for PVE ($M_w = 5.6$ kg/mol) containing a trace amount of PI ($M_w = 80$ kg/mol), demonstrating the breakdown of time–temperature superpositioning in miscible blends, even in the absence of effects due to concentration fluctuations. Data from [13].

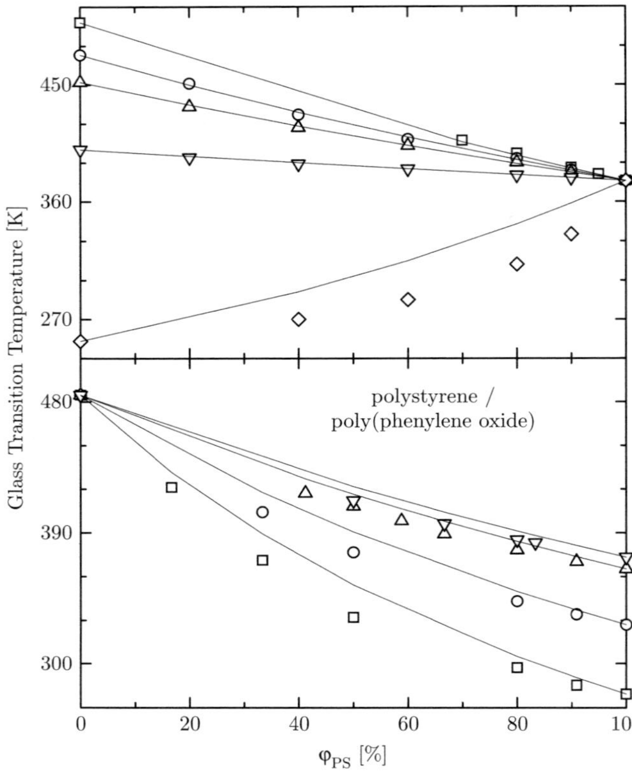

Figure 7.6 Calorimetric T_g for blends of polystyrene and polyphenylene oxide. (top) PS: $M_n = 1.44\times10^5$ g/mol; PPO: $362 \leq M\,(\text{g/mol}) \leq 4.40\times10^4$ g/mol [15]. (bottom) PPO: $M_n = 1.10\times10^4$ g/mol; PS: $800 \leq M\,(\text{g/mol}) \leq 3.70\times10^4$ g/mol [16]. The lines represent eqn (7.1).

vitrification; these employ additional parameters, which of course improves their ability to fit experimental data [17–22].

There are many polymer mixtures that exhibit anomalous concentration dependences for T_g and τ_α. For example, the addition of PS substantially increases τ_α of an oligomeric polystyrene, as expected since the former has a 130 K higher T_g [23]. However, addition of the same high-M PS to a polychlorinated biphenyl (PCB) has almost no effect on the latter's dynamics (Figure 7.7), even though T_g and τ_α for the PCB are the same as those for the oligomeric PS. Another example is blends of PEO and PMMA [24]. High-resolution calorimetry measurements on this blend reveal two glass transitions, corresponding to each of the components [25]. Obviously such behavior cannot be captured by simple mixing rules like eqn (7.1). More interesting is the composition dependence of the dynamics. Segmental relaxation of the PEO at lower temperatures is 12 orders of magnitude faster than that of the PMMA, due to the disparity in the neat component T_g values (>200 K). Nevertheless, the segmental dynamics of the PEO are nearly independent of blend composition. Up through about 10% PEO, $\tau_\alpha(T)$ for PEO remains the same and changes only weakly at higher concentrations (Figure 7.8). An explanation for this behavior is given in Section 7.2.5.

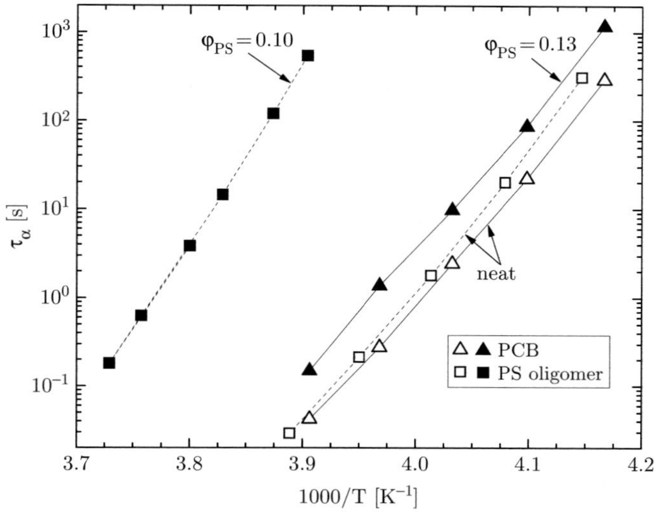

Figure 7.7 Mechanical relaxation times for an oligomeric PS (M_w = 0.59 kg/mol; T_g = 244 K; open squares) and PCB (54% chlorine; T_g = 246 K; open triangles). Addition of about 10% high molecular weight (M_w = 90 kg/mol) PS has a small effect on τ_α of PCB (filled triangles), but increases the relaxation time of the oligomeric PS by 4 orders of magnitude. At these temperatures, the high M, neat PS is in the glassy state. Data from ref. [23].

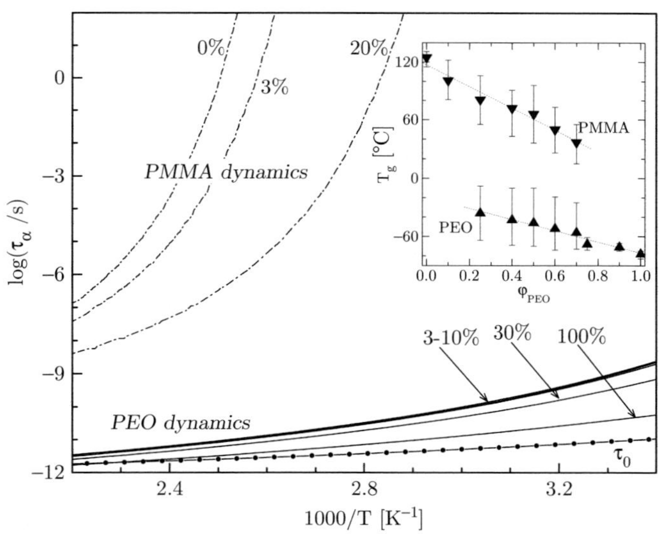

Figure 7.8 Segmental relaxation times for PMMA (dash-dotted lines) measured by dielectric spectroscopy and PEO (solid lines) measured by ^{2}H-NMR, in blends containing the indicated amount of PEO. The variation of τ_α for the PEO component is very weak with respect to T and negligible with respect to composition for low PEO contents (the curves for $\varphi_{PEO} \leq 10\%$ are indistinguishable). The lowest curve (line with solid circles) is the non-cooperative relaxation time calculated for PEO (eqn (7.7)). Since τ_0 falls close to the characteristic time, $t_c \sim 2$ ps, of the coupling model, τ_α for PEO becomes independent of n; that is, independent of its local environment. Data from ref. [24]. The inset shows the calorimetric T_g measured for the components; $dT_g/d\varphi$ for PMMA is twice that for the PEO. Data from ref. [25].

The dynamics in Figure 7.7 and Figure 7.8 have unusual composition-dependences, but τ_α and T_g are always intermediate between the values for the neat components. However, some blends exhibit qualitative deviations from mixing rules such as eqn (7.1). Examples of both minima and maxima in T_g versus concentration curves are shown in Figure 7.9 [20,26–28]. The segmental relaxation times for these polymer blends can be orders of magnitude shorter or longer than τ_α of the neat components. Similar effects are seen in concentrated polymer solutions [23,29,30]. For example, the addition of PCB to polyvinylethylene decreases τ_α a hundred-fold, making it lower than the relaxation times of either neat component (Figure 7.10 [29]). There are various mechanisms that can give rise to this behavior, including excess mixing volumes, specific interactions, and steric coupling of the components, all having no counterpart in neat materials.

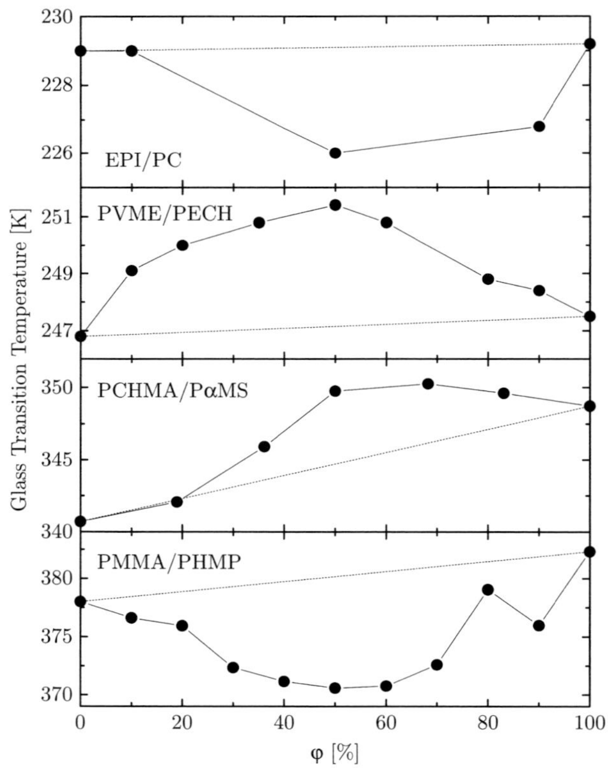

Figure 7.9 Glass-transition temperature as a function of the weight per cent of the lower T_g component for blends of (top to bottom): epoxidized polyisoprene and polychloroprene [26], polyvinylmethylether and polyepichlorhydrin [27], polycylohexylmethacrylate and poly-α-methylstyrene [28], and polymethylmethacrylate with poly(1-hydroxy-2,6-phenylene)-methylene phenolic resin [20]. The dotted lines are the linear interpolation of T_g of the pure components.

7.1.3 Confinement effect

The intermolecularly cooperative nature of the segmental dynamics means that the local environment has a profound effect on the relaxation of a polymer chain. A striking example of this is when the neighboring segments are in the glassy state; that is, on the timescale of a chain's motion, surrounding segments from primarily the other component are frozen. This is different from the glassy state of a neat material because in the blend, chains from one component retain orientational mobility. However, this mobility is severely restricted, and the effect of temperature becomes weaker, since changes in T only affect the thermal energy, not the density or local packing.

When PVME is blended with PS, the more polar PVME dominates the dielectric response, enabling its dynamics to be characterized (Figure 7.11). At high temperatures both components are above T_g, and the PVME dynamics are normal. However, at lower temperatures the PS phase becomes glassy, causing a change in the T dependence of τ_α for PVME. This phenomenon has been interpreted as confinement of the more mobile PVME by the glassy component [31]. The Arrhenius behavior at lower temperatures is analogous to the JG process, as under these conditions structural relaxation is restricted to local conformational transitions, characteristic of the weakly cooperative, secondary relaxations [32].

Blends of PVME with P2CS exhibit similar behavior (Figure 7.12) [9], although in this case both components have sufficient dielectric strength to be observed in the spectrum. When τ_α for the P2CS becomes long, exceeding the value associated with the glassy state ($\tau_\alpha(T_g) \approx 100$ s), the apparent activation energy for segmental relaxation of the PVME component is much reduced; the T sensitivity of τ_α for PVME is five times smaller upon vitrification of the P2CS.

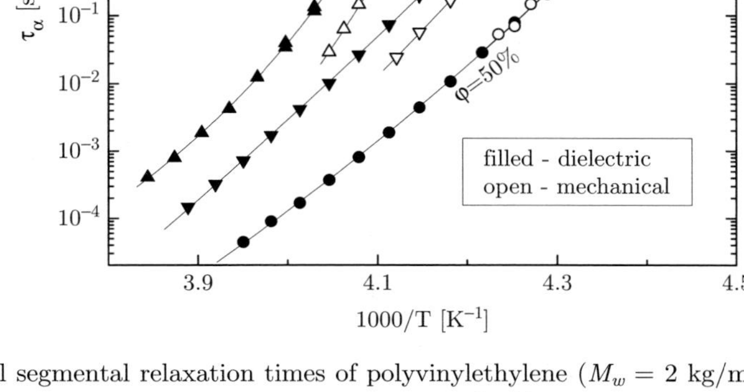

Figure 7.10 Local segmental relaxation times of polyvinylethylene (M_w = 2 kg/mol; triangles), reorientational relaxation times of polychlorinated biphenyl (48% chlorine; inverted triangles), and τ_α measured for their equal parts mixture (circles). The more polar PCB dominates the dielectric relaxation (filled symbols) and, by inference, also the mechanical response. Data from ref. [29].

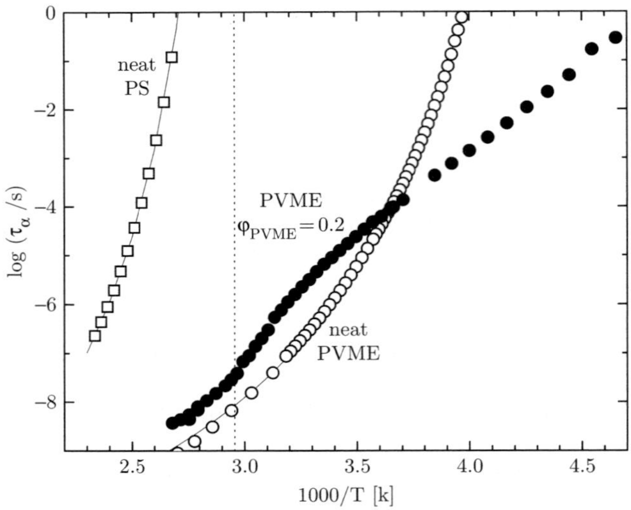

Figure 7.11 Segmental relaxation times measured by dielectric spectroscopy for PS (open squares), PVME (open circles) and a blend containing 20% PVME (filled circles). The dielectric strength of the PVME is about 80 times larger than $\Delta\varepsilon$ of PS, so that τ_α for the blend corresponds to motion of the PVME. At temperatures below the calorimetric T_g of PS (indicated by the vertical dotted line), the T dependence of the PVME becomes weaker and essentially non-Arrhenius. (*Data courtesy of Juan Colmenero of the Universidad del País Vasco.*)

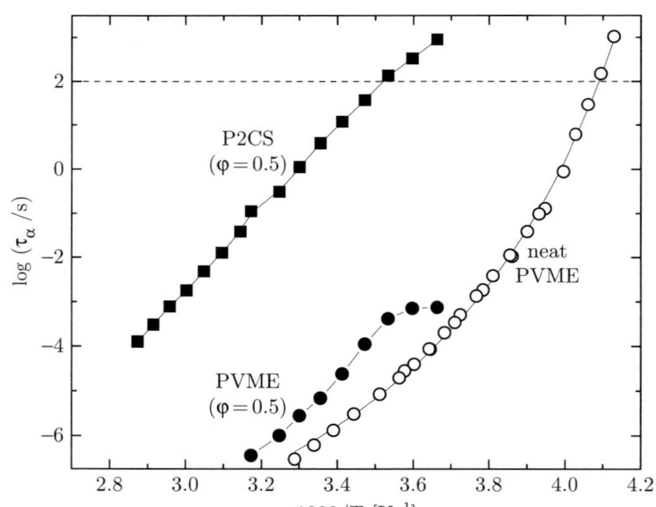

Figure 7.12 Dielectric segmental relaxation times for the components of an equal parts blend of poly-2-chlorostyrene (filled squares) and polyvinylmethylether (filled circles). When τ_α for the former becomes very long, τ_α for PVME becomes nearly T invariant. Also shown are the corresponding segmental relaxation times of neat PVME (open circles), which maintain the usual VFTH (eqn (2.4)) T dependence throughout the measured range. Data from ref. [9].

7.1.4 Relative role of volume and temperature

Chapter 2 describes how volume and thermal energy govern the segmental dynamics near T_g, with their relative influence varying among polymers. In favorable cases the ratio of the isochoric and isobaric activation energies, which quantifies the effect of V and T, can be determined for both components of a blend; the requirement is that the relaxation peak of each can be resolved. Results for PVME/P2CS ($\varphi = 0.5$) are shown in Figure 7.13 [9]. The general trend is that as the dynamics are enhanced (smaller τ_α) by either higher temperature or lower pressure, volume exerts a stronger effect (smaller E_V/E_a). Comparing the two neat polymers, volume is more dominant for P2CS than PVME. However, blending affects the two materials in opposite fashion; the role of volume is diminished for P2CS, but becomes stronger for PVME. Thus, there is a reversal in order, relaxation of P2CS depending less on V for P2CS than for PVME. Such results reveal how the inherent chemical structure, in combination with intermolecular effects arising from constraints imposed by the local environment, determine the response of the segmental dynamics to changes in thermodynamic variables.

Since PS is only weakly polar, dielectric relaxation probes only the dynamics of the polar component in PS blends. Results are shown in Figure 7.14 for PS/PPO and for PS/PVME [33]. The very low value of E_V/E_a for neat PPO, indicative of volume-dominated dynamics, increases with addition of PS. This diminution of the influence of volume is due to alleviation of steric constraints on segmental motion of the PPO, due to the presence of more mobile (lower T_g) PS segments. However, the opposite situation prevails when PS is added to PVME. PS has a higher T_g, so that local relaxation of PVME becomes more constrained; that is,

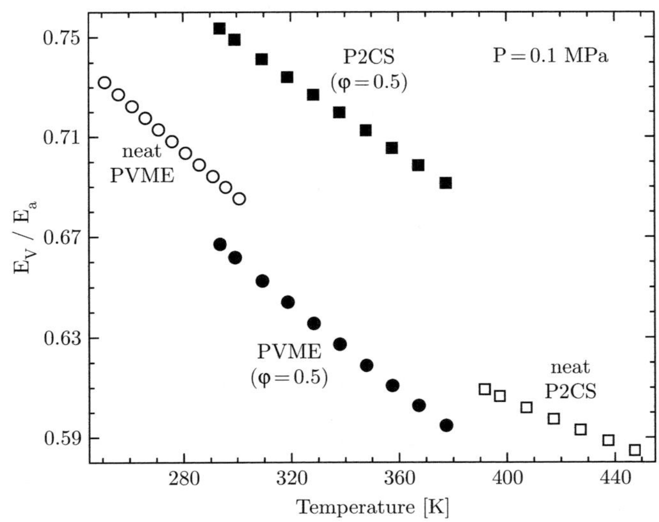

Figure 7.13 Ratio of isochoric and isobaric activation energies for PVME neat (open circles) and blended with P2CS (filled circles) and for P2CS neat (open squares) and in the same blend with PVME (filled squares). Volume exerts a stronger effect on the segmental dynamics of P2CS than of PVME for the neat polymers. In the blends the situation is reversed, with the PVME dynamics having the more volume-dominated segmental relaxation. Data from ref. [9].

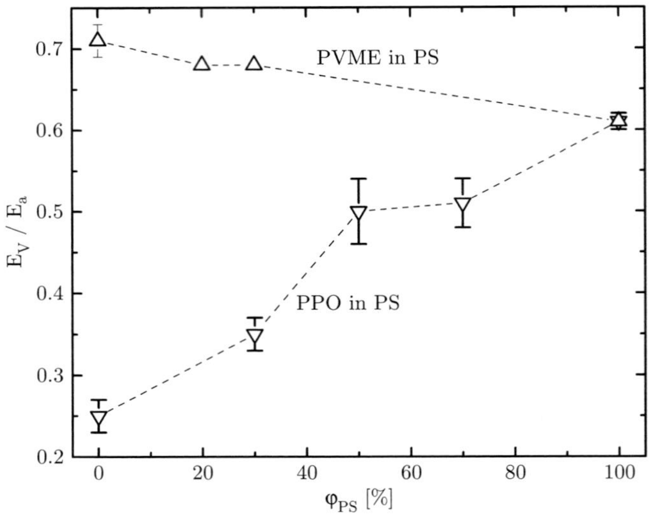

Figure 7.14 Effect of added polystyrene on the relative effect of temperature and volume on the segmental dynamics, as measured by the ratio of the isochoric to isobaric activation energies, for polyvinylmethylether (triangles) and polyphenylene oxide (inverted triangles), respectively. An increasing concentration of PS makes the α-relaxation of the lower T_g PVME more dependent on density, while PS has the opposite effect on the higher T_g PPO. Data from ref. [33].

motion of PVME segments requires greater coordination with the relatively sluggish neighbors. This strong steric coupling makes volume the more dominant control parameter.

The influence of blending cannot always be predicted from the relative T_g of the components, especially when the neat T_g values are similar. For example, the addition of PαMS (T_g = 341 K) to PCHMA (T_g = 349 K) enhances the effect volume has on the latter's segmental dynamics [28]. This is because T_g depends not only on the rate of conformational changes of a chain's backbone, but also on local steric coupling and enthalpic interactions. Both the latter change upon blending, making their effect on the neat material's T_g irrelevant.

7.1.5 Emergence of secondary peaks

Blending has a relatively small effect on the frequency of secondary relaxations because the smaller amplitude of secondary motions makes them less sensitive to the local environment. This is illustrated for PCHMA in Figure 7.15 [28]. Addition of lower T_g PαMS causes a significant change in τ_α for PCHMA. The two higher-frequency secondary relaxations of PCHMA, measurable in the glassy state, are essentially unaffected by the presence of the PαMS. Figure 7.16 [34] shows a similar result for PBD, only in this case the additive, a low molecular weight paraffin oil, has a higher T_g. Thus, segmental relaxation of the PBD slows down, but there is no corresponding change in the secondary dynamics.

The secondary relaxations in Figure 7.15 and Figure 7.16 are observed in the glassy state and thus are well removed from the segmental (α-) process. At higher temperatures the JG secondary relaxation merges with the α-peak, and, as described in Section 2.4, their different response to pressure can be used to resolve the peaks in the region of overlap near T_g. The

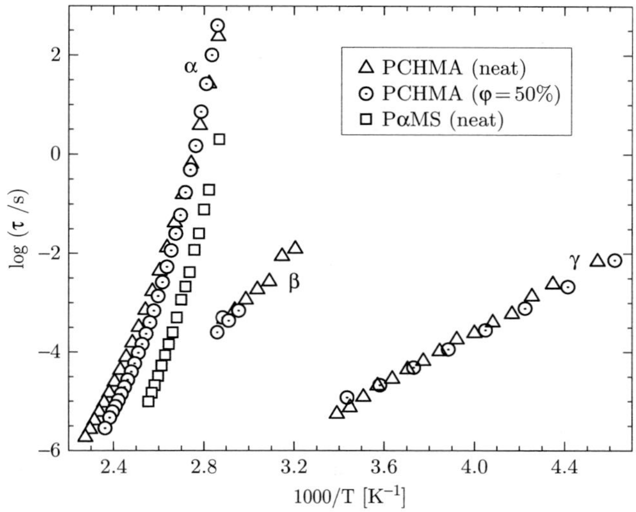

Figure 7.15 Relaxation times for local segmental relaxation and two secondary processes in polycyclohexylmethacrylate neat (triangles) and blended with an equal weight of poly-α-methylestyrene (circles). τ_α is also shown for the neat PαMS (squares) but it is too non-polar to be measured in the blend. Data from ref. [28].

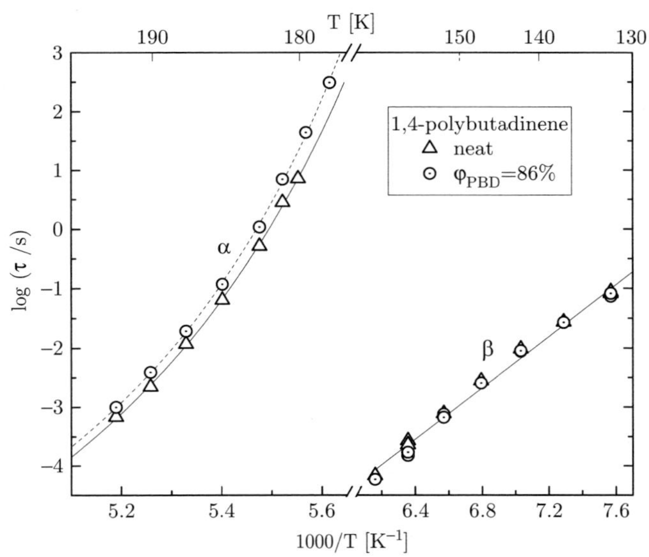

Figure 7.16 Relaxation times for local segmental motion and the JG secondary process (circles) in PBD neat (triangles) and blended with 14% paraffin oil (circles). (Note the scale break separating the higher and lower temperatures.) Data from ref. [34].

greater sensitivity of the α-process to the local environment, and hence to local composition, can likewise be utilized to deconvolute overlapping α- and β-peaks. This is shown in Figure 7.17 [35]. In the dielectric spectra of the neat molecular liquids *t*-butylpyridine (TBP) and quinaldine (QN), the primary α-peak exhibits an excess wing (extra intensity on the high-frequency side). In both cases, addition of a higher T_g, non-polar component shifts the α-peak toward lower frequency, with the secondary dispersion, being less sensitive to local composition, evolving into a distinct maximum in the spectra. Thus, by taking advantage of the relative invariance of secondary relaxations to changes in local environment, a consequence of their very local character, blending offers a means to deconvolute overlapping peaks.

7.1.6 Terminal dynamics

Since the friction factor for the chain dynamics is related to, if not identified with, the friction factor for local segmental motions (Chapter 6), the terminal relaxation of polymers is affected

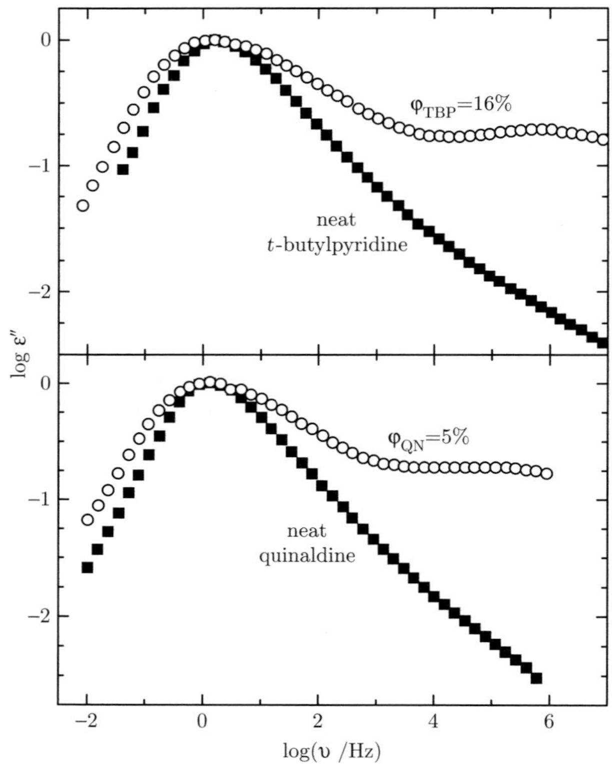

Figure 7.17 Normalized dielectric loss for (top) neat *tert*-butylpyridine at 169 K (solid squares) and its mixture with an excess of non-polar, tristyrene at 223 K (open circles); (bottom) neat quinaldine at 188 K (solid squares) and its mixture with tristyrene at 236 K (open circles). The temperatures were chosen so that τ_α was constant. The higher T_g tristyrene shifts the α-peak more than the JG-secondary relaxation shifts; thus, the excess wing in the neat spectra is resolved as a distinct maximum in the mixture. Data from ref. [35].

by miscible blending. However, because local concentration fluctuations are averaged over the spatial extent of chain molecules, local variations in friction will usually not cause distinct terminal relaxation peaks. Likewise, the plateau modulus of a blend is just the composition-weighted mean of G_N for the pure components [36]. Nevertheless, if the components have very large differences in their respective degrees of entanglement, their terminal relaxation times can be sufficiently different to yield heterogeneous terminal dynamics. Blends of PEO and PMMA in which M_w/M_e was about 15 times larger for the PEO show a breakdown of time-temperature superpositioning in the terminal zone [37], with a suggestion of two maxima in the loss modulus spectra (Figure 7.18). Similar thermorheolgoical complexity of the terminal dynamics was found when low molecular weight poly(4-*tert*-butylstyrene) was mixed with PI that had a 7 times larger value of M_w/M_e [38].

A powerful method to resolve the component dynamics is by combining experimental methods having different sensitivity to the components; examples include infrared dichroism with optical birefringence [39–41] and dielectric spectroscopy with mechanical rheometry [8]. This is illustrated by results in Figure 7.19 for the component terminal relaxations in a blend of PI and PVE [39]. Mechanical and birefringence measurements were combined, with the analysis assuming the component stresses were additive (rather than, for example, the strains; see Section 5.2.2).

7.2 Relaxation models for miscible blends

The potential for new physics in the dynamics of miscible blends has aroused considerable theoretical interest and the development of a variety of models. Aside from any intuitive appeal, the criterion for a model's success is consistency with experimental results. This is not

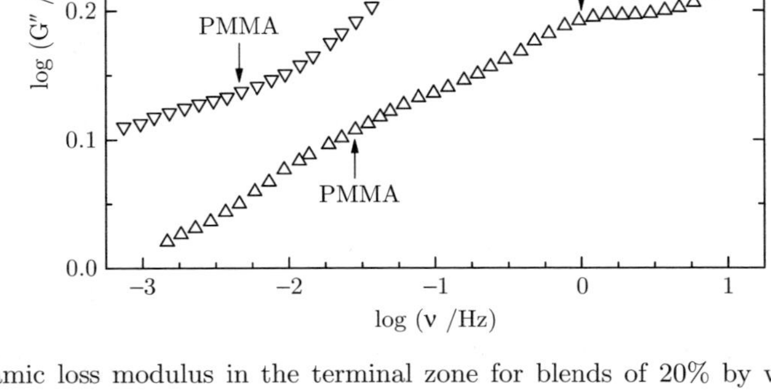

Figure 7.18 Dynamic loss modulus in the terminal zone for blends of 20% by weight PEO (M_w = 235 kg/mol; M_e = 2 kg/mol) in PMMA (M_w = 107 kg/mol; M_e = 13.6 kg/mol). The thermorheological complexity is a result of the large difference in degree of entanglement of the components. The estimated frequencies at which the contribution of each is centered are indicated by the arrows. Data from ref. [37].

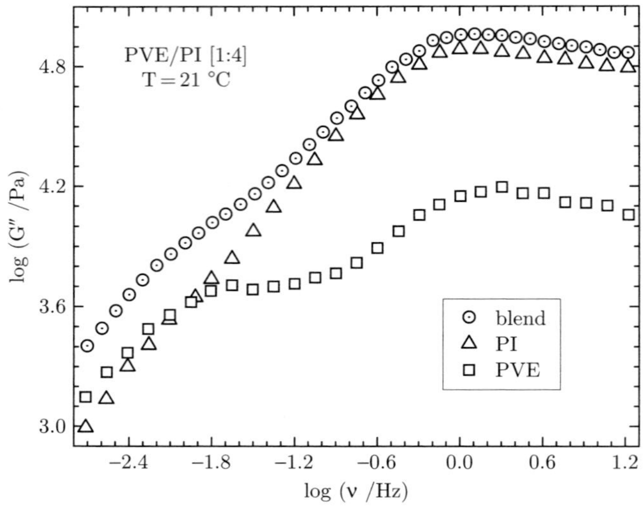

Figure 7.19 Dynamic loss modulus in the terminal zone of a 20% by weight PVE in PI (circles), along with the respective contributions from the PVE (squares) and PI (triangles) components, deduced from birefringence measurements with the assumption of stress additivity. Data from ref. [39].

straightforward, however, since some models address the common features of the component dynamics, while others attempt to explain anomalous aspects that are unique to certain blends. Since no model is completely successfully, agreement with experimental data alone cannot be used to choose among them. Moreover, if recourse is made to adjustable parameters that cannot be determined independently, a model's utility is limited to curve fitting. Some popular models for relaxation in blends are described briefly:

7.2.1 Zetsche and Fischer [42,43]

This model focuses on the effect of local composition on the glass-transition temperature and thus on the dynamics of segments comprising a given local environment. There is no assumption about different intrinsic component mobilities, but since the higher T_g component on average will be associated with higher T_g domains and vice versa for the lower T_g component, their mean relaxation times are predicted to differ, consistent with experiments. Each local volume has a concentration $\varphi_i(r)$ of the ith component, which fluctuates as governed by the mixing enthalpy. Figure 7.20 shows the computed mean square fluctuation in local composition, $\langle \delta\varphi_{PS}^2 \rangle$, for PS in blends with PVME. The magnitude of $\langle \delta\varphi_{PS}^2 \rangle$ is substantial and grows with increasing temperature, since this particular blend has a lower critical solution temperature (LCST). The problem in applying the model is lack of knowledge of the local T_g and its dependence on φ. Zetsche and Fischer [42,43] assumed that the local T_g for any subvolume is equal to the bulk T_g associated with the same macroscopic composition.

The characteristic broadening of the relaxation function towards lower frequency seen in blends at lower temperatures can be reproduced using the Zetsche–Fischer model. For example, it reproduces the main features of the segmental peak shapes in blends of PI and PVE [44]. However, predictions for the relaxation time distribution in this blend were not in quantitative agreement with results from deuterium NMR experiments [45]. This failing was

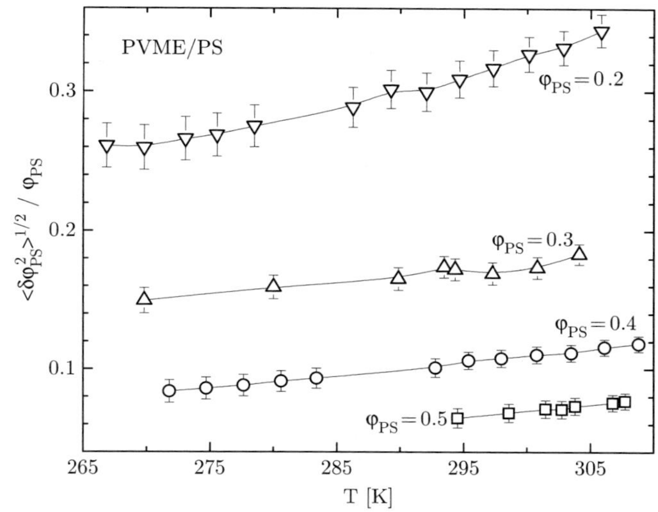

Figure 7.20 Relative root-mean-square fluctuation in the concentration of PS in blends with PVME having the indicated bulk volume fraction of PS. The concentration fluctuations were deduced from the breadth of the α-dispersion in the dielectric loss using the model of Zetsche and Fischer [42].

ascribed to the model's neglect of intrinsic mobility differences between the components; i.e. the model expects τ_α to depend only on local composition.

7.2.2 Kumar and coworkers [46,47]

Kumar and coworkers extended the Zetsche–Fischer concentration fluctuation model with subvolumes similarly defined by concentration fluctuations giving rise to a local glass temperature. This local T_g, or parameters describing $\tau_\alpha(T)$, are the weighted average of those for the pure components. The important distinction made by Kumar *et al.* is the assumption that experimental probes of the dynamics only sense composition fluctuations occurring over a certain cooperative range. The length scale for the relevant cooperative volume is not constant, but governed by the local composition. If this volume is identified with the cooperatively rearranging region of glass transition models such as that of Adam and Gibbs [48], the approach of Kumar *et al.* [47] fails to describe experimental data. Rather than growing to the point of divergence near T_g, the subvolumes are smaller and roughly independent of T and composition [49].

7.2.3 Jones and coworkers [50,51]

Jones *et al.* [50,51] developed a model based on a Flory–Huggins-type lattice. Nearest-neighbor contacts are enumerated and their influence on local dynamics assessed by assuming that the latter is described by a distribution of relaxation times, obtained by direct mapping of the compositional distribution. The model has been applied to both polymer blends [50] and to polymer/diluent mixtures [52–54]. The analysis yields information concerning local populations and their mobilities. A lattice approach is limited in its ability to capture the interactions in polymer blends, and the problem remains of how variations in composition of a local volume affect the associated relaxation time.

7.2.4 Lodge and McLeish [55–57]

According to this model, the segmental relaxation rate of a component in a binary blend is determined by the composition of a local volume whose length scale is assumed to be the Kuhn length of the polymer chain. Chain connectivity causes this local volume to be enriched in "like contacts"; that is, the average concentration of component A in one of its subvolumes is larger than the bulk concentration ϕ_A. This "self-concentration" leads to an effective local concentration of A given by the sum of the segments from a given chain and those from other A chains

$$\varphi_A^{eff} = \varphi_A^{self} + (1 - \varphi_A^{self})\varphi_A \tag{7.2}$$

The self-concentration of A, φ_A^{self}, is determined by the volume fraction of A units in a volume having a size sim l_K^3. The model has no adjustable parameters because φ_A^{self} can be calculated from the chain parameters

$$\varphi_A^{self} = \frac{C_n m_0 N}{n\rho N_A l_K^3} \tag{7.3}$$

where n/N is the number of bonds per repeat unit and m_0/ρ is the volume of a repeat unit.[1] The size of the subvolume according to this model is constant with respect to temperature, although experiments indicate that the relevant length scale increases with decreasing temperature.

Note that unless φ^{self} is the same for the two components, the model predicts distinct component T_g values. If there is a large difference between the respective self-concentrations (i.e. the components differ significantly in chain flexibility), eqn (7.2) must be modified to ensure that the composition averaged over all subvolumes equals the bulk φ_A [58]

$$\varphi_A^{eff} = \varphi_A^{self} + (1 - \varphi_A^{self})\hat{p}_A \tag{7.4}$$

where

$$\hat{p}_A = \frac{(1 - \varphi_A^{self})\varphi_A}{\varphi_A(1 - \varphi_A^{self}) + \varphi_B(1 - \varphi_B^{self})} \tag{7.5}$$

If the chains are similar so that $\varphi_A^{self} \sim \varphi_B^{self}$, then $\hat{p}_A$ is just ϕ_A.

The assumption that the blend dynamics are the cumulative result of a distribution of "local" glass-transition temperatures is common to the Zetsche–Fischer and Kumar models; the unique feature of the Lodge–McLeish model is identification of a chain property, l_k, as the length scale determining the relevant subvolume. There remains the problem that the dependence of T_g on ϕ_{eff} is unknown a priori. Assuming eqn (7.1) gives

$$T_g(\phi_A) = \left(\frac{\phi_{eff}}{T_{g,A}} + \frac{1 - \phi_{eff}}{T_{g,B}}\right)^{-1} \tag{7.6}$$

[1] $\phi^{self} = l_p/l_k$, where l_p is the packing length defined in Chapter 3.

for a blend of A and B polymers. Reflecting the limitations of the Fox equation, eqn (7.6) cannot describe a T_g that does not vary linearly with composition, such as in Figure 7.9.

Although in principle the model has no adjustable parameters, usually φ^{self} must be varied to fit experimental data, leading to larger values than calculated from eqn (7.3). Some of this discrepancy has its origin in the inadequacy of eqn (7.6). Also, the assumption that the relevant subvolume is temperature (and composition) invariant must be abandoned; the cooperative volume increases with decreasing temperature [59,60]. A more fundamental question is whether the Kuhn length ($\approx$ 1 nm) is the proper size to characterize the segmental relaxation of polymers. Although the rotational motions frozen out below T_g involve changes in orientation of a Kuhn segment, the properties of vitrifying polymers are qualitatively indistinguishable from those of molecular liquids, for which l_K has no meaning. Note also that concentration fluctuations (due to Brownian motion) are neglected in predicting the component relaxation times; for this reason the Lodge-McLeish model can only predict mean component relaxation times, without addressing the shape of the relaxation dispersion.

7.2.5 Roland and Ngai [61–64]

This model for blend dynamics uses as its starting point the coupling model (Chapter 4) to describe the segmental dynamics, with effects specific to blending treated by incorporating dynamic heterogeneity, arising from both intrinsic mobility differences between the components (reflected in their different T_g and coupling parameters) and from local compositional heterogeneity due to concentration fluctuations. The dynamics of a blend component is determined by its chemical structure (intrinsic mobility) and by its local environment, which governs the degree of intermolecular cooperativity. The component whose segmental motion is more intermolecularly cooperative will have both longer relaxation times and a broader relaxation function; thus, there is a characteristic skewing of the measured relaxation peak toward lower frequencies.

In this model, which was the first to account for dynamic heterogeneity of blends [61], concentration fluctuations inherent to miscible blends cause the degree of intermolecular cooperativity for each component to differ from one another and from that associated with the neat state. Each component has its own most probable relaxation time, reflecting its most probable local composition. The corresponding relaxation function of each component has a distribution, determined both by the nature of the intermolecular cooperativity it experiences and by the manner in which local composition alters this cooperativity. The shape of the relaxation function and the temperature dependence of the relaxation time for either component are determined by its coupling parameter, which has a distribution due to concentration fluctuations. Since the intermolecular coupling also depends on chemical structure [65,66], the blend components have different intermolecular coupling, even if present in the same local environment. This means that a component present at a trace concentration still differs in its dynamics from the host; thus, the model predicts the thermorheological complexity seen even at extremes of concentrations (Figure 7.5).

The model has been applied to mechanical and dielectric spectra for blends of PVE and PI (some of these data are in Figures 7.1 and 7.2), with the assumption that concentration fluctuations produce a Gaussian distribution of coupling parameters for each component, reflecting the influence of the local environment on the strength of the intermolecular coupling.[2] The mean values of n (Figure 7.21) were obtained by fitting the segmental relaxation peaks (*cf.* eqn (4.59)), with the usual assumption that the stresses are additive among the components. PVE has a larger coupling parameter and higher T_g than PI in their respective neat states, so that dilution with PI reduces the cooperativity of the PVE segmental relaxation; thus, n for PVE becomes smaller and the temperature dependence of its τ_α becomes weaker. Results for τ_α normalized by the component T_g are shown in Figure 7.21, where the fragility is seen to decrease with increasing concentration of the PI. The analysis of the combined mechanical and dielectric results yield $\tau_\alpha(T)$ for the components that are consistent with the valued obtained from deuterium NMR measurements on the same blend [3,10].

The peculiar NMR results in Figure 7.8 [24], showing the relaxation times for PEO in blends with PMMA that are essentially independent of composition, can be explained by this blend

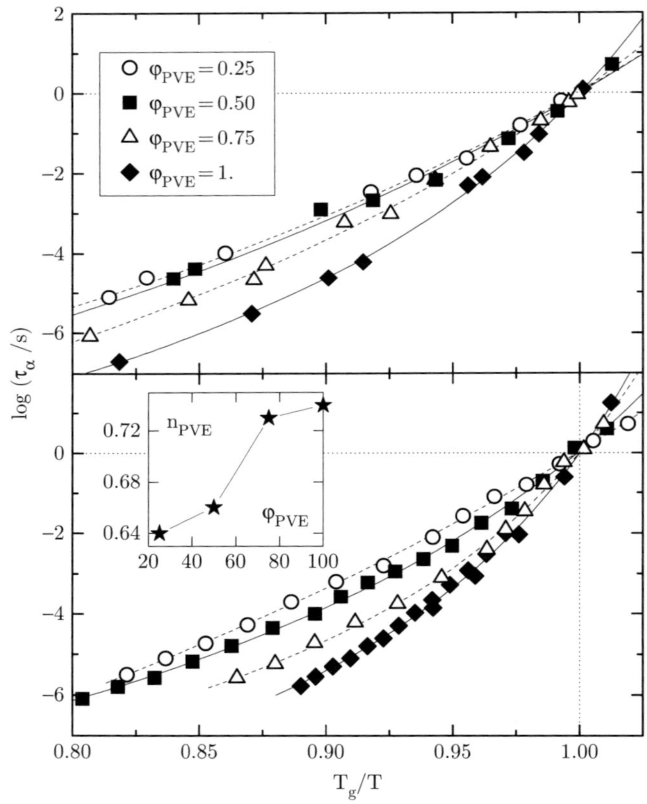

Figure 7.21 Fragility plots of the most probable segmental relaxation times of PVE in blends with PI having the indicated composition. τ_α was determined (upper) by deuterium NMR [11] and (lower) by combined mechanical and dielectric spectroscopies [8]. T_g is taken as the temperature at which $\tau_\alpha = 1$ s. The fragility decreases with increasing PI content, consistent with the value of the coupling parameter (shown in the inset) obtained by fitting the experimental loss modulus. Data from ref. [10].

[2] The Zetsche–Fischer model assumes the concentration fluctuations have a Gaussian distribution, whereas this model assumes the coupling parameters, as governed by the local composition, are Gaussian distributed.

model [67]. At lower PEO compositions, PEO chains are surrounded entirely by PMMA segments, which eliminates any effect of concentration fluctuations on the PEO dynamics. The different behavior of PEO and PMMA are due to differences in their intrinsic mobilities, which can be assessed from the magnitude of their respective non-cooperative relaxation times, τ_0. Rearranging eqn (2.18) (*cf.* eqn (4.60)) and dropping the slowing varying prefactor gives

$$\tau_\alpha/\tau_0 = [\tau_o/t_c]^{n/(1-n)} \tag{7.7}$$

In the temperature range of the NMR measurements, τ_0 for PEO is very short, ranging from 10^{-12} to 2×10^{-11} s. This is of the same magnitude as the crossover time of the coupling model, $t_c \sim 2\times10^{-12}$ s (Section 2.4). Since the ratio τ_0/t_c is not large, $[\tau_0/t_c]^{n/(1-n)}$ and thus τ_α do not vary much within the range of n engendered by the range of PMMA concentrations. If $\tau_0 = t_c$, $\tau_\alpha(n)$ becomes independent of n; that is, the PEO relaxation time becomes independent of PMMA concentration. The dynamics of PMMA remain composition dependent, since for PMMA $\tau_\alpha \gg t_c$.

The difficulty in applying this model is in quantifying the manner in which the local environment alters intermolecular cooperativity. A higher local T_g will increase the coupling, but chemical structure also exerts an effect through changes in steric interactions, etc. Nevertheless, analyses of blend dynamics based on the coupling model have been successful in describing the shape of segmental dispersions and the T dependences of τ_α for blends of PIP/PVE [8], PVME/PS [63], and TMPC/PS [64]. It has the additional virtue of making successful predictions in the limit $\varphi \approx 1$; that is, neat polymers.

7.3 Phase-separated blends

In most polymer blends the components are segregated into distinct domains having sizes of a few hundred nm and larger. Except at the interface of these phases, the relaxation properties of the components are unaltered from those of the pure materials. However, unique combinations of properties can be achieved with immiscible mixtures, and for this reason many commercial elastomers are based on such compositions. Technical issues include the effect of the phase morphology and the spatial distribution of crosslinks. Consideration here is limited to prominent features of phase-separated blends that influence their viscoelastic properties.

7.3.1 Morphology

The phases of an immiscible blend can be co-continuous, or one component can be dispersed within a continuous matrix of the other. The former is favored by equal concentrations and equal viscosities. This is seen in Figure 7.22 in data for a blend of PBD and EPDM [68]. In most blends, a dispersed particle morphology is obtained, with the continuous phase invariably the lower-viscosity phase, provided it is present at a sufficient level. This morphology arises during mechanical mixing, as domains of the lower-viscosity material deform more and thereby encapsulate phases of the other, relatively less deformable component.

During mixing of polymer blends, the dispersed phase undergoes continual breakup and coalescence (Section 3.3.2), leading to a final particle size distribution that depends both on the properties of the polymers and the severity of the mixing. Generally, the particle-size distribution has negligible effect on most bulk properties, as seen in empirical mixing rules for

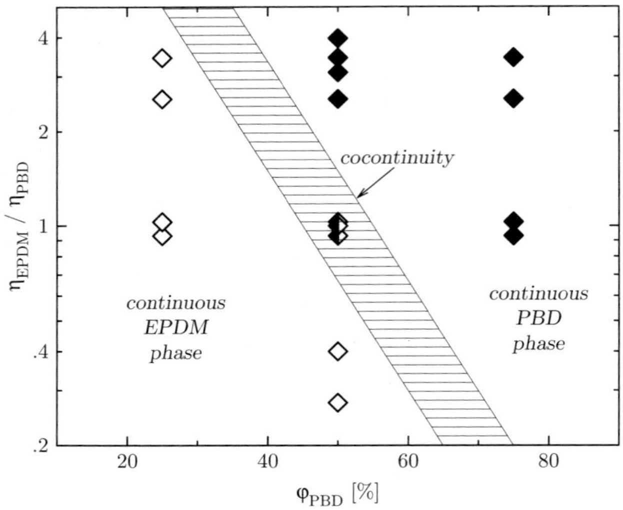

Figure 7.22 Dependence of the phase morphology on the viscosity ratio and composition for blends of EPDM and PBD. Dispersed PBD and EPDM particles respectively are indicated by open and filled symbols; half-filled symbols indicate a co-continuous morphology. Data from ref. [68].

blend properties that consider only the relative amounts of the components. Examples [69] include series

$$\eta_{AB} = \varphi_A \eta_A + \varphi_B \eta_B \tag{7.8}$$

and parallel representations

$$\eta_{AB}^{-1} = \varphi_A \eta_A^{-1} + \varphi_B \eta_B^{-1} \tag{7.9}$$

and a log-additivity rule [70]

$$\eta_{AB} = \eta_A^{\varphi_A} \eta_B^{\varphi_B} \tag{7.10}$$

The limited utility of any of these is evident from the fact that there is more than one rule, as well as various extensions utilizing adjustable parameters. Blend viscosities can exhibit complex composition dependences, as illustrated in Figure 7.23 [71] for two immiscible blends of natural rubber, which cannot be described by any mixing rule.

Similar relations have been proposed for the modulus of rubber blends, the mechanical response expected to be a superposition of that of the components. However, interesting effects of phase morphology arise when the temperature is reduced below T_g of one component. The blend modulus is no longer additive in the component moduli because of the different stiffnesses of the phases. The distribution of strain becomes very non-linear as one component undergoes its glass transition and the other remains rubbery. If the glassy phase is present as discrete particles, its influence on the overall modulus is relatively small, acting in the manner of a filler to amplify strain in the continuous phase. On the other hand, if the higher T_g component forms the continuous phase, stresses are transmitted through the glassy domains, with consequent large increase of the blend modulus. These varying behaviors are illustrated for blends of NR and PBD in Figure 7.24 [72]. Unlike the modulus, failure

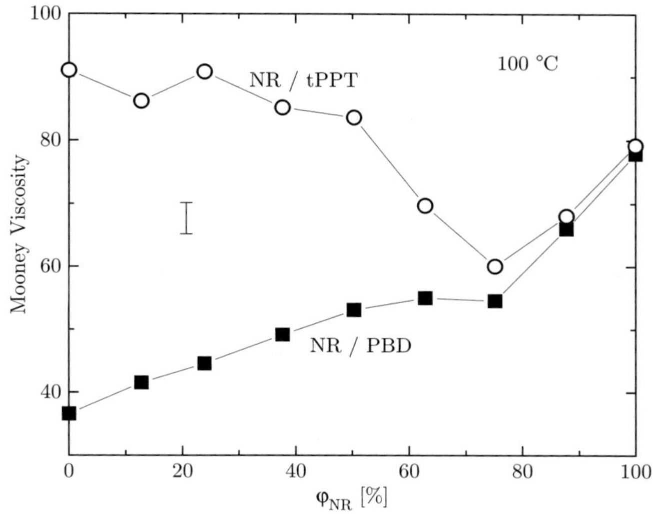

Figure 7.23 Mooney viscosity of blends of natural rubber with 1,4-polybutadiene (filled squares) and with *trans*-polypentenamer (open circles). Data from ref. [71].

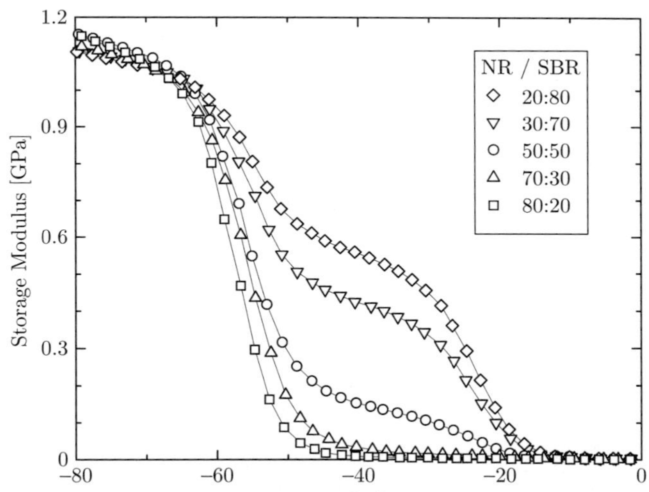

Figure 7.24 Dynamic storage modulus of blends of natural rubber and styrene–butadiene copolymer having the indicated compositions, at temperatures encompassing their respective T_g. Vitrification of the SBR has a large effect on G' only for the blends enriched in SBR. Data from ref. [72].

properties are very sensitive to the detailed structure of the dispersed domains. When the continuous phase is glassy, the dispersed particles very effectively reduce brittleness; this is the morphology of rubber-toughened plastics [73,74].

7.3.2 Surface segregation

Moving machine parts are almost always lubricated to reduce friction and wear. Lubricants can also decrease the resistance to flow of polymers during processing operations by inducing wall slip. This slippage suppresses shear flow without affecting the stretching (extensional) deformation that is most effective for dispersing[3] domains or particles. However, lubricants are degraded by mechanical stresses and elevated temperatures and thus have to be replenished. An alternative is to incorporate the lubricant into the polymer, and low molecular weight liquids (referred to as plasticizers or processing aids) are commonly used this way to facilitate polymer flow through dies, extruders, roll mills, etc. These monomeric or oligomeric additives are especially effective if they migrate to the surface because of their lower viscosity and propensity to promote interfacial slippage.

An alternative to liquid additives is to introduce a small amount of a lower-viscosity polymer. When one component of a phase-segregated blend has lower viscosity, there is a driving force during flow for its accumulation at the interface (the region of highest shear rate) [75,76]. This propensity for rearrangement of the morphology to better accommodate the applied stresses is known as the principle of minimum viscous dissipation [77]. Figure 7.25 [78] shows the viscosity measured dynamically and during continuous shear flow of an SBR containing low levels of a PDMS, the latter ten-fold less viscous. The oscillatory measurement imposes a homogeneous strain that is the same throughout the material. The PDMS remains uniformly distributed, and the small PDMS content has a negligible effect on $\eta^*(\omega)$. However, resistance to flow through a capillary die is reduced by diffusion of the PDMS to the interface with the wall (this surface enrichment confirmed by post-factum surface analysis [78]). The

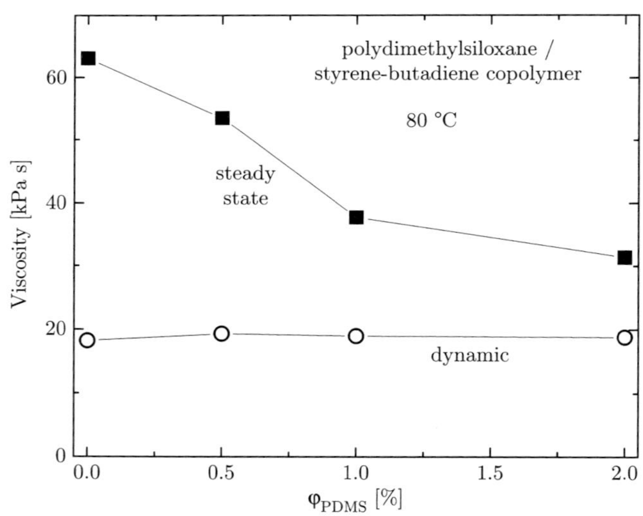

Figure 7.25 Viscosity of an SBR with 50 phr N326 carbon black as a function of added PDMS measured by oscillatory (open circles) and continuous (filled squares) shearing at a rate equal to 4.5 s^{-1}. Data from ref. [78].

[3] Dispersive mixing involves particle breakup, whereas distributive mixing refers to spatial homogenization.

velocity profile becomes very non-uniform,[4] and the "apparent" viscosity is reduced as deformation of the lower viscosity component becomes dominant. Since there remains a strong extensional component of the flow, good dispersion of carbon black in the SBR was attained, independent of the presence of the PDMS [78].

The advantage of a polymeric component as a plasticizer is that it functions specifically at the surface, with bulk properties relatively unaffected. In fact, if the components are not too chemically incompatible, the surface segregation transpires only during viscous flow. This allows the phase morphology to be altered by subsequent processing, to yield a final material with more randomly distributed components.

A variation on this method that does not rely on surface enrichment takes advantage of the incompatibility of the component polymers. During flow through a die or extruder, often contamination of the wall occurs, and a small amount of adhered polymer can cause the other component to slip due to their incompatibility. The reduction in wall adhesion can cause a dramatic reduction in the apparent viscosity, as shown for a blend of EPDM with a fluorelastomer in Figure 7.26) [79]. Note that the lower stress is a consequence of the lack of adhesion between the two polymers, not the lower viscosity of either.

7.3.3 Crosslink and filler distribution

The distribution between phases of crosslinks, plasticizers, antioxidants, fillers, etc. can have a role in the properties of the blend. Especially if the components differ in unsaturation, achieving uniformity of the crosslink density can be very difficult, due both to reactivity differences and the potential for partitioning of the curatives (i.e. their migration to the more

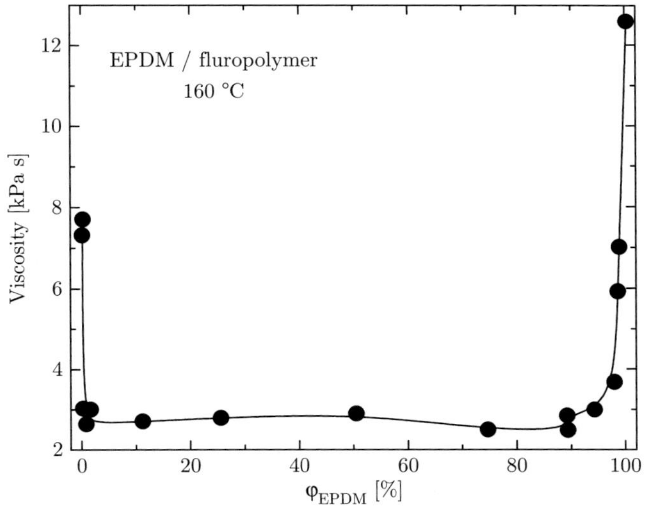

Figure 7.26 Viscosity of a blend of EPDM and a Viton® fluoroelastomer measured in a capillary rheometer at a nominal shear rate equal to 14 s^{-1}. Data from ref. [79].

[4] For homogeneous polymer melts, the velocity profile through a pipe departs from the parabolic flow curve of Newtonian fluids under stick conditions. The flow rate decreases more strongly with radial distance from the wall and filled polymers with yield stresses can even exhibit "plug flow" in the center region.

polar component). For strength properties there is an additional requirement of achieving interfacial crosslinking to adhere the domains.

As discussed in Chapter 4, crosslinking affects the segmental relaxation and T_g of a polymer. Thus, a disparity in crosslink densities for the components can affect how their respective T_g values are altered by blending. Figure 7.27 [80] shows component T_g values measured for a blend of NR and NBR as a function of their respective crosslink densities. The latter were determined from the breadth of NMR lines (which reflects the increasing anisotropy of the motions with increased crosslinking [81]), with the components resolved from the differences in their chemical shifts.

The affinity for fillers varies among polymers, which can lead to a non-uniform distribution in blends. For example, carbon black preferentially resides in the more unsaturated or polar component. The consequent viscosity difference of the phases in turn affects the blend morphology (e.g., Figure 7.22). Since filler particles generally do not migrate between phases, a non-uniform distribution can be avoided by incorporating the filler prior to mixing the polymers. In fact, this can be done with a single material, to obtain different phases of the same polymer. Since certain properties (elasticity and hysteresis) depend non-linearly on the filler content, the behavior for a non-uniform distribution can differ from that for filler that is equally partitioned between components. Studies in which carbon black is selectively incorporated into one or the other of the components show systematic changes in mechanical hysteresis and elastic properties such as die swell and rebound [82,83]; however, bulk properties such as the viscosity are affected only by the amount of filler, not by its phase distribution.

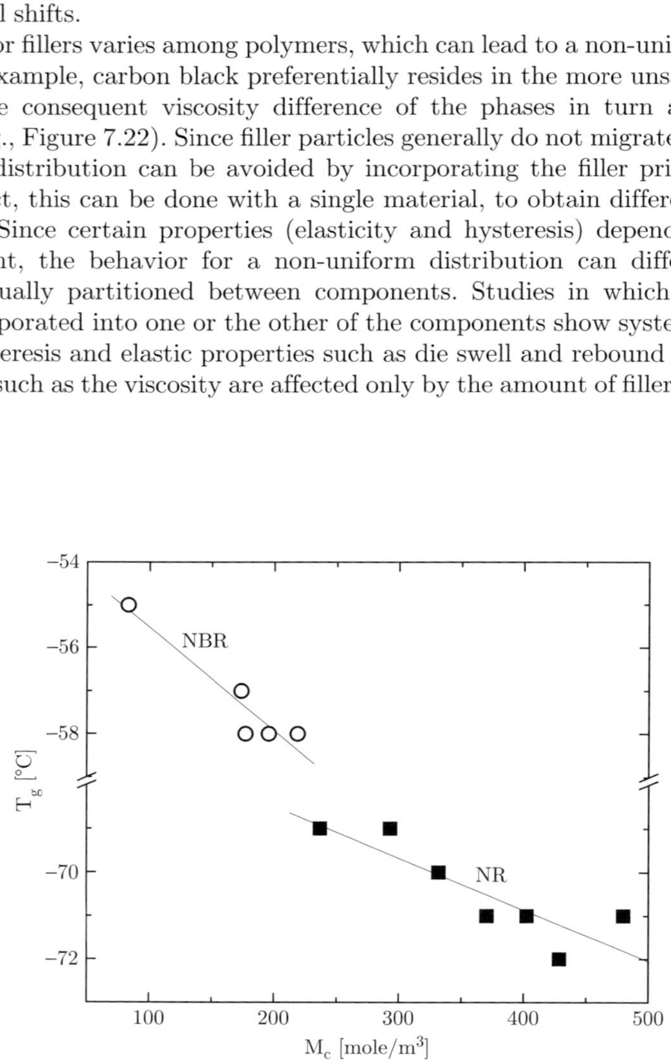

Figure 7.27 Glass-transition temperature of each component in blends of natural and nitrile rubbers as a function of the number of repeat units between crosslink junctions. The cure system, tetramethylthiuram monosulfide with either *bis*-alkylphenol disulfide or free sulfur, preferentially resides in the more polar NBR phase, causing the latter to be two to five times more crosslinked. Data from ref. [80].

References

1. C.M. Kuo and S.J. Clarson, Polymer **41**, 5993 (2000).
2. D.J. Lohse, R.T. Garner, and W.W. Graessley, *Rubber Chem. Technol.* **72**, 569 (1999).
3. G.C. Reichart, W.W. Graessley, R.A. Register, and D.J. Lohse, *Macromolecules* **31**, 7886 (1998).
4. B. Crist and M. J. Hill, *J. Polym. Sci., Polym. Phys. Ed.* **35**, 2329 (1997).
5. T.R. Maier, A.M. Jamieson, and R. Simha, *J. Appl. Polym. Sci.* **51**, 1053 (1994).
6. S. Sakurai, T. Izumitani, H. Hasegawa, T. Hashimoto, and C.C. Han, *Macromolecules* **24**, 4844 (1991).
7. J.B. Miller, K.J. McGrath, C.M. Roland, C.A. Trask, and A.N. Garroway, *Macromolecules* **23**, 4543 (1990).
8. A. Alegria, J. Colmenero, K.L. Ngai, and C.M. Roland, *Macromolecules* **27**, 4486 (1994).
9. C.M. Roland, K.J. McGrath, and R. Casalini, *Macromolecules* **39**, 3581 (2006).
10. K.L. Ngai and C.M. Roland, *Macromolecules* **28**, 4033 (1995).
11. G.C. Chung, J.A. Kornfield, and S.D. Smith, *Macromolecules* **27**, 5729 (1994).
12. C.M. Roland and R. Casalini, *J. Non-Cryst. Solids* **353**, 3996 (2007).
13. J.C. Haley and T.P. Lodge, *Coll. Polym. Sci.* **282**, 793 (2004).
14. T.G. Fox, *Bull. Am. Phys. Soc.* **1**, 123 (1956).
15. T.K. Kwei and H.L. Frisch, *Macromolecules* **11**, 1267 (1978).
16. M.A. de Araujo, R. Stadler, and H.J. Cantow, *Polymer* **29**, 2235 (1988).
17. P.R. Couchman and F.E. Karasz, *Macromolecules* **11**, 177 (1978).
18. M. Gordon and J.S. Taylor, *J. Appl. Chem.* **2**, 493 (1962).
19. E.A. Di Marzio, *Polymer* **31**, 2294 (1990).
20. T.K. Kwei, E.M. Pearce, J.R. Pennacchia, and M. Charton, *Macromolecules* **20**, 1174 (1987).
21. M.J. Brekner, H.A. Schneider, and H.J. Cantow, *Polymer* **29**, 78 (1988).
22. E. Leroy, A. Alegria, and J. Colmenero, *Macromolecules* **35**, 5587 (2002).
23. R. Casalini, P.G. Santangelo, and C.M. Roland, *J. Chem. Phys.* **117**, 4585 (2002).
24. T.R. Lutz, Y. He, M.D. Ediger, H. Cao, G. Lin, and A.A. Jones, *Macromolecules* **36**, 1724 (2003).
25. T.P. Lodge, E.R. Wood, and J.C. Haley, *J. Polym. Sci. Polym. Phys. Ed.* **44**, 756 (2006).
26. C.M. Roland, P.G. Santangelo, Z. Baram, and J. Runt, *Macromolecules* **27**, 5382 (1994).
27. A. Alegria, C. Elizetxea, I. Cendoya, and J. Colmenero, *Macromolecules* **28**, 8819 (1995).
28. C.M. Roland and R. Casalini, *Macromolecules* **40**, 3631 (2007).
29. C.M. Roland, *Macromolecules* **28**, 3463 (1995).
30. C.M. Roland, P.G. Santangelo, K.L. Ngai, and G. Meier, *Macromolecules* **26**, 6164 (1993).
31. C. Lorthioir, A. Alegria, and J. Colmenero, *Phys. Rev. E* **68**, 031805 (2003).
32. K.L. Ngai and Capaccioli, *J. Phys. Condens. Matter* **19**, 205114 (2007).
33. C.M. Roland and R. Casalini, *Macromolecules* **38**, 8729 (2005).
34. R. Casalini, K.L. Ngai, C.G. Robertson, and C.M. Roland, *J. Poly. Sci. Poly. Phys. Ed.* **38**, 1841 (2000).
35. D. Prevosto, K. Kessair, S. Capaccioli, M. Lucchesi, and P.A. Rolla, *Philos. Mag.* **87**, 643 (2007).
36. C.M. Roland, *J. Polym. Sci. Pt. B-Polym. Phys.* **26**, 839 (1988).

37. R.H. Colby, *Polymer* **30**, 1275 (1989).
38. Q. Chen, Y. Matsumiya, Y. Masubuchi, H. Watanabe, and T. Inoue, *Macromolecules* **41**, 8694 (2008).
39. K. Osaki, E. Takatori, M. Ueda, M. Kurata, T. Kotaka, and H. Ohnuma, *Macromolecules* **22**, 2457 (1989).
40. Y. Zhao, B. Jasse, and L. Monnerie, *Polymer* **30**, 1643 (1989).
41. J.A. Zawada, G.G. Fuller, R.H. Colby, L.J. Fetters, and J. Roovers, *Macromolecules* **27**, 6851 (1994); *ibid.* **27**, 6861 (1994).
42. A. Zetsche and E.W. Fischer, *Acta Polym.* **45**, 168 (1994).
43. G. Katana, E.W. Fischer, Th. Hack, V. Abetz, and F. Kremer, *Macromolecules* **28**, 2714 (1995).
44. F. Alvarez, A. Alegria, and Colmenero, *Macromolecules* **30**, 597 (1997).
45. G.C. Chung, J.A. Kornfield, and S.D. Smith, *Macromolecules* **27**, 964 (1994).
46. S.K. Kumar, R.H. Colby, S.H. Anastasiadis, and G. Fytas, *J. Chem. Phys.* **105**, 3777 (1996).
47. S. Kamath, R.H. Colby, S.K. Kumar, K. Karatasos, G. Floudas, G. Fytas, and J.E.L. Roovers, *J. Chem. Phys.* **111**, 6121 (1999).
48. G. Adam and J.H. Gibbs, *J. Chem. Phys.* **43**, 139 (1965).
49. S.K. Kumar, S. Shenogin, and R.H. Colby, *Macromolecules* **40**, 5759 (2007).
50. A.A. Jones, P.T. Inglefield, Y. Liu, A.K. Roy, and B.J. Cauley, *J. Non-Cryst. Solids* **131–133**, 556 (1991).
51. Y.H. Chin, C. Zhang, P. Wang, P.T Inglefield, A.A. Jones, R.P. Kambour, J.T. Bendler, and D.M. White, *Macromoleucles* **25**, 3031 (1992).
52. C. Zhang, P. Wang, A.A. Jones, and P.T Inglefield, *Macromolecules* **24**, 338 (1991).
53. B.J. Cauley, C. Cipriana, K. Ellis, A.K. Roy, A.A. Jones, P.T. Inglefield, B.J. McKinley, and R.P. Kambour, *Macromolecules* **24**, 403 (1991).
54. Y. Liu, A.K. Roy, A.A. Jones, P.T. Inglefield, and P. Ogden, *Macromolecules* **23**, 968 (1990).
55. T. Lodge and T.C.B. McLeish, *Macromolecules* **33**, 5278 (2000).
56. J.C. Haley, T.P. Lodge, Y. He, M.D. Ediger, E.D. von Meerwall, and J. Mijovic, *Macromolecules* **36**, 6142 (2003).
57. Y. He, T.R. Lutz, M.D. Ediger, and T.P. Lodge, *Macromolecules* **36**, 9170 (2003).
58. J.E.G. Lipson, and S.T. Milner, *J. Polym. Sci. Polym. Phys. Ed.* **44**, 3528 (2006).
59. R. Kant, S.K. Kumar, and R.H. Colby, *Macromolecules* **36**, 10087 (2003).
60. E. Krygier, G. Lin, J. Mendes, G. Mukandela, D. Azar, A.A. Jones, J.A. Pathak, R.H. Colby, S.K. Kumar, G. Floudas, R. Krishnamoorti, and R. Faust, *Macromolecules* **38**, 7721 (2005).
61. C.M. Roland and K.L. Ngai, *Macromolecules* **24**, 2261 (1991).
62. C.M. Roland and K.L. Ngai, *J. Rheology* **36**, 1691 (1992).
63. C.M. Roland and K.L. Ngai, *Macromolecules* **25**, 363 (1992); **33**, 3184 (2000).
64. C.M. Roland, K.L. Ngai, J.M. O'Reilly, and J.S. Sedita, *Macromolecules* **25**, 3906 (1992).
65. C.M. Roland and K.L. Ngai, *Macromolecules* **24**, 5315 (1991).
66. K.L. Ngai and C.M. Roland, *Macromolecules* **26**, 6824 (1993).
67. K.L. Ngai and C.M. Roland, *Macromolecules* **37**, 2817 (2004).
68. G.N. Avgeropoulos, F.C. Weissert, P.H. Biddison, and G.G.A. Bohm, *Rubber Chem. Technol.* **49**, 93 (1976).

69. L.E. Nielsen, *Predicting the Properties of Mixtures*, Marcel Dekker, New York (1978).

70. L.A. Utracki and M.R. Kamal, *Polym. Eng. Sci.* **22**, 96 (1982).

71. V.L. Folt and R.W. Smith, *Rubber Chem. Technol.* **46**, 1193 (1973).

72. M. Kluppel, R.H. Schuster, and J. Schaper, *Rubber Chem. Technol.* **72**, 91 (1999).

73. *Rubber Toughened Plastics* (C.K. Riew, ed.), Adv. Chem. Ser., Amer. Chem. Soc. Washington, DC, Vol. 222 (1989).

74. *Rubber Toughened Engineering Plastics* (A.A. Collyer, ed.), Chapman and Hall, London (1994).

75. J.L. White, R.C. Ufford, K.R. Dharod, and R.L. Price, *J. Appl. Polym. Sci.* **16**, 1313 (1972).

76. A.E. Everage, *Trans. Soc. Rheol.* **17**, 629 (1973).

77. D.L. MacLean, *Trans. Soc. Rheol.* **17**, 385 (1973).

78. C.M. Roland and M. Nguyen, *J. Appl. Polym. Sci.* **35**, 2141 (1988).

79. C.K. Shih, *Polym. Eng. Sci.* **16**, 742 (1976).

80. A.J. Tinker, *Rubber Chem. Technol.* **63**, 503 (1990).

81. M.J.R. Loadman and A.J. Tinker, *Rubber Chem. Technol.* **62**, 234 (1989).

82. B.L. Lee, *Polym. Eng. Sci.* **21**, 294 (1981).

83. W.M. Hess, *Rubber Chem. Technol.* **64**, 386 (1991).

8
Liquid-crystalline materials

A liquid crystal is a compound that exhibits simultaneously both solid and fluid properties. This unusual behavior confers unique optical and electrical properties exploited in applications, most notably the multi-billion dollar liquid-crystal display (LCD) industry. This and other areas of engineering and device technology exploit the ability to induce phase changes in liquid crystals with thermal, mechanical, electric, or magnetic fields. Liquid crystallinity [1,2] involves arrest of the orientational, and to a lesser extent positional, degrees of freedom, yielding a fluid with molecular order. This is in contrast to glasses, the molecules of which cannot rotate or translate but are disordered. Generally, liquid crystals are organic molecules that self-assemble; however, liquid crystals lack the degree of three-dimensional positional order of crystalline solids, and therefore have some mobility, in the manner of isotropic liquids. The propensity to form an ordered phase is due to an anisotropic intermolecular potential, usually resulting from an anisotropic molecular shape, or in the case of polymers composed of rigid units, to maximize packing of the highly extended chains.

For rigid-rod polymers the ordered state may be the only stable one below the decomposition temperature, so their processing is often done in solution (lyotropic liquid crystallinity, in which the phase behavior depends on concentration). Rigid-rod polymers such as poly-*p*-phenyleneterphthalamide (e.g. Kevlar®), poly-*p*-phenylene-*cis*-benzobisoxazole (e.g. Zylon®), and aromatic ester copolymers (Vectran®) are used in the form of fibers or, less often, films. The high chain orientation in combination with lateral connectivity (for example through hydrogen bonding) confers exceptional stiffness and strength. The mechanical properties can be limited by defects resulting from frustration in the ordering due to the limited mobility of large molecules. However, the modulus is much higher than for flexible chains because the conformational changes governing the elasticity of flexible chains are unavailable. There have been many efforts to obtain the mechanical properties of rigid-rod polymers by "extended-chain crystallization" of flexible-chain polymers [3,4]. The idea is that prior to crystallization, an ordered "nematic" state is induced [5,6] by stretching the chains to primarily an all *trans* conformation; this is achieved by carrying out the deformation at high strain rates (to avoid back-folding) [7,8], high temperatures (for which the gauche rotamers necessary for back-folding are less stable) [9], low temperatures [10] (at which only highly extended chains have sufficient crystallization rates [11]), or high pressures [12,13] (to exploit the better packing of extended chains).

Liquid-crystal-forming moieties can be grafted onto or copolymerized with flexible chains to form a liquid crystal elastomer (LCE). LCE are viscoelastic, with the coupling of rubber elasticity and liquid crystallinity leading to unique properties; for example, the phase diagram is affected by mechanical strain and spontaneous shape changes can occur in single-domain LCE when switched between the ordered and isotropic phases. The time constants for LCE are usually long, hindering their application, in contrast molecular liquid crystals used in

LCD. Nevertheless, the use of LCE as reversible shape-memory materials for actuators has attracted significant interest [14–18].

Liquid crystallinity was first discovered[1] in derivatives of cholesterol, a lipid[2] found in blood. Many biological systems exist as liquid-crystalline materials, and there is a natural connection, in that the collective orientational ordering that defines liquid crystallinity is central to the functioning of cell membranes and muscles, embryonic development, and other biological processes [19]. Biological self-assembly usually involves lyotropic liquid crystals, which have hydrophobic and hydrophilic groups at different locations. Examples of lyotropic phases include cell membranes, enveloped viruses, and DNA. Ordering of lyotropics is governed by their concentration in a solvent, whereas thermotropic liquid crystallinity is driven primarily by temperature.

8.1 Liquid crystals

Thermotropic liquid crystals are either rod-like (calamitic) or disk-like (discotic). At least in principle, there are 18 possible mesomorphic[3] arrangements between the extremes of crystalline molecular registry and the isotropic amorphous state [20,21]. Figure 8.1 shows several of the phases of rod-shaped liquid crystals, which can be liquid-like or solid-like, depending on the degree of positional order. Discotic liquid crystals [22] arrange into two main structures: In columnar phases, the discs are stacked to form two-dimensional arrays of columns; the columns may be tilted rather than strictly parallel. In nematic phases the disks are mutually oriented but without translational order. There can be a systematic twist of the molecules to give rise to a cholesteric phase.

8.1.1 Phase transitions in thermotropic liquid crystals

Temperature enhances molecular disorder, so higher temperature mesophases are closer to the liquid state. The phase transitions of liquid crystals are accompanied by discontinuities in the specific heat and volume; these changes are generally small compared to the latent heat and volume expansion associated with the melting of crystals. At low pressures the phase transition temperature, T_c, increases linearly with pressure. Some pressure coefficients of T_c for different liquid crystals are listed in Table 8.1; generally these are somewhat higher than $\mathrm{d}T_g/\mathrm{d}P$ (Chapter 2). The effect of pressure on the phase transitions in 4-(4'-hexylcyclohexyl)-isothiocyanato-benzene is shown in plots of volume versus temperature in Figure 8.2 [23]. The temperatures for the nematic to isotropic transition from these data are the same as measured by calorimetry (Figure 8.3). The equivalence for the different experiments is due to the relative insensitivity of the clearing transition[4] to supercooling for many nematics. This differs from the behavior of transitions involving phases with significant positional order, which are

[1] The Austrian biologist F. Reinitzer is generally credited with the discovery of liquid crystallinity. In 1888 he reported two melting points and birefringence in esters of cholesterol, properties resulting from their ordered, helical structure.

[2] Lipids are water-insoluble organic compounds that comprise along with carbohydrates and proteins the main structural material of cells.

[3] Forms of a liquid crystal are called mesophases because they are intermediate between the crystalline and liquid states (Greek *mesos*, middle). The structural units causing the liquid-crystalline ordering are called mesogens.

[4] The temperature of the transition to the isotropic liquid is referred to as the clearing temperature because it's associated with a loss of opacity and birefringence.

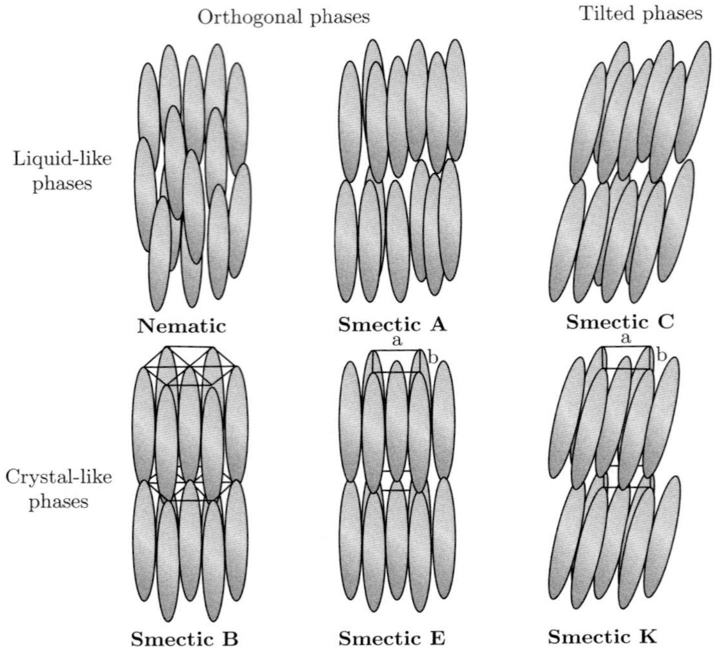

Figure 8.1 Liquid-crystal phases of rod-shaped molecules, all possessing long-range orientational order. Nematic molecules have randomly distributed centers and three degrees of translational freedom; smectic phases are layered (and tilted in the smectic C and K phases) and one less degree of translational freedom. The lack of positional disorder in the upper structures permits a degree of liquid-like mobility that is lacking in the crystal-like smectic phases. Molecules of the latter are regularly spaced within a layer, giving some long-range positional order; however, between layers this ordering is weak compared to a crystalline solid. (*Diagram courtesy of Joanna Czub of the AGH University of Science and Technology*).

more strongly affected by thermal history (Figure 8.4 [24]). (Because of the problem with supercooling, temperature calibration of instruments using crystalline materials must be based on melting points, not crystallization temperatures. Many liquid crystals, however, can be used for temperature calibration during cooling and heating [25,26].) The range of the nematic–isotropic transition is fairly narrow, usually less than one degree, whereas transitions involving smectic and other more ordered phases can be a few degrees in breadth.

There are various theories of the thermodynamics of liquid crystallinity, but these address the nematic–isotropic transition, involving only changes in molecular orientations. Other transitions, such as the nematic-to-smectic, which require rearranging the position of molecular centers, are too complex for a realistic model. The Onsager [27] and Flory [28] theories consider only steric effects (repulsive interactions), with a phase transition reflecting the competition between combinatorial entropy (rods pack more efficiently when aligned) and orientational entropy (which is reduced by alignment). The mean field theory of Maier and Saupe [29] assumes stability of the nematic phase is due to anisotropic dispersion forces. The orientational energy of a molecule in an average field generated by its neighbours is

Table 8.1 Relaxation parameters for liquid crystals in the ordered state. [51]

Liquid crystal	Phase	$\frac{T_c}{\text{(K)}}$	$\frac{\mathrm{d}T_c/\mathrm{d}P}{\text{(K/GPa)}}$	E_V/E_a	γ	Γ
5CB	nematic	308.3	424	0.61	4.1	5.3
6CB	nematic	301.2	390	0.44	4.1	6.3
7CB	nematic	314.6	370	0.47	3.3	4.7
8CB	nematic/smectic A	313.8	370	0.53/0.57	4.2	4.0
5PCH	nematic	328.1	440	0.57	3.5	5.2
7PCH	nematic	331.0	420	0.50	3.9	3.3
8PCH	nematic	328.3	412	0.51	3.6	3.4
6CHBT	nematic	316.7	419	0.52	5.0	5.0
6DBT	smectic A	350.1	259	0.40	4.0	2.9
6OPB8	smectic A/smectic C	339.2	234	0.66/0.66	2.7	2.5
5BT	smectic E	347.1	284	0.72	2.3	2.2
6BT	smectic E	344.7	285	0.57	3.1	2.4
7BT	smectic E	345.6	241	0.71	2.3	2.2
8BT	smectic E	341.1	237	0.45	4.1	2.7

*n*CB: alkyl-cyanobiphenyl
*n*PCH: *trans*-4-*n*-alkyl-(4'-cyanophenyl)hexane
6CHBT: 4-(*trans*-4'-hexylcyclohexyl)isothiocyanatobenzene;
6DBT: 5-*trans*-hexyl-2-(4'-isothiocyanatophenyl)-1,3-dioxane
6OPB8: 2-(4-hexyloxyphenyl)-5-octyl-pyrimidine
*n*BT: alkyl-isothiocyanato-biphenyls
Note: *n* is number of carbons in alkyl substituent.

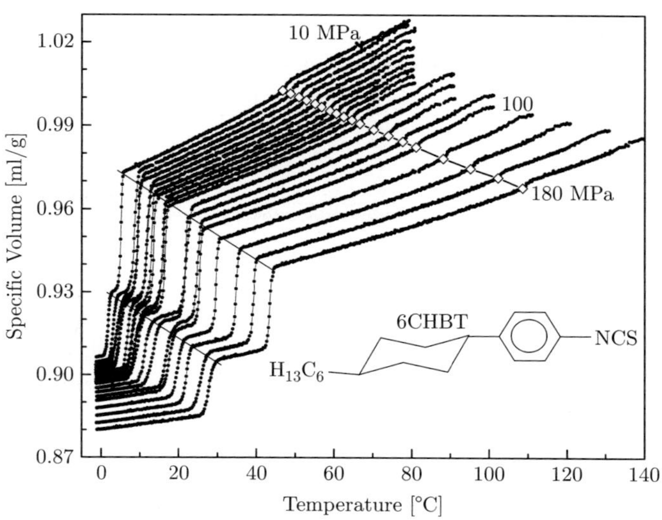

Figure 8.2 Specific volume *versus* temperature for 4-(4'-hexylcyclohexyl)-isothiocyanato-benzene at pressures from 10 MPa to 50 MPa in steps of 5 MPa, from 50 MPa to 100 MPa in steps of 10 MPa, and from 100 MPa to 180 MPa in steps of 20 MPa. Abrupt volume changes signify transitions between the various phases, which consist of two crystalline forms, the nematic state, and the isotropic liquid. The clearing line ($T_c(P) > 40$ °C) is denoted by the diamond symbols. Data from ref. [23].

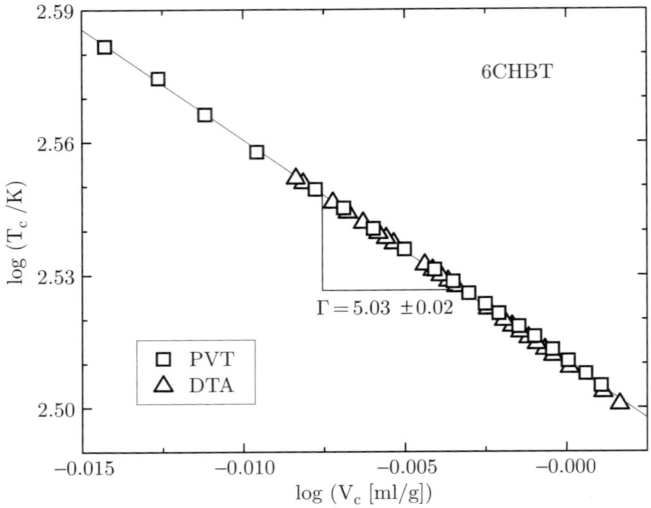

Figure 8.3 Double-logarithmic plot of the clearing temperature *versus* volume at T_c for 6CHBT from measurements by PVT (squares) and differential thermal analysis (triangles) at pressures over the range from atmospheric to 180 MPa. The negative slope yields the indicated value for the thermodynamic ratio. Data from ref. [23].

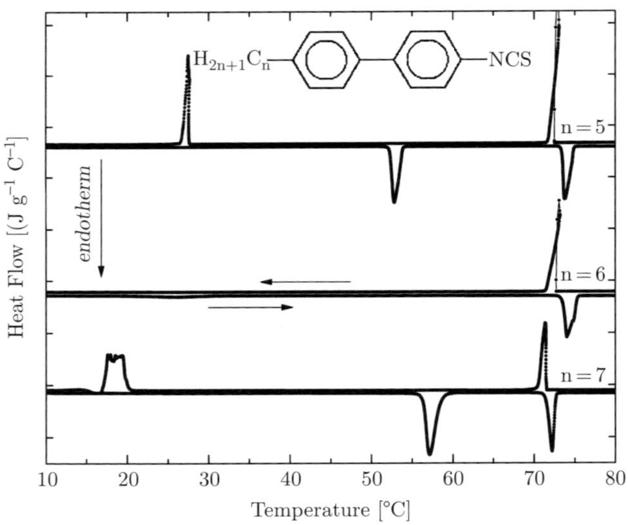

Figure 8.4 DSC curves for homologous series of alkyl-isothiocyanato-biphenyls (data displaced on the ordinate scale for clarity with vertical tic marks every 40 J/g C). The higher temperature transition is the smectic E–isotropic liquid phase change and the lower is the crystal–smectic E transition (the latter is below 0 °C for $n = 6$ and thus not shown). The strong exotherm during cooling through the liquid to smectic liquid transition for the pentyl- and hexyl-substituted compounds causes a brief loss of temperature control, evident in the peaks. Data from ref. [24].

$$U_i(\theta_i) = -\hat{A}f_2(\theta)\left(\frac{3\cos^2\theta_i - 1}{2}\right)V^{-2} \tag{8.1}$$

where $\hat{A}$ is a material constant, $f_2(\theta)$ the order parameter defined in eqn (5.18), and θ is the angle between the molecular long axis and the axis of the mesophase (the director). It follows from eqn (8.1) that the temperature of the nematic–isotropic transition, T_c, is given by

$$T_c = \hat{A}/(4.541 k_B V_c^2) \tag{8.2}$$

where V_c is the specific volume at the transition. From eqn (8.2) it follows that the product $T_c V_c^2$ should be constant along the pressure-dependent phase boundary; however, high-pressure experimental data for nematics deviate from this prediction of the Maier–Saupe model. The obvious shortcoming is neglect of the repulsive part of the interaction potential [30–34]. By introducing a repulsive term, the calculated change in the clearing temperature with pressure becomes closer to experimental values [35]. The details of the interactions are not crucial for the prediction of phase boundaries; the main requirement is an anisotropic potential having a reasonable dependence on molecular orientation [30,36].

The presence of anisotropic repulsive interactions makes the potential depend more strongly on intermolecular distances than predicted by Maier and Saupe. This can be accounted for by replacing the quadratic volume dependence in eqn (8.2) with the general form

$$T_c(P) \sim V_c^{-\Gamma}(P) \tag{8.3}$$

The parameter Γ, known as the thermodynamic ratio, appears in the orientation-dependent part of eqn (8.1), $U_i \propto V^{-\Gamma}$. Γ is a measure of the strength of the steric repulsions relative to the attractive part of the potential. Chandrasekhar and Madhusudana [37] modified the Maier–Saupe model to obtain $\Gamma = 3$, but more usefully it is determined from experimental data as

$$\Gamma = -\left(\frac{d\log T_c}{d\log V_c}\right)_P \tag{8.4}$$

This definition implies the product $T_c V_c^{\Gamma}$ is constant (independent of pressure) at a phase transition. Experimentally Γ can be determined from PVT measurements or calorimetry (Figure 8.3 [23]), with values found to lie in the range 2 to 8 [38–40], depending on the molecular structure [41–44] (Table 8.1). Although the Maier–Saupe theory [29] only considered the nematic–isotropic transition, this analysis has been extended to other liquid crystal phase changes, with the obtained Γ smaller than for the clearing line [39,41,43, 45–47].

8.1.2 Molecular motions in liquid crystals

Notwithstanding their long-range order, liquid crystals exhibit fluidity. For nematic phases, longitudinal correlation of the molecules is maintained during translational diffusion. Smectic phases are more viscous; the molecules move randomly within the layers, but their long axes remain perpendicular to the layer and some positional order is retained. The forces between smectic layers are usually weak enough to also allow layers to slide past one another.

Molecules in both nematic and smectic phases reorient, and these rotations are generally much faster than translational motions in liquid crystals. Rotations about the long axes (i.e. lower inertial moment) are fast and relatively insensitive to phase changes [48,49]. Longitudinal rotations ("flip-flop" motion about the short axis) have large activation barriers [50,51] and, since the distance between the molecules is typically less than half their length [52,53], the motion is coupled to fluctuations in the lateral position of neighboring molecules. This coupling slows these reorientations, with time constants in the mesogenic phase typically 2 to 20 times larger than the corresponding τ_a in the isotropic phase (the ratio of these rotational relaxation times is called the retardation factor). Data are shown for a homologous series of liquid crystals in Figure 8.5 [54]. In this example the clearing temperature increases strongly with size of the alkyl side group (more stable ordered state), with consequently enormous increases (8 orders of magnitude) in the mobility at T_c. Notwithstanding the cooperative aspect of longitudinal reorientational motions, the relaxation peaks of liquid crystals are almost always Debye (Figure 8.6), with a temperature dependence that is usually Arrhenius (Figure 8.7).

Since hydrostatic pressure reduces intermolecular distances, it retards the dynamics of liquid crystals. In the manner described for vitrifying liquids and polymers in Chapter 2, the

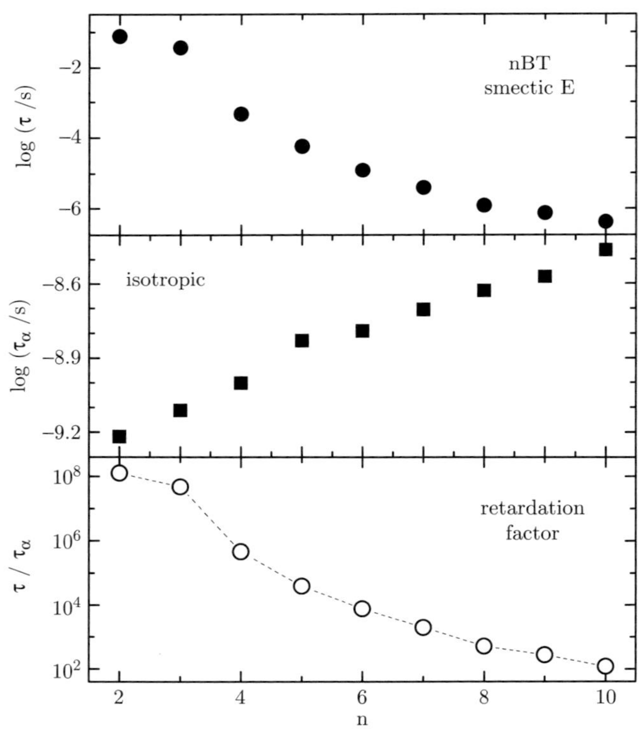

Figure 8.5 Relaxation times for alkyl-isothiocyanato-biphenyls (n represents the number of alkyl carbons) at the clearing point (top) in the smectic E phase and (middle) in the isotropic liquid. The corresponding retardation factor is plotted in the lower panel. (*Data courtesy of Stanislaw Urban of Jagiellonian University*) [54].

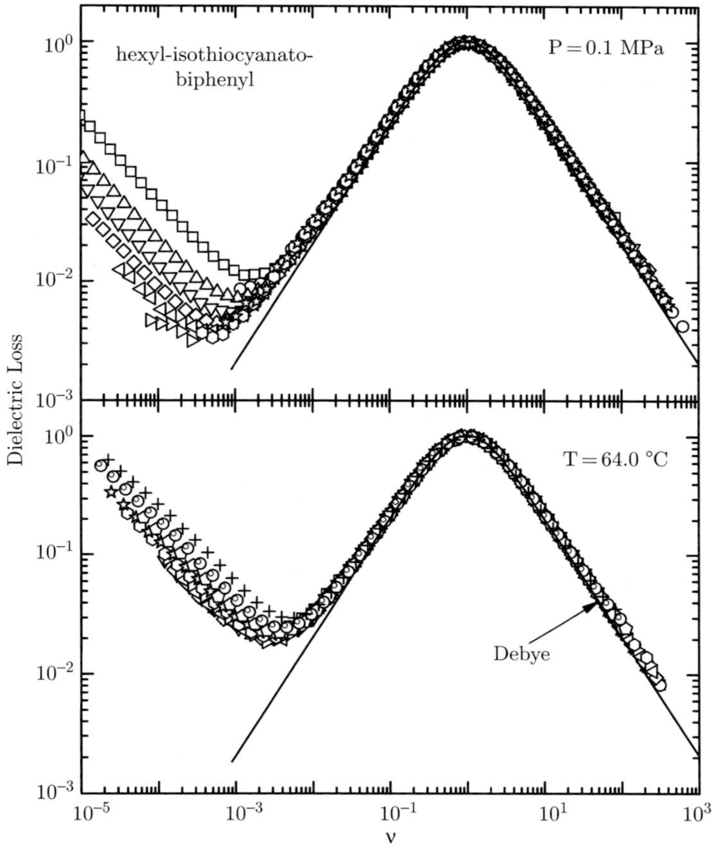

Figure 8.6 Dielectric loss curves of a liquid crystal in the smectic phase. The curves were shifted to bring the peaks into coincidence. The rise at lower frequencies is due to DC conductivity. (upper) from top to bottom: $T = 63.6$, 54.5, 45.5, 36.9, 28.1, 19.5, 10.4, 0.7, and −5.6 °C at ambient pressure; (lower) from top to bottom: $P = 7.7$, 27.6, 49.7, 72.6, 95.9, 118.0, 143.7, 169.7, 196.5, 220.2, 245.4, and 269.2 MPa at 64.0 °C. Solid lines represent the Debye function (Kohlrausch exponent = 1). Data from ref. [24].

ratio of the isochoric and isobaric activation energies can be used to assess the relative effect that temperature and pressure have on the longitudinal relaxation time. Listed in Table 8.1 are values of E_V/E_a for a number of liquid crystals. E_V, which includes only thermal effects, is roughly half the constant pressure value, indicating a comparable influence of thermal energy and volume on reorientation in the ordered state. In this regard there is no qualitative difference between ordered and isotropic liquids.

The density scaling of structural relaxation times (eqn (6.7)) can also be applied to τ in the liquid-crystalline state. Typical results are shown in Figure 8.8 for several alkyl-isothiocyanato-biphenyls. These data also illustrate the odd–even effect in a homologous series of mesogens [55,56]. The range of stability of the ordered state is larger when there is an even number of carbon atoms in the alkyl side chain (e.g. $n = 6$ or 8 in nBT) because the end chain in the *all-trans* orientation lies mainly parallel to the major molecular axis, thus enhancing

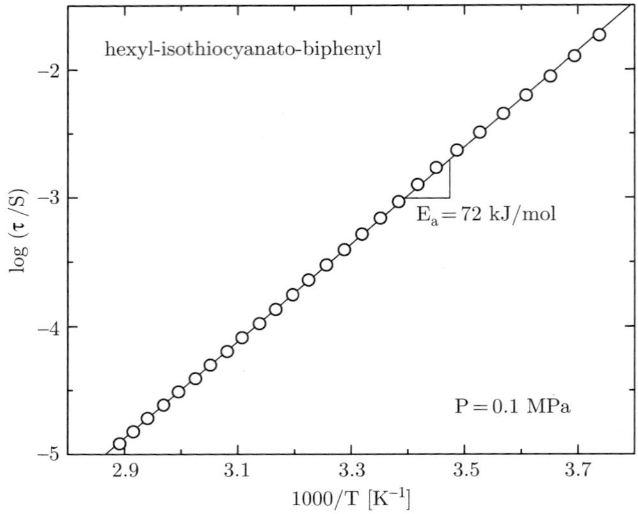

Figure 8.7 Longitudinal relaxation times for a liquid crystal in the smectic phase at ambient pressure (the corresponding spectra are in the top half of Figure 8.6). A linear fit gives the indicated activation energy. Data from ref. [24].

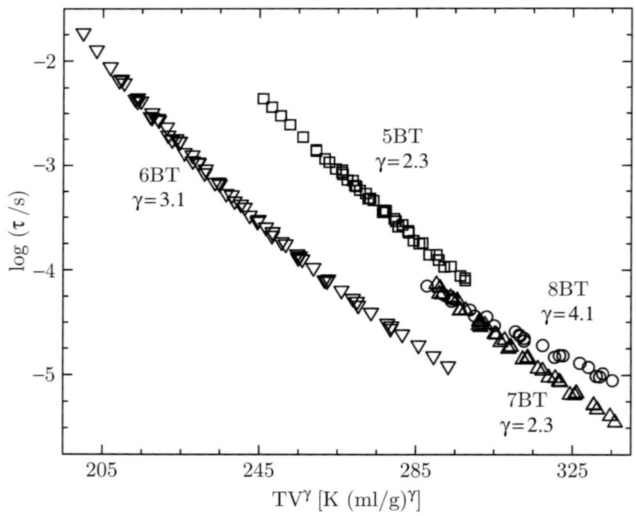

Figure 8.8 Longitudinal relaxation times measured at various temperatures and pressures for alkyl-isothiocyanato-biphenyls in the smectic E phase plotted as a function of TV^{γ}, with the respective values of the scaling exponent indicated. Data from refs. [24,54].

ordering, whereas for odd values of n the terminal carbon is out of registry. For this reason the crystal to smectic transition temperatures in hexyl- and octyl-isothiocyanato-biphenyls are lower and, as seen in the figure, the longitudinal relaxation times do not rank order with the size of the alkyl group.

The density scaling of longitudinal relaxation times has been verified for a large number of liquid crystals (Table 8.1); moreover, there is a rough correspondence between the exponent that superposes longitudinal relaxation times and the thermodynamic ratio, Γ. Their equivalence implies an interesting phenomenon concerning the magnitude of τ at a phase transition. Since the quantity $T_c V_c^{\Gamma}$ is fixed at a phase transition and τ is uniquely determined by the product variable TV^{γ}, it follows that $\tau(T_c V_c)$ must be a constant; that is, changes in pressure will change the transition temperature, but the relaxation time remains constant along a phase transition line, $T_c(P)$. Experimental results are shown in Figure 8.9, with the value of $\tau(T_c)$ and its variation with pressure tabulated in Table 8.2 [57]. Some data exhibit a weak

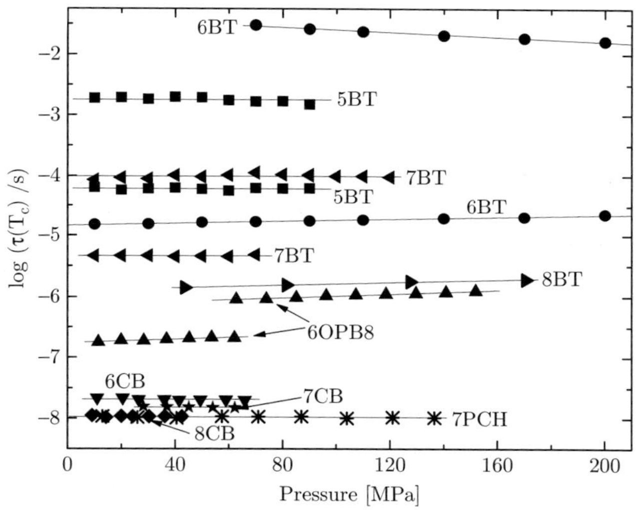

Figure 8.9 Pressure variation of the longitudinal relaxation times at the phase transition for various liquid crystals (identified in Table 8.2). Data from ref. [57].

Table 8.2 Relaxation times at transitions of various liquid crystals. [57]

Liquid crystal	Transition	log $(\tau(T_c)/\text{s})$
5BT	crystal – smectic E	–2.75 ± 0.04
	smectic E – isotropic	–4.21 ± 0.01
6BT	crystal – smectic E	–1.65 ± 0.10
	smectic E – isotropic	–4.75 ± 0.06
7BT	crystal – smectic E	–4.00 ± 0.01
	smectic E – isotropic	–5.34 ± 0.02
8BT	smectic E – isotropic	–5.78 ± 0.06
6OPB8	smectic C – smectic A	–5.97 ± 0.05
	smectic A – nematic	–6.70 ± 0.03
5CB	nematic – isotropic	–7.85 ± 0.20
6CB	nematic – isotropic	–7.68 ± 0.01
7CB	nematic – isotropic	–7.81 ± 0.01
8CB	nematic – isotropic	–7.97 ± 0.01
7PCH	nematic – isotropic	–7.97 ± 0.01

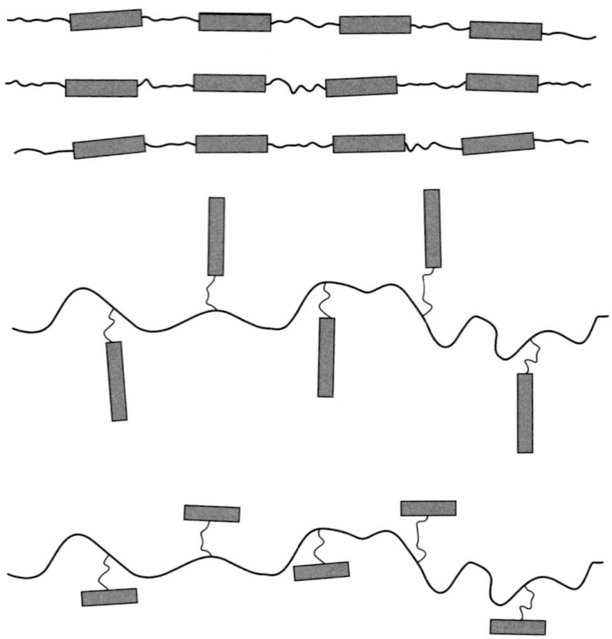

Figure 8.10 Illustration of liquid crystal elastomers with main-chain (top) and side-chain (middle and bottom) mesogens.

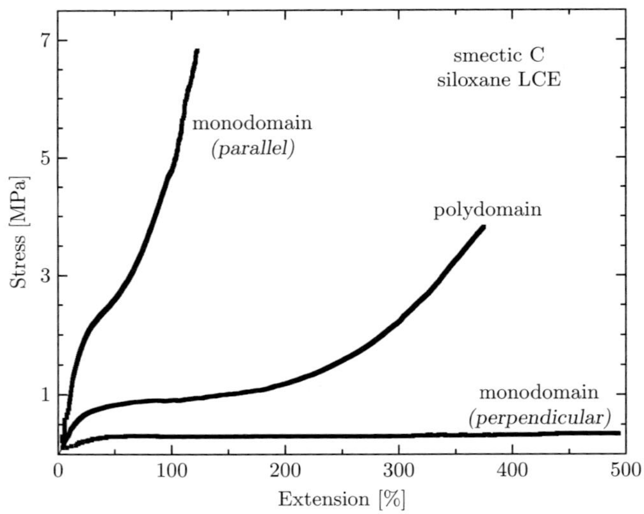

Figure 8.11 Mechanical response of an oriented monodomain LCE stretched parallel and perpendicular to the director, and the same material with unoriented polydomains. The Young's modulus equals 11.9 and 1.5 MPa, respectively, for parallel and perpendicular extension, and 3.8 MPa for the polydomain morphology. Data from ref. [61].

slope, although this might arise from systematic error in the measurements; notwithstanding, $\tau(T_c)$ varies less than 10% for a given material over the range of P.

There is no a priori reason to expect the thermodynamic conditions associated with a phase change to bear any relationship to the timescale of molecular rotations; this result is not predicted by any model. Thermodynamic phase changes reflect competition between the anisotropic interaction energy and the orientational entropy, and evidently the factors minimizing the Gibbs free energy are also the control parameters governing the liquid-crystal reorientational dynamics.

8.2 Liquid-crystal elastomers

Mesogens can be attached as pendant groups to the polymer backbone or copolymerized into the main chain [58–60] (Figure 8.10). To achieve elastomeric properties, side groups must be connected by a flexible "spacer," otherwise a rigid-rod polymer results. The backbone itself must be flexible, and invariably it is crosslinked. The three main types of LCE phases are

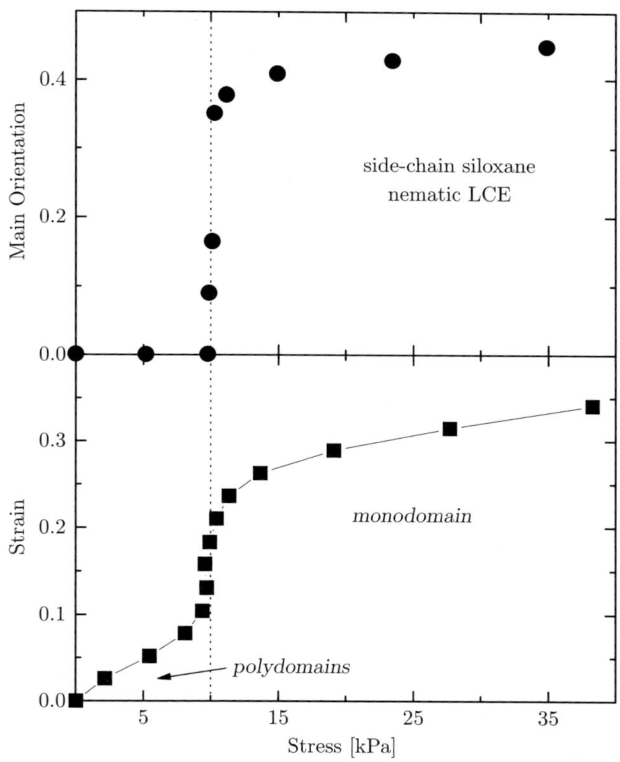

Figure 8.12 The average orientation determined from X-ray scattering (top) and the corresponding macroscopic strain imposed transverse to the director (bottom) as a function of the engineering stress in a side-chain substituted polydimethylsiloxane nematic liquid crystal. The change in behavior at *ca.* 10 kPa is due to the onset of rotation of the polydomains into the stretch direction to form an oriented monodomain. Data from ref. [62].

nematic, smectic, and cholesteric. For main chain LCE, the phase transitions tend to be very broad, reflecting polydispersity of the sequence lengths. The mesogenic units also have a distribution of orientations about the director, with the coherence range of the director defining a domain. LCE can exist as either one ordered domain or as polydomains, each having its own director. The usual method of preparing a monodomain LCE is by crosslinking in the ordered state. (Network junctions are mechanically stable at the orientation present during their formation, as described for double networks in Section 4.5.2). The mechanical properties of multi- and monodomain LCE are quite different, with the latter highly anisotropic (Figure 8.11) [61].

Since the mesogens are mechanically coupled to the elastomeric network, stretching can convert randomly oriented domains into a single monodomain, as illustrated in Figure 8.12 [62]. Stretching transverse to the director can give rise to the phenomenon of "soft elasticity" – at a certain threshold strain, deformation of the polymer chains is avoided by domain rotation without appreciable stress. However, to avoid a bulk shear strain, the domain rotation does not encompass macroscopic length scales; rather the morphology breaks up into "stripe domains," which are bands, several micrometers wide, having an alternating

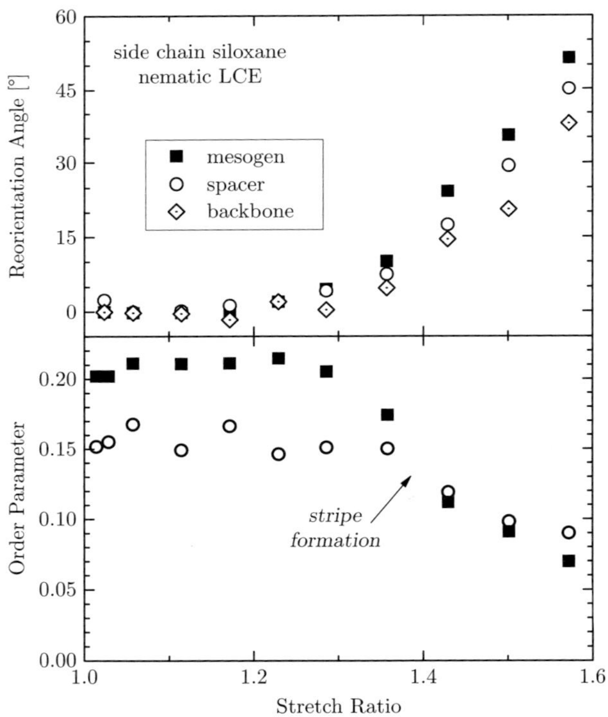

Figure 8.13 Effect of stretching perpendicular to the initial director (top) on the angular orientation of the domain and (bottom) on the order parameter determined from infrared dichroism for: the backbone (diamonds), flexible space (open circles), and pendant mesogen (filled squares) of a polydimethysiloxane nematic side-chain liquid crystal elastomer. The rotation of the domains introduces some dispersion into the orientation, reducing the degree of order. Data from ref. [65].

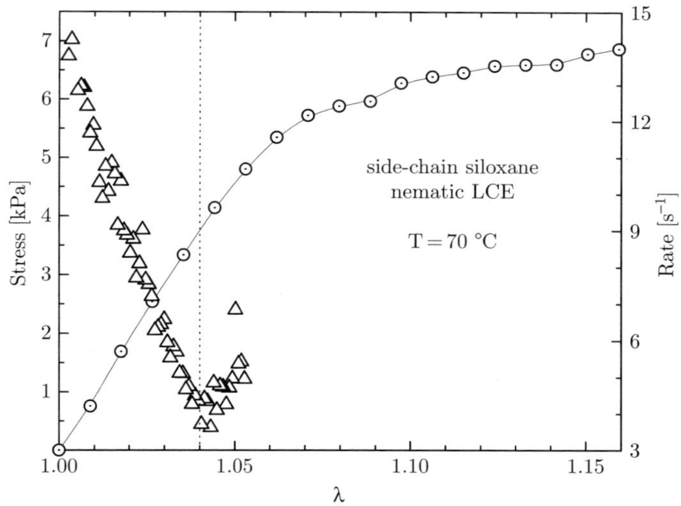

Figure 8.14 The stress (circles) and nematic director fluctuation rate (triangles) as a function of the extension perpendicular to the director. With increasing strain, the relaxation rate decreases, reaching a minimum value at the threshold ($\lambda \sim 1.04$) for onset of the soft elastic response. The director then begins to rotate, so that the rotational fluctuations are no longer governed by the internal elasticity. Some residual nematic order remains above the clearing temperature ($T_c = 78$ °C). Data from ref. [66].

direction of orientation in a semi-regular pattern [63,64]. Figure 8.13 [65] shows the change in angular orientation of the backbone, spacer, and nematogen of a side-chain siloxane LCE. Domain rotations only occur beyond a threshold strain (about 30% in this example) and are accompanied by some loss of orientation, as seen in the decrease of the order parameter in Figure 8.13. When the director and the long axis of the polymer chains become parallel to the stretching direction, the behavior transitions to the usual rubbery stress–strain response.

Since fluctuations of the nematic director are coupled to the shear modulus of the LCE, the soft elasticity effected by domain rotation affects these fluctuations. This is seen in the relaxation rate for nematic fluctuations measured by depolarized light scattering (Figure 8.14 [66]). In the vicinity of the threshold strain, there is a minimum in the frequency of these fluctuations [66,67] as the internal aligning field approaches zero. This critical strain corresponds to the onset of director rotation and stripe formation.

The attachment of mesogens to the backbone in side-chain LCE precludes rotation about their short axis, the most important motion in molecular liquid crystals (Figures 8.5 through 8.9). The dynamics are limited to oscillations about the long axis. This librational motion is not intermolecularly cooperative; thus, its frequency is insensitive to the molecular weight of the polymer chain and depends only weakly on the length of the spacer, provided the latter is sufficiently long to decouple the mesogen from the backbone. For example, in polymethylmethacrylate LCE with phenyl benzoate mesogens [68], increasing the spacer from 6 to 11 methylene units increased the librational relaxation time by less than one decade; the activation energy also showed only a small (~10%) increase. Consistent with the non-cooperative nature, the librational frequencies are unaffected by the phase structure (Figure 8.15 [68]).

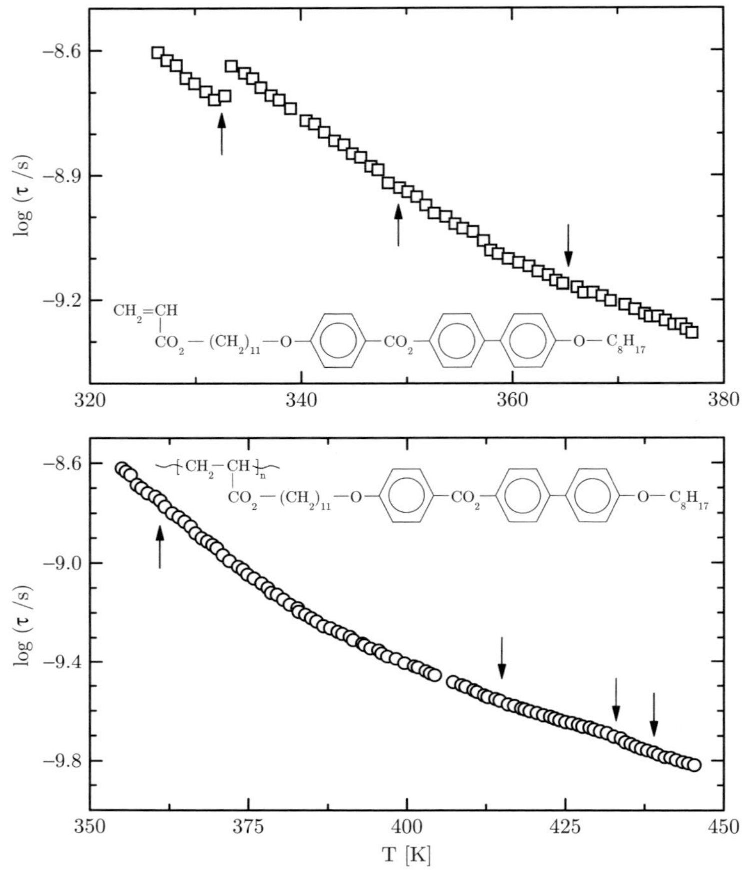

Figure 8.15 Librational relaxation times of polymethacrylate with pendant phenyl benzoate mesogens (lower) and the corresponding liquid crystal monomer (upper). The arrows denote the location of phase changes, with the highest temperature transition being the clearing point. Data from ref. [68].

The chain-segmental dynamics, however, show some influence of the pendant mesogen. For short spacers, e.g. an ethyl group, the fragility (defined in Chapter 2) increases, whereas long spacers "plasticize" structural relaxation, reducing the fragility [69].

Since the viscoelastic response of global motions of polymers is generally slow (» seconds), LCE are not used for display applications. The fact that LCE mechanical properties can couple to external fields offers promise for applications such as actuators or muscle tissue. Substantial changes in shape of LCE can be brought about by reversible phase transitions induced by variations of temperature or solvent [70,71]. The effects are stronger for main-chain than for side-chain LCE [72]. For single-domain, smectic materials, an electric field can be employed to induce tilting of the phase and consequent shape changes [73,74]. Although LCE are thermorheologically complex in their ordered phase [75,76], the dynamic mechanical properties of nematic and smectic LCE exhibit only very small changes on transitioning to the isotropic state [77–79] (Figure 8.16).

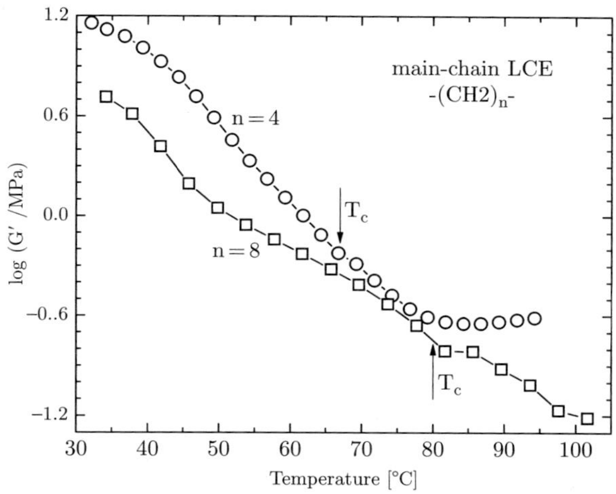

Figure 8.16 Storage modulus *versus* temperature for an unoriented main-chain LCE having an aromatic epoxy mesogen copolymerized with flexible segments having the indicated number of methylene carbons. There is only a very small change in G' at the transition of the smectic A phase to the isotropic elastomer (clearing points indicated by the vertical arrows). Data from ref. [78].

References

1. S. Chandrasekhar, *Liquid Crystals*, 2nd edn, Cambridge University Press, New York (1992).
2. S. Singh, *Phys. Rep.* **324**, 107 (2000).
3. B. Wunderlich, *J. Polym. Sci. Polym. Phys. Ed.* **46**, 2647 (2008).
4. G.K. Elyashevich, *Adv. Polym. Sci.* **43**, 205 (1982).
5. S. Frenkel, *J. Polym. Sci. Polym. Symp.* **58**, 195 (1977).
6. R.S. Zurabian, V.G. Baranov, and S. Frenkel, *J. Polym. Sci. Sump. Ser.* **44**, 163 (1974).
7. *High-speed Fiber Spinning – Science and Engineering Aspects*, A. Ziabicki and H. Kawai, eds., John Wiley and Sons, New York (1985).
8. D.W. Tomlin, C.M. Roland, and L.I. Slutsker, *J. Polym. Sci. Polym. Phys. Ed.* **31**, 1331 (1993).
9. C.M. Roland, *Polym. Eng. Sci.* **31**, 849 (1991).
10. C.M. Roland and M.F. Sonnenschein, *Polym. Eng. Sci.* **31**, 1434 (1991).
11. E.B. Gowd, C. Ramesh, M.S. Byrne, N.S. Murthy, and J. Radhakrishanan, *Polymer* **45**, 6707 (2004).
12. D.C. Bassett, *Polymer* **17**, 460 (1976).
13. L.B. Li, L. Zhang, R. Huang, W.Y. Fan, S.M. Hong, G. Wang, and Y.M. Zhong, *J Polym. Sci. Polym. Phys. Ed.* **38**, 1612 (2000).
14. K. Hiraoka, W. Sagano, T. Nose, and H. Finkelmann, *Macromolecules* **38**, 7352 (2005).
15. D.L. Thomsen, P. Keller, J. Naciri, R. Pink, H. Jeon, D. Shenoy, and B.R. Ratna, *Macromolecules* **34**, 5868 (2001).
16. S. Serak, N. Tabiryan, R. Vergara, T.J. White, R.A. Vaia, and T.J. Bunning, *Soft Matter* **6**, 779 (2010).
17. D. Corbett and M. Warner, *Liq. Cryst.* **36**, 1263 (2009).

18. S.J. Woltman, G.D. Jay, and G.P. Crawford, *Nature Mater.* **6**, 929 (2007).
19. G.T. Stewart, *Liq. Cryst.* **30**, 541 (2003); *ibid.* **31**, 443 (2004).
20. K. Herrmann, *Z. Krist.* **92**, 49 (1935).
21. N. Boccara, *Ann. Phys.* **76**, 72 (1973).
22. H.K. Bisoyi and S. Kumar, *Chem. Soc. Rev.* **39**, 264 (2010).
23. R.B. Bogoslovov, C.M. Roland, J. Czub, and S. Urban, *J. Phys. Chem. B* **112**, 16008 (2008).
24. C.M. Roland, R.B. Bogoslovov, R. Casalini, A.R. Ellis, S. Bair, S.J. Rzoska, K. Czuprynski, and S. Urban, *J. Chem. Phys.* **128**, 224506 (2008).
25. J. D. Menczel and T.M. Leslie, *Thermochimica Acta* **166**, 309 (1990).
26. S. Neuenfeld and C. Schick,, *Thermochimica Acta* **446**, 55 (2006).
27. L. Onsager, *Ann. N.Y. Acad. Sci.* **51**, 627 (1949).
28. P.J. Flory, *Proc. Roy. Soc. A* **234**, 73 (1956).
29. W. Maier and A. Saupe, *Z. Naturforsch.* ***13a***, 564 (1958); **14a**, 882 (1959); **15a**, 287 (1960).
30. G.R. Luckhurst and C. Zannoni, *Nature* **267**, 412 (1977).
31. R. Tao, P. Sheng, and Z.F. Lin, *Phys. Rev. Lett.* **70**, 1271 (1993).
32. J.G. Gay and B.J. Berne, *J. Chem. Phys.* **74**, 3316 (1981).
33. L. Longa, G. Cholewiak, and J. Stelzer, *Acta. Phys. Polon. B* **31**, 801 (2000).
34. L. Longa, H.-R. Trebin, and G. Cholewiak, "Computer Simulations of Polar Liquid Crystals," in *Relaxation Phenomena*, W. Haase and S. Wróbel, ed., Springer-Verlag, Berlin, ch. 4.3, p. 204 (2003).
35. M.A. Bates and G.R. Luckhurst, *Chem. Phys. Lett.* **281**, 193 (1997).
36. R. L. Humphries, G. James, and G. R. Luckhurst, *J. Chem. Soc. Faraday Trans. II* **68**, 1031 (1972).
37. S. Chandrasekhar and N.W. Madhusudana, *Acta Cryst.* **A27**, 303 (1971).
38. T. Shirakawa, T. Inoue, and T. Tokuda, *J. Phys. Chem.* **87**, 1406 (1983).
39. C.S. Johnson and P.J. Collings *J. Chem. Phys.* **79**, 4056 (1983).
40. M.A. Bates and G.R. Luckhurst Chem. *Phys. Lett.* **281**, 193 (1997).
41. J.R. McColl and C.S. Shih, *Phys. Rev. Lett.* **29**, 85 (1972).
42. S. Urban and A. Würflinger, *Adv. Chem. Phys.* **98**, 143 (1997).
43. A. Würflinger and M. Sandmann, in *Physical Properties of Liquid Crystals. Nematics*, D. A. Dunmur, A. Fukuda, and G.R. Luckhurst (ed.), EMIS Data Review Series, 25, IEE, London (2001).
44. T. Shirakawa, H. Eura, H. Ichimura, T. Ito, K. Toi, and T. Seimiya, *Thermochim. Acta* **105**, 251 (1986).
45. A. Würflinger and S. Urban, *Liq. Cryst.* **29**, 799 (2002).
46. M.W. Lampe and P.J. Collings Phys. *Rev. A* **34**, 524 (1986).
47. McColl, J.R.; Shih, C.S. *Phys. Rev. Lett.* **29**, 85 (1972).
48. A.J. Dianoux and F. Volino, *J. Phys.*, **40**, 181 (1979).
49. V. Domenici, J. Czub, M. Geppi, B. Gestblom, S. Urban, and C.-A. Veracini, *Liq. Cryst.* **1**, 91 (2004).
50. H. Kresse, *Adv. Liq. Cryst.* **6**, 109 (1983).
51. S. Urban and C.M. Roland, *J. Non-Cryst. Solids* **357**, 740 (2011).
52. S. Urban, J. Czub, R. Dąbrowski, and H. Kresse, *Liq. Cryst.* **32**, 119 (2005).

53. S. Urban, J. Przedmojski, and J. Czub, *Liq. Cryst.* **32**, 619 (2005).
54. S. Urban and A. Würflinger, *Phys. Rev. E* **72**, 021707 (2005).
55. R.G. Horn and T.E. Faber, Proc. *Roy. Soc. Lond. Ser. A* **368**, 199 (1979).
56. M. Tokita and J. Watanabe, *Polym. J.* **38**, 611 (2006).
57. C.M. Roland, *Soft Matter* **4**, 2316 (2008).
58. S. Mayer and R. Zentel, *Curr. Opin. Solid State Mater. Sci.* **6**, 545 (2002).
59. S. M. Kelly, *Liq. Cryst.* **24**, 71 (1998).
60. G.G. Barclay and C.K. Ober, *Prog. Polym. Sci.* **18**, 899 (1993).
61. W. Ren, P.J. McMullan, and A.C. Griffin, *Macromol. Chem. Phys.* **209**, 1896 (2008).
62. E.M. Terentjev, *J. Phys. Condens. Matter* **11**, R239 (1999).
63. I. Kundler and H. Finkelmann, *Macro. Chem. Rap. Commun.* **16**, 79 (1995).
64. I. Kundler and H. Finkelmann, *Macromol. Chem. Phys.* **199**, 677 (1998).
65. M. Tammer, J. Li, A. Komp, H. Finkelmann, and F. Kremer, *Macro. Chem. Phys.* **206**, 709 (2005).
66. A. Petelin and M. Copic, *Phys. Rev. Lett.* **103**, 077801 (2009).
67. A. Petelin and M. Copic, *Phys. Rev. E* **82**, 011703 (2010).
68. A. Schonfeld, F. Kremer, A. Hofmann, K. Kuhnpast, J. Springer, and G. Scherowsky, G. *Makromol. Chem.* **194**, 1149 (1993).
69. K.L. Ngai and A. Schonhals, *J. Polym. Sci. Polym. Phys. Ed.* **36**, 1927 (1998).
70. H. Finkelmann, E. Nishikawa, G.G. Pereira, and M. Warner, *Phys. Rev. Lett.* **87**, 015501 (2001).
71. J. Naciri, A. Srinivasan, H. Jeon, N. Nikolov, P. Keller, and B.R. Ratna, *Macromolecules* **36**, 8499 (2003).
72. P.G. de Gennes, M. Hebert, and R. Kant, *Macromol. Symp.* **113**, 399 (1997).
73. M. Brehmer, R. Zentel, F. Gieselmann, R. Germer, and P. Zugenmaier, *Liq. Cryst.* **21**, 589 (1996).
74. H. Poths, A. Schonfeld, R. Zentel, F. Kremer, and K. Siemensmeyer, *Adv. Mater.* **4**, 351 (1992).
75. M. Giamberini, V. Ambrogi, P. Cerruti, and C. Carfagna, *Polymer* **47**, 4490 (2006).
76. P. Stein, N. Assfalg, H. Finkelmann, and P. Martinoty, *Eur. Phys. J. E* **4**, 255 (2001).
77. J. Weilepp, J.J. Zanna, N. Assfalg, P. Stein, L. Hilliou, M. Mauzac, H. Finkelmann, H.R. Brand, and P. Martinoty, *Macromolecules* **32**, 4566 (1999).
78. M. Giamberini, P. Cerruti, V. Ambrogi, C. Vestito, F. Covino, and C. Carfagn, *Polymer* **46**, 9113 (2005).
79. J.J. Zanna, P. Stein, J.D. Marty, M. Mauzac, and P. Martinoty, *Macromolecules* **35**, 5459 (2002).

9 Bioelastomers

9.1 Proteins

Biological macromolecules include nucleic acids, polysaccharides (polymeric carbohydrates), and proteins, all essential to life. Proteins are linear polymers made up of one or more chains of α-amino acids linked by amide bonds (peptide bonds). Amino acids (repeat units of which are called residues) contain an amine attached to the carbon adjacent to a carboxylic acid group, having the formula NH_2-CHR-COOH (Figure 9.1). The pendant "R" differentiates the structure of the 20 amino acids found in proteins, for which M ranges from 75 (glycine) to 204 g/mol (tryptophan). Except for glycine, the α-carbon is asymmetric and all amino acids in nature have the L-chiral form. The particular sequence of amino acids defining a protein is determined by the genetic code. Internal flexibility of polypeptide chains is essential to protein function. These dynamics involve changes in torsional angles of the α-carbon with the adjacent carbon or nitrogen atoms. There is a hierarchy of protein structures having increasing complexity (Figure 9.2 [1]). The native state of a protein refers to the folded, three-dimensional structure in which there is biological activity [2]. Protein folding involves the transition from a random coil to a complex helical form by a series of cooperative conformational changes [3]. The folded state is the thermodynamically stable configuration determined by the myriad interactions among amino acid sequences in a particular environment. It minimizes the Gibbs free energy by maximizing hydrogen-bond formation and suppressing hydrophobic interactions through positioning non-polar residues within the protein core. This coil to helix phase change occurs well above the glass transition of a protein. There has been longstanding interest in understanding the pathways of protein folding because of the insights obtained on how amino acid sequences specify the functioning structure. For shorter proteins the process can be described by a simple two-state model of folded and unfolded configurations [4–6]. Kinetic measurements of the transition are analyzed by construction of chevron plots (the term refers to the characteristic V-shape of the curve); see Figure 9.3 [7]. The folded structure is destabilized by addition of a denaturant, usually urea or guanidinium chloride, or by raising the temperature [8]. The relative concentrations of the two forms are measured and the sum of the rate constants for folding and unfolding is plotted versus denaturant concentration (or inverse T). By extrapolation to zero denaturant, the kinetics are determined. By selectively changing certain amino acids (mutagenesis), the folding pathway can be changed [9–11], and in this manner, folding mechanism and protein structure can be related to the amino acid sequence, with the ultimate goal being their prediction [12].

Biological molecules share characteristics of liquid-crystalline polymers. They exhibit molecular self-assembly, the best-known example of which is the helical mesophase of

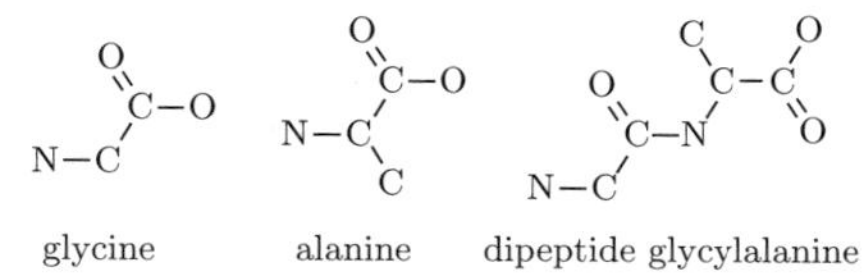

Figure 9.1 Two smallest amino acids and the dipeptide formed by their condensation reaction. By convention, 2–50 linked amino acids are called peptides, 50–100 amino acids polypeptides, and over 100 amino acids proteins (although the latter term is sometimes reserved for polypeptides in a stable conformation). The α-carbons are adjacent to a carbonyl carbon.

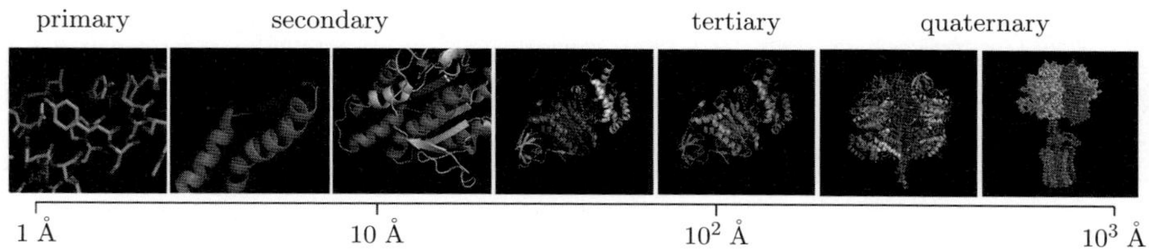

Figure 9.2 Illustration of protein structures of increasing complexity. Primary structures are isolated peptides; secondary structures consist of helical and sheet configurations with hydrogen bonding; tertiary structures are three-dimensional, folded, and involving side-chain interactions; quaternary structure is an aggregate of polypeptides having biological function. [1] *(Courtesy of Gustavo Caetano-Anolles of the Univ. Illinois Urbana; reproduced with permission)*

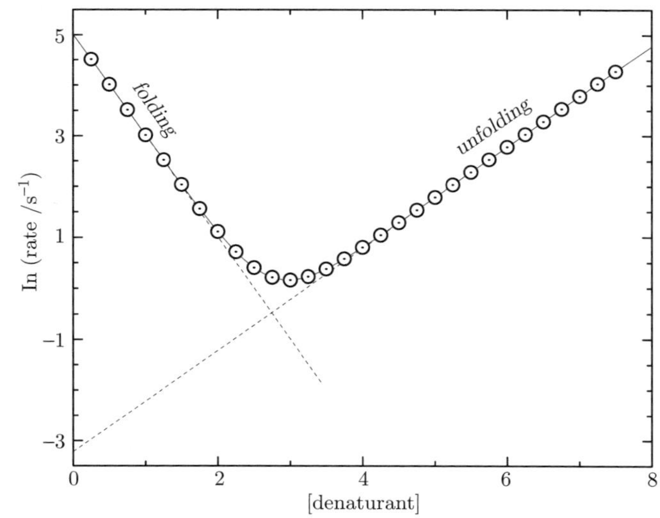

Figure 9.3 Representative chevron plot of the natural logarithm of the observed rate of protein folding as a function of denaturant concentration. Linear extrapolation of the folding and unfolding arms of the data yield the respective rate constants in water absent denaturant. The minimum corresponds to equal concentrations of expanded and folded proteins. Data from ref. [7].

polypeptides (chains of amino acids) and nucleic acids (e.g. RNA and DNA[1]). Transitions of lyotropic liquid crystals have even been invoked as the mechanism for spontaneous formation of elementary life forms [13,14]. Elastic proteins exhibit rubber (high) elasticity and viscoelasticity, and generally have the form of amphiphilic[2] block-copolymers, with hydrophilic and hydrophobic sequences of amino acids that segregate into organized structures [15,16]. Such proteins have characteristics associated with LCE [17]. An example is fibrillar collagen, a widely occurring animal protein prominent in connective tissue that is comprised of long polypeptide chains having axial periodicity in combination with liquid-like regions. Reduction in the water content of collagen induces tilting of the ordered domains, consistent with a lyotropic phase change (e.g. smectic A to C or E to K; see Figure 8.1) [18]. In the protein elastin, small polypeptides assemble to form hydrophobic domains that bring the chains into position for covalent crosslinking of adjacent lysine amino acid units [19]. Lamellar LCE properties have also been found in dragline silk, used to form the frame of spider webs. Silk from the silkworm (*Bombyx mori*) is formed from two proteins spun to produce a nematic liquid crystal fiber. Spider silk can be modeled as nematic discotic units produced during extensional flow of solution from the glands producing the silk [19].

9.2 Bioelastomers

Rubber-like elasticity is central to the utility of many proteins in animals (but the only elastomeric protein in plants is gluten[3]). Figure 9.4 [39] shows stress–strain curves for various bioelastomers, all polypeptides with different amino acid side groups. In Table 9.1 the mechanical properties corresponding to these data are tabulated [39,23]. Note the four orders of magnitude difference in the modulus even between two forms of spider silk. The origin of elasticity varies among different proteins in accord with their function [24,25]. Extension of the amorphous regions of polypeptide chains with consequent reduction in the number of conformations (i.e. entropic rubber elasticity) is the mechanism for high fatigue life materials such as elastin and resilin. In bioelastomers whose function is mainly energy dissipation, high strain recovery involves unfolding of domains with disruption of H bonds or other associated structures.

The elasticity of biological macromolecules reflects the semiflexible nature of their chain structure. The presence of β-strands (rigid-rod segments connected laterally by backbone H bonds and analogous to the mesogenic sequences in the backbone of liquid-crystal polymers) results in large persistence lengths. Since random packing of extended chains is inefficient, local order develops (as in rigid-rod polymers, Chapter 8), causing highly anisotropic and non-affine mechanical behavior [26]. The contribution of bending modes causes further deviations

[1] Nucleic acids are linear polymers that encode the information required for construction of each cell and tissue in an organism. The information is stored serially, with a given monomer sequence, known as a gene, responsible for a particular trait.

[2] Amphiphilic molecules have distinct hydrophilic and hydrophobic regions, usually in the form of a polar moiety attached to an alkyl chain. This structure enables formation of micelles in aqueous solutions; thus, amphiphiles can function as surfactants (substances that reduce the surface tension).

[3] We only consider protein-based bioelastomers in this chapter. The commodity rubber *cis*-1,4-plyisoprene is obtained commercially by coagulation of latex from the *Hevea brasiliensis* tree (and also made by about one thousand other plants [20]); the latex is produced to protect wounds from insects. The proteins present in natural rubber are low in concentration (<0.1%) and non-essential (although they improve mechanical properties by enhancing crystallization [21]). Biocompatibility is a characteristic associated with bioelastomers; however, allergic reactions to proteins in NR are common among chronic users of "latex rubber" products, particularly gloves [22].

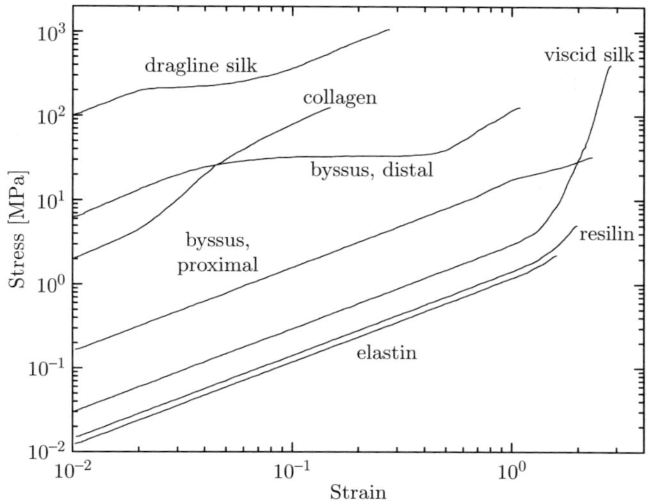

Figure 9.4 Double-logarithmic plot of stress–strain data for seven bioelastomers. Measurements were made on macroscopic samples at temperatures and water contents corresponding to *in vivo* conditions. "Viscid silk" is found in the capture spiral of a spider web, viscid a reference to the aqueous glue coating the threads. "Byssus" refers to filaments used by certain bivalve mollusks for attachment to fixed surfaces such as rocks. Data from ref. [39].

Table 9.1 Mechanical properties of various elastic proteins. [39,23]

Protein	Young's modulus (MPa)	elongation at break (%)
silk, viscid	0.3–3	270
elastin	0.41–1.1	150
resilin	0.65–2	190
fibrillin	0.2–100	>185
fibrin (uncrosslinked)	2	230
fibrin (crosslinked)	15	330
titin	1	200
byssus, distal	10–500	200
byssus, proximal	870	109
actin	2000	≤15
keratin (wet)	2000	45
tendon collagen	250–400	10–50
silk, dragline	10 000	30

from Gaussian behavior. The shear modulus is very non-linear (Figure 9.5) [27], as the chain bending at small strains transitions to stretching of the chains at higher strains. However, prior to the onset of a nematic phase transition, the stress–strain relationship for longitudinal strains (tension or compression) is qualitatively independent of degree of chain rigidity [28,29].

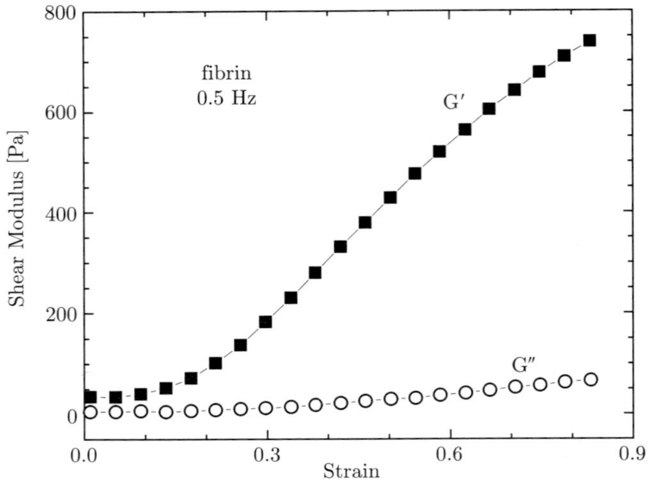

Figure 9.5 Strain dependence of the storage and loss modulus of fibrin gels formed under physiological conditions. The twenty-fold increase in stiffness reflects a change from bending to chain extension mechanisms. Data from ref. [27].

The mechanical properties of proteins are of interest given their role in biological function and thus their relation to aging and disease. LCE behavior in biopolymers has led to efforts to use crosslinked synthetic liquid-crystal polymers as substitutes for muscle and other biological materials [16,30–33]. The interpretation of mechanical data for biological materials is difficult, however, because of their complexity; thus, modeling and computer simulations are often used. A current limitation of numerical modeling is that biologically relevant timescales (ms – s) are beyond the limits for systems with any appreciable number of atoms. Since even a single protein can have as many as 10^3 covalent bonds or residues, with more than 10^5 in a protein assembly, recourse is often made to coarse-grained methods.[4]

Beyond the difficulties in interpreting experimental results, problems unique to the mechanical characterization of proteins *ex vivo* are manifest, including securing sufficient test material and carrying out measurements on small quantities. Most mechanical testing is done on naturally occurring samples obtained by extraction, although biocompatible proteins can be chemically synthesized (*in vitro* biosynthesis). The interpretation of data on proteins is more complex than for studies of conventional rubber, for several reasons: (i) The properties vary with the size-scale of the test, reflecting compositional variations. Generally, the modulus of a protein is larger than measured for a mesoscopic sample, which in turn is stiffer than the tissue itself. For example, one study found $E = 1$ to 10 000 MPa for single elastin fibrils from blood vessels; 7 to 30 MPa from nanoindentation of the tissue itself; and 0.2 to 4 MPa from tests on the 10-μm thick whole tissue [36]. (ii) When fiber bundles are tested, the individual fibers

[4] The complexity of proteins leads to two paradoxes: The Levinthal paradox [34] concerns the improbability of a randomly coiled protein finding its native (metastable) state, since the latter is one among an astronomical number of possible configurations, yet the coil–helix transition requires only a few seconds. The Hoyle paradox [35] describes the difficulty in assembling an optimized (life-sustaining) amino acid sequence by a random (Darwinian) process, given the finite (*ca.* 14 billion years) age of the universe. The resolution of both problems is that the searches are directed, so that all possibilities need not be sampled.

experience different strains [37]. This means the results are not necessarily an inherent property of the protein, but rather an ill-defined average that depends on fiber orientation and structure. (iii) The properties *in vitro* differ from *in vivo*. The latter usually corresponds to an equilibrium extent of hydration, but this water content (both intra- and extrafibrillar) changes with temperature and mechanical stress [38] and is not easily reproduced during laboratory testing. The effect of degree of hydration on the viscoelasticity of elastin is illustrated in Figure 9.6 [39,40]. Besides water content, other components of living tissue, such as salts and lipids, can affect the conformation of biopolymers and hence their viscoelastic behavior [41,42].

The strong effect of water on properties is a consequence of the hydration shell surrounding proteins. A few layers of hydrogen-bonded water, which are fixed in position, confer biological activity to the protein [43]. There are also more loosely bound secondary layers of water that have some diffusional mobility. The hydration shell governs the conformational dynamics, involving large-scale shape changes of the protein that are directly related to folding, the passage of ligands,[5] and other essential processes. Dehydrated proteins lack the oscillatory and rotational mobilities required to function. Structural relaxation of a protein involves some translational displacement of bonded water molecules [44]. This coupling of the protein dynamics to motion of the surrounding water is illustrated in Figure 9.7 [45]. The water dynamics become non-Arrhenius above *ca.* 200 K, coinciding with the onset of protein conformational dynamics over a measurable timescale ($\tau \leq 100$ s) [46,47]. The temperature demarcating the non-Arrhenius behavior can be identified with the glass-transition

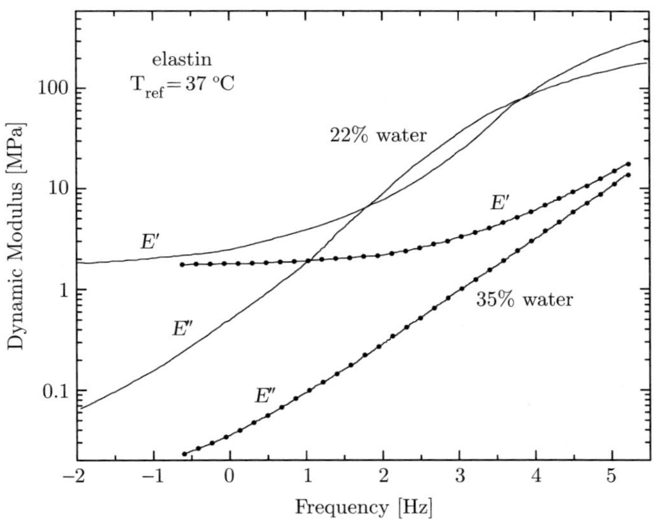

Figure 9.6 Master curves of the storage and loss moduli of elastin fully hydrated (–•–) and conditioned at 97% relative humidity (—). At equilibrium swelling, the loss tangent is near its maximum (∼ unity) at *ca.* 5 Hz. At lower water content, the curves shift to lower frequency due to less plasticization. Data from refs. [39 and 40].

[5] Ligands are chemical moieties bound to a central atom, usually a metal, to form a coordination compound. In biology, the term refers to signal-triggering molecules that bind at specific sites on proteins.

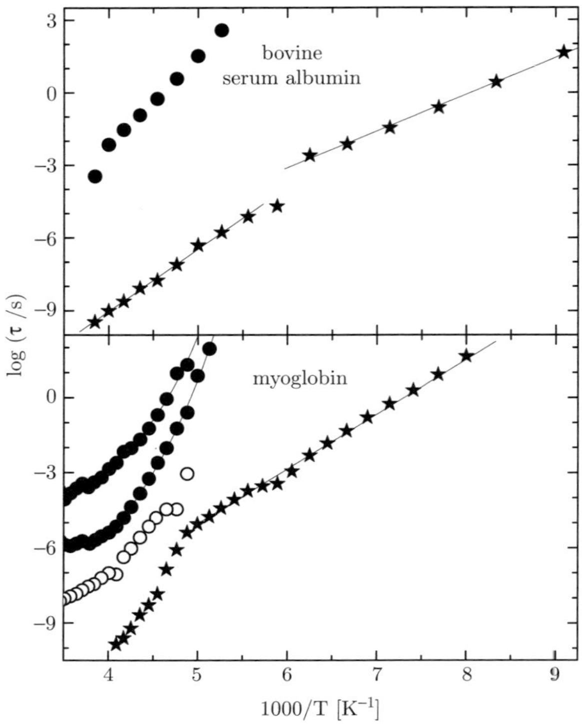

Figure 9.7 Dielectric relaxation times for: (top) conformational fluctuations of bovine serum albumin (solid circles) and reorientation of water in the hydration cell (solid stars) for a 40% solution of the protein; (bottom) conformational fluctuations of myoglobin protein (solid circles – two dielectrically active processes), side-group motion in the protein (open circles), and reorientation of molecules in the hydration shell (solid stars) for an equal weight mixture of the protein, water, and glycerol. At temperatures above the structural relaxation temperature of the proteins, the water dynamics become more temperature sensitive. Data from ref. [45].

temperature, below which fluctuations of the protein are frozen and biological activity ceases [48–50]. Figure 9.8 [51] illustrates this connection in myoglobin, a muscle protein. The conformational changes of the protein involved in association and dissociation of oxygen are slaved to the secondary relaxation of the surface-bound solvent molecules [49,52].

9.2.1 Elastin

A particularly well-studied bioelastomer is elastin, an extracellular protein that provides tissues with the ability to deform to large strains and completely recover with negligible energy loss. Elastin is mechanically active in large blood vessels, lungs, ligaments, skin, etc., with its network structure specific to the type of tissue. The dried fibers are stiff and almost inextensible; hydrophilic solutions (aqueous *in vivo*) are required for elastomeric behavior.

At equilibrium water content, the storage modulus of elastin is relatively flat up to frequencies of about 5 Hz, beyond which there is an upturn due to the onset of the glass transition (Figure 9.6). This rubbery range coincides with the heart rate of mammals, enabling

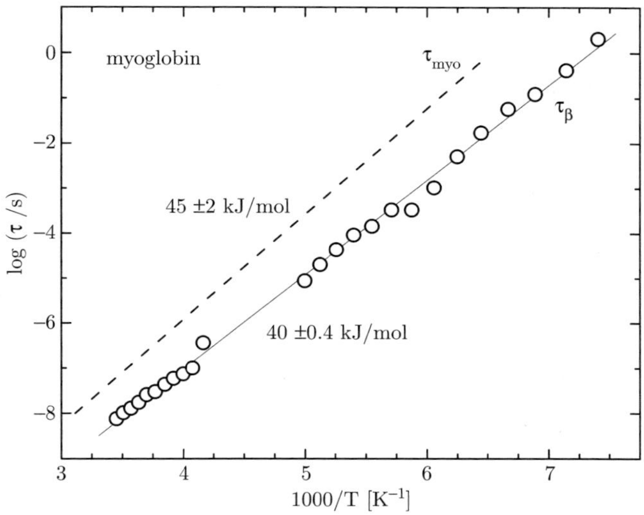

Figure 9.8 Secondary relaxation time of solvent molecules in the hydration cell of myoglobin (circles) and time constant for internal motions (ligand binding and structural relaxation) of the protein (dashed line). Data from ref. [51].

elastin to elastically restore arterial walls during the relaxation (diastolic) phase of the blood pressure cycle. At higher rates the mechanical loss increases due to the onset of the viscoelastic glass transition. It has been speculated that the greater damping in the transition zone serves to attenuate high-frequency perturbations induced by turbulent flow of blood around obstacles such as atherosclerotic plaque [53]. This same mechanism is operative in wet-skid resistant tires and other applications (Chapter 2). Even though elastin in the body is operating near its T_g, it has the remarkable property of a near invariance of its viscoelastic properties to temperature. This is due to the increased water absorption as temperature is lowered, so that plasticization compensates for the reduced thermal mobility.

With the exception of elastin in the uterus, this protein does not get replaced over the lifetime of a mammal [54]. Thus, for example, elastin in the human aorta evidently retains its resiliency without rupture for more than 10^9 cycles.[6] From extrapolation of fatigue measurements and assuming a dynamic stress of 0.55 MPa *in vivo*, a mechanical lifetime of 18 years was predicted for elastin obtained from pig arteries [39]. This compares to the 10–15 year life span of the animal, and indicates that the arterial walls operate close to the highest stress compatible with their required duration. Such behavior differs from that of striated muscle tissue (involving the elastomeric protein titin; see below), in which stress levels remain well below the fibers' tensile strength even during extreme exertion. The elastic fibers found in connective tissue are a composite network of elastin and collagen; the tenderness of cooked meat correlates with the degree of crosslinking of this network [55].

[6] This assumes a human heart beats about once per second. Pulse rates vary in proportion to an animal's size, ranging from 0.1 Hz for the great whale to 13 Hz for the shrew.

9.2.2 Resilin

Resilin in insects serves a role analogous to that of elastin in vertebrates, and generally the low-strain dynamic mechanical properties of the two proteins are similar (Figure 9.4) [53]. As implied by its name, the outstanding characteristic of resilin is its resilience. Resilin has the smallest mechanical damping of any bioelastomer (~0.4% energy loss) [56]. This can be critical for minimizing temperature rise during dynamic applications, as illustrated in Figure 9.9, showing the energy dissipation measured in the prealar arm[7] of locusts [57]. The hinge and main wing structure in locusts, dragonflies, etc., are composed almost entirely of resilin and chitin, a hard polymer common to exoskeletons [58,59]. The wing structure is a laminate with anisotropic stiffness. The modulus transverse to the laminate layers, in the direction of the wing hinge, is about 2 MPa, essentially the modulus of the resilin protein itself [57]. The layer under the wing undergoes alternate compression and extension during flight. Interestingly, changes in the degree of swelling of the protein *in vivo* alter the shape of the wing, which greatly affects its mechanical properties [60]. Resilin is found in numerous other insect structures, including leg muscles and joint membranes [56,59], antennae and setae base plates [61,62], and food pumps (insect sucking mechanism) [63].

9.2.3 Abductin

The rubbery protein abductin serves as the inner hinge ligament in the shell of bivalve mollusks (clams, oysters, mussels, scallops, etc.). Whereas elastin and resilin operate in

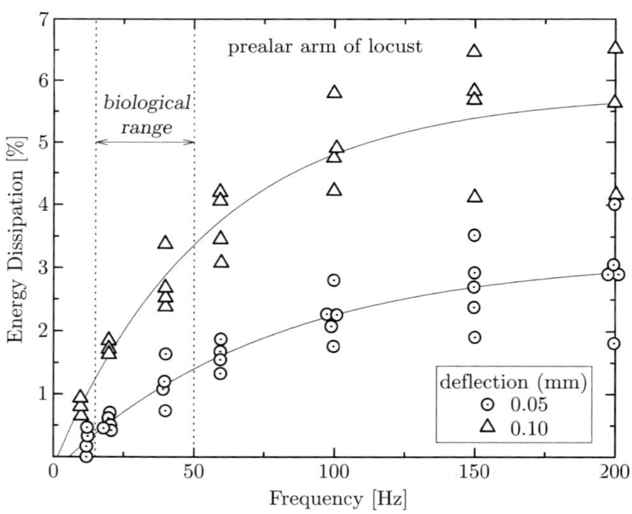

Figure 9.9 The energy dissipated as heat in the attachment arm of a locust during reversible lateral deflections of 0.05 (circles) and 0.10 (triangles) mm. The material is about 98% resilient at the frequencies 15–50 Hz, the biologically relevant range. Data from ref. [57].

[7] The prealar arm consists of about 80% resilin. It connects the insect's thorax to the plate between the forewings, deforming during wing motion.

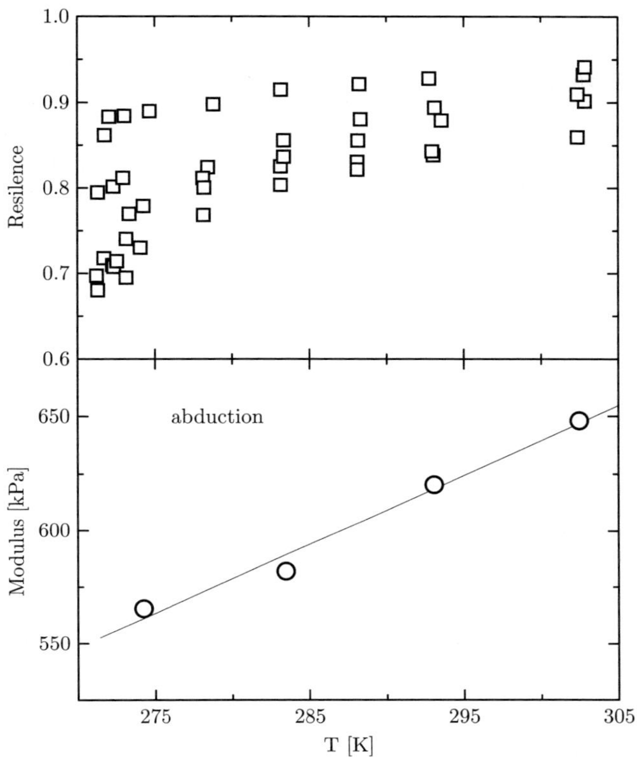

Figure 9.10 Temperature variation in (top) stored energy and (bottom) secant modulus of abductin. The scatter in the resilience reflects variations among samples; the modulus is for 15% compressive strain measured at 3–4 Hz. Data from ref. [65].

tension, abductin stores mechanical energy by compressing when the shell is closed. When the adductor muscles relax, abductin forces the shell open. In scallops rapid ($\sim$ 4 Hz) opening and closing of the shell is used as a remarkably resilient (<10% energy dissipation) propulsion mechanism [64]. Power outputs of 0.1–0.2 W/g are attained, comparable on a weight basis to that of small gasoline engines. The elasticity of abductin is due to rearrangements of the chain configurations; that is, the mechanism is primarily entropic and thus the modulus increases in proportion to temperature (Figure 9.10) [65].

9.2.4 Titin

Also referred to as connectin, titin is filamentous protein, responsible for the elasticity of striated muscles.[8] It is the largest known protein, with molecular weights of a few million g/mol, corresponding to contour lengths exceeding one micrometer. From measurements on muscle tissue, the extension of titin has been shown to produce forces as large as a few tens of

[8] Striated tissues, constructed of parallel fibers and having a striped appearance, comprise the voluntary muscles in the heart and skeletal system. Involuntary muscles are smooth muscles found in blood vessels, the digestive and respiratory tracts, etc.

pN per molecule, corresponding to a molecular modulus of several MPa [66]. During muscle stretching, initially contracted portions of the titin chain elastically extend, followed at higher stresses by unfolding of other regions of the molecule [67]. The details of the stretching mechanism depend on the fiber type (e.g. skeletal versus cardiac muscle) and its location in the muscle tissue. Titin is responsible both for force generation and the rapid straightening of muscles when the tissue becomes slack. Many experimental studies of stresses and recoil velocities of titin fibers have been carried out [68–70]. The measured recoil velocities are independent of temperature over the biologically relevant range, consistent with an entropic mechanism for the elasticity of titin. However, titin interacts with actin, a globular protein found in muscle and other tissues, resulting in energy dissipation and a significant change in the stiffness of titin fibers [68,70]. Generally, biological force-generation mechanisms are extremely complex compared to conventional elastomers. Depending on the activity, the muscle force can change or remain constant as the muscle fibers contract. Moreover, the number of fibers contracting will vary over time and with the required level of force. These factors make it difficult to interpret data obtained on extracted fibers in terms of their *in vivo* functioning.

9.2.5 Silk

Silks consist of filaments spun from various protein feedstocks by silkworms and spiders. It is this spinning[9] (extrusion of aqueous solution to form solid monofilaments or bundled fibers) that differentiates silk from other bioelastomers, since their chemical compositions are similar. Unlike silk, other elastic proteins are formed by the sequential combining of components, whose final assembly relies on the evolution of properties during the process. There are thousands of different silk compositions, with functions ranging from the capture and storage of prey to protection of eggs and pupae, in climates that vary from underwater to deserts [72]. Accordingly, silks display a broad range of mechanical properties (Figure 9.4), with a tensile modulus ranging from 10^4 Pa for well-hydrated silk to more than 10^{10} Pa for spider-web draglines. Figure 9.11 [73] shows the various silk glands in the *Nephila* golden spider, along with the corresponding function and stress–strain response of each silk. The mechanical behavior of dragline silk is believed to reflect an isotropic–nematic phase transition occurring at the yield strain ($\cong 0.02$ in Figure 9.4) [74]. The modulus of dragline silk is about the same as for textile silk fibers (which are obtained from the cocoon of silkmoth larvae). The range of stiffness of different silks reflects variations in their heterogeneous structure of crystalline hard domains embedded in an amorphous, soft matrix. For example, in spider silk alanine-rich segments form the ordered phase, while those having mainly glycine residues comprise the soft domains [75]. Generally, stiffer silk fibers have higher crystalline contents. The morphology of silk proteins is stabilized by hydrogen bonding, which makes the mechanical properties sensitive to water content. Water molecules can bind to amide groups, thus reducing the concentration of H bonds. This plasticizing effect is seen in the change in the modulus of silk fibers as a function of the humidity (Figure 9.12) [76]. At any given temperature, there is a significant decrease with water content, but this hydration level is reduced as temperature rises. The behavior reflects the combined effects of water and thermal energy on hydrogen bonding.

[9] A fossil of a spider spinneret was discovered that dates to *ca.* 380 million BC [71].

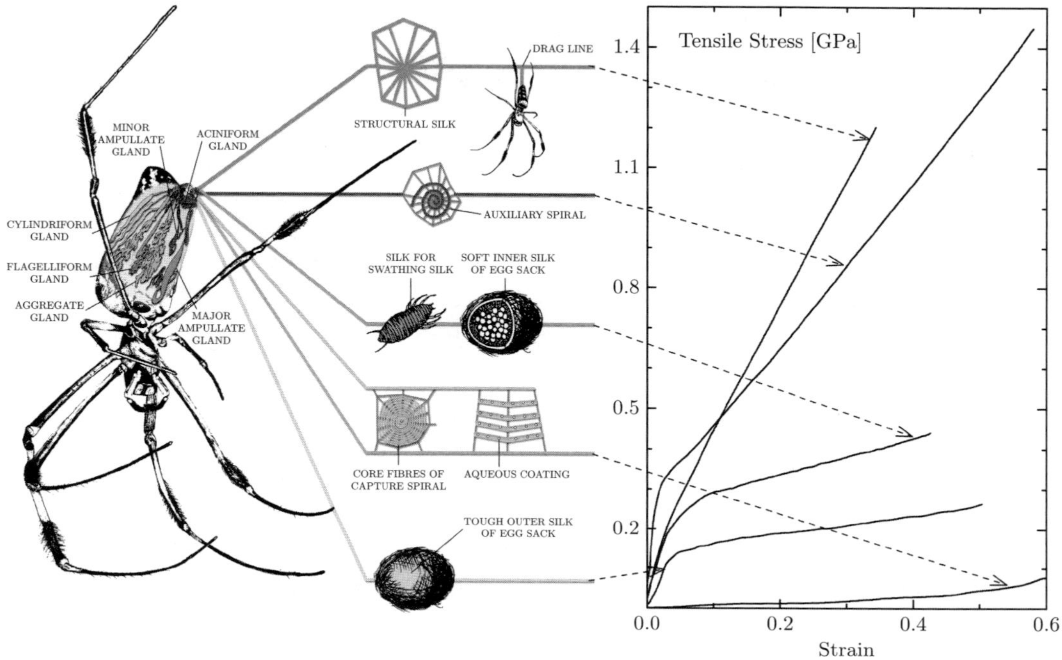

Figure 9.11 Golden silk orb-weaver spider (genus *Nephila*, from the Greek *nen*, to spin, and *philos*, love). The various glands and their silk are depicted, with the corresponding mechanical responses indicated. The stress–strain data for flagelliform silk (the lowest curve) is for the dry form; the extension is much larger when the protein is wet. (*Illustration courtesy of Chris Holland and Fritz Vollrath of Oxford University and reproduced with permission.)* Data from ref. [73].

9.2.7 Fibrin

Fibrinogen is a water-soluble protein, ~45 nm long and composed of six polypeptide chains. It is cleaved by thrombin, the final enzyme in the cascade of blood-coagulation processes, with the cleaved segments reacting to form fibrin fibers. The fibers polymerize into a highly branched, crosslinked mesh, with lateral aggregation extending the diameter to as much as 200 nm [77]. This fibrin network binds with activated platelets to form an insoluble, hydrated assembly that traps red blood cells, yielding a blood clot. The platelets serve as an initial plug to slow bleeding, and enhance the stiffness of the subsequent fibrin network [78]. The modulus of a clot is orders of magnitude smaller than the longitudinal modulus of a single fibrin fiber [79], although fibrin networks exhibit substantial strain hardening (Figure 9.5) [27,80,81]. The dynamic mechanical response of coagulating blood is shown in Figure 9.13 [82]. The formation of a network increases the shear modulus, as the liquid transitions to a solid state. Traditionally, the time for clot formation is quantified by a rigidity criterion (thromboelastography [83]). As seen in Figure 9.13, this empirical clotting time falls in the region where G' begins to strongly increase, which is at substantially longer times than the gel point defined by a frequency-independent loss tangent (Section 3.4.5).

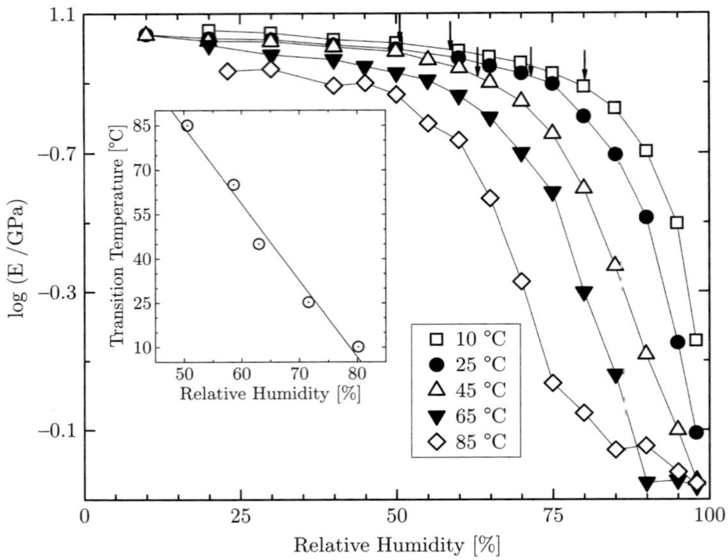

Figure 9.12 Effect of relative humidity on the tensile modulus of silk fibers (obtained from silkworm larvae). The intersection of the low- and high-modulus regions (indicated by arrows) yields the transition temperatures plotted in the inset. This transition behavior is analogous to the glass transition. Data from ref. [76].

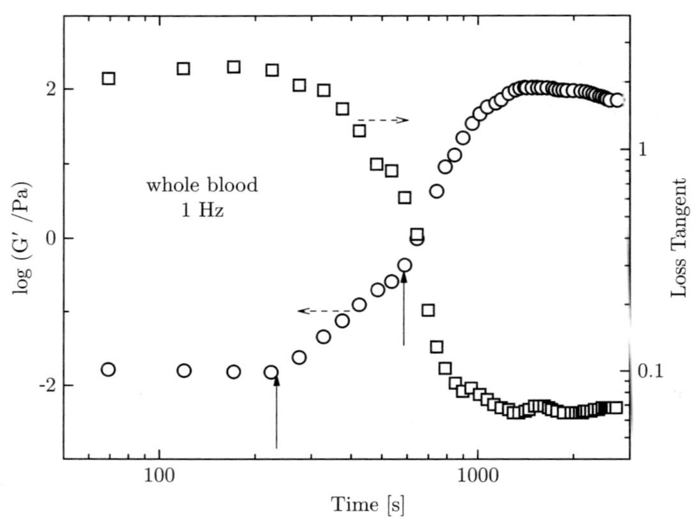

Figure 9.13 Storage modulus (circles) and loss tangent (squares) measured on coagulating blood. The vertical arrows indicate the gel point (frequency-independent dynamic response) and the subsequent time of clot formation as determined by thromboelastography. Data from ref. [82].

9.2.6 Gluten

Gluten, a main component of flour, is the only elastomeric protein found in plants (such as wheat, rye, and barley), where it is used for seed storage [84]. Gluten is a mixture of polypeptides, varying from single chains having molecular weights as low as 15 000 g/mol to chains linked by disulfide bonds to form extended structures with M as high as 10^6 g/mol. In the presence of water these proteins aggregate to form a network ("dough"). Cooking or other processing extends the network, with the elasticity of gluten derived mainly from the aggregation. For example, the modulus of gluten has been found to vary as the cube of the number of aggregated proteins [85]. This can be compared to the prediction of entropy theories of rubber elasticity of a direct proportionality between modulus and the number of crosslinks (Chapter 4). The inference is a significant enthalpic contribution to the elasticity of wheat protein, which would also account for the decrease of its modulus with temperature [84].

References

1. G. Caetano-Anolles, M. Wang, D. Caetano-Anolles, and J.E. Mittenthal, *Biochem. J.* **417**, 621 (2009).
2. K. Linderstrøm-Lang and J.A. Schellman in *The Enzymes*, 2nd edn (H. Lardy and K. Myrback, ed.), Academic Press, New York, p. 443 (1959).
3. D.J. Brockwell and S.E. Radford, *Curr. Opin. Struct. Biol.* **17**, 30 (2007).
4. S.E. Jackson and A.R. Fersht, *Biochemistry* **30**, 10428 (1991).
5. R. Zwanzig, *Proc. Natl. Acad. Sci.* **94**, 148 (1997).
6. H. Kaya and H.S. Chan, *Phys. Rev. Lett.* **90**, 258104 (2003).
7. A. Zarrine-Afsar and A.R. Davidson, *Methods* **34**, 41 (2004).
8. B. Kuhlman, D.L. Luisi, P.A. Evans, and D.P. Raleigh, *J. Mol. Biol.* **284**, 1661 (1998).
9. A. Matouschek, J.T. Kellis, L. Serrano, and A.R. Fersht, *Nature* **340**, 122 (1989).
10. A.R. Viguera, C. Vega, and L. Serrano, *Proc. Natl. Acad. Sci.* **99**, 5349 (2002).
11. T. Ternstrom, U. Mayor, M. Akke, and M. Oliveberg, *Proc. Natl. Acad. Sci.* **96** 14854 (1999).
12. D. Baker, *Nature* **405**, 39 (2000).
13. G.T. Stewart, *Liq. Cryst.* **30**, 541 (2003); *ibid.* **31**, 443 (2004).
14. G. Zanchetta, M. Nakata, M. Buscaglia, T. Bellini, and N.A. Clark, *Proc. Natl. Acad. Sci.* **105**, 1111 (2008).
15. S. Zhang, *Biotechnol Adv.* **20**, 321 (2002).
16. W.F. Daamen, J.H. Veerkamp, J.C.M. van Hest, and T.H. van Kuppevelt, *Biomaterials* **28**, 4378 (2007).
17. D.P. Knight and F. Vollrath, *Philos. Trans. R. Soc. Lond. B* **357**, 155 (2002).
18. D.W.L. Hukins and J. Woodhead-Galloway, *Mol. Cryst. Liq. Cryst.* **41**, 33 (1977).
19. F.W. Keeley, C.M. Bellingham, and K.A. Woodhouse, *Philos. Trans. Roy. Soc. Lond. B* **357**, 185 (2002).
20. J.B. Gomez, *Anatomy of Hevea and Its Influence on Latex Production*, Malaysian Rub. Res. Dev. Board, Kuala Lumpur (1982).
21. I.S. Choi and C.M. Roland, *Rubber Chem. Technol.* **70**, 202 (1997)

22. R. Deval, V. Ramesh, G.B.K.S. Prasad, and A.K. Jain, *Ind. J. Dermatol., Venereol., Leprol.* **74**, 304 (2008).
23. M. Guthold, W. Liu, E.A. Sparks, L.M. Jawerth, L. Peng, M. Falvo, R. Superfine, R.R. Hantgan, and S.T. Lord, *Cell. Biochem. Biophys.* **49**, 165 (2007).
24. Y. Liu, Z. Shao, and F. Vollrath, *Biomacromolecules* **9**, 1782 (2008).
25. A.S. Tatham and P.R. Shewry, *Trends Biochem. Sci.* **25**, 567 (2000).
26. C. Heussinger, B. Schaefer, and E. Frey, *Phys. Rev. E* **76**, 031906 (2007).
27. Q. Wen, A. Basu, J.P. Winer, A. Yodh, and P.A. Janmey, *New J. Phys.* **9**, 428 (2007).
28. F.C. MacKintosh, J. Kas, and P.A. Jamney, *Phys. Rev. Lett.* **76**, 4425 (1995).
29. K. Kroy and E. Frey, *Phys. Rev. Lett.* **77**, 306 (1996).
30. D.L. Thomsen, P. Keller, J. Naciri, R. Pink, H. Jeon, D. Shenoy, and B.R. Ratna, *Macromolecules* **34**, 5868 (2001).
31. P.G. de Gennes, *Comptes Rendus Acad. Sci. II B* **324**, 343 (1997).
32. H. Finkelmann, W. Kaufhold, L. Noirez, A. ten bosch, and P. Sixou, *J. Physique II* **4**, 1363 (1994).
33. S. Lv, D.M. Dudek, Y. Cao, M.M. Balamurali, J. Gosline, and H. Li, *Nature* **465**, 69 (2010).
34. C. Levinthal, *J. Chim. Phys. Phys.-Chim. Biol.* **65**, 44 (1968).
35. F. Hoyle, *Eng. Sci.*, November, p. 8 (1981).
36. R. Akhtar, N. Schwarzer, M.J. Sherratt, R.E.B. Watson, H.K. Graham, A.W. Trafford, P.M. Mummery, and B. Derby, *J. Mater. Res.* **24**, 638 (2009).
37. E. Green, R. Ellis, and P. Winlove, *Biopolymers* **89**, 931 (2008).
38. P.D. Weinberg, C.P. Winlove, and K.H. Parker, *Biopolymers* **35**, 161 (1995).
39. J. Gosline, M. Lillie, E. Carrington, P. Guerette, C. Ortlepp, and K. Savage, *Philos. Trans. Roy. Soc. Lond. B* **357**, 121 (2002).
40. M.A. Lillie and J.M. Gosline, *Biopolymers* **29**, 1147 (1990).
41. J. Reguera, D.W. Urry, T.M. Parker, D.T. McPherson, and J.C. Rodriguez-Cabello, *Biomacromolecules* **8**, 354 (2007).
42. M.A. Lillie and J.M. Gosline, *Biopolymers* **39**, 641 (1996); *ibid.* **64**, 127 (2002).
43. K. Kotitschke, R. Kimmich, E. Rommel, and F. Parak, *Prog. Coll. Polym. Sci.* **83**, 211 (1990).
44. D. Ringe and G.A. Petsko, *Biophys. Chem.* **105**, 667 (2003).
45. J. Swenson, H. Jansson, J. Hedstrom, and R. Bergman, *J. Phys. Condens. Matter* **19**, 205109 (2007).
46. K.L. Ngai, S. Capaccioli, and N. Shinyashiki, *J. Phys. Chem. B* **112**, 3826 (2008).
47. K.L. Ngai, S. Capaccioli, M.S. Thayyil, and N. Shinyashiki, *J. Therm. Anal. Calor.* **99**, 123 (2010).
48. K. Kawai, T. Suzuki, and M. Oguni, *Biophys. J.* **90**, 3732 (2006).
49. H. Frauenfelder, G. Chen, J. Berendzen, P.W. Fenimore, H. Jansson, B.HY. McMahon, I.R. Stroe, J. Swenson, and R.D. Young, *Proc. Natl. Acad. Sci.* **106**, 5129 (2009).
50. W. Doster, *Biochim. Biophys. Acta* **1804**, 3 (2010).
51. T. Kleinert, W. Doster, H. Leyser, W. Petry, V. Schwarz, and M. Settles, *Biochemistry* **37**, 717 (1998).
52. L. Olivares-Quiroz and L.S. Garcia-Colin, *Biophys. Chem.* **144**,123 (2009).
53. J.M. Gosline, *Rubber Chem. Technol.* **60**, 417 (1987).
54. E. Davis, *J. Histochem.* **100**, 17 (1993).
55. J. Lepetit, *Meat Sci.* **76**, 147 (2007).

56. S. Vogel, *J. Biosciences* **30**, 303 (2005).
57. M. Jensen and T. Weis-Fogh, *Philos. Trans. Roy. Soc. Lond. B* **245**, 137 (1962.)
58. A.C. Neville, *Quart. J. Micr. Sci.* **106**, 269 (1965).
59. S.O. Andersen, *Insect Biochem. Molec. Biol.* 34, 459 (2004).
60. R.C. Herbert, P.G. Young, C.W. Smith, R.J. Wootton, and K.E. Evans, *J. Exp. Biol.* **203**, 2945 (2000).
61. J. Okada and Y. Toh, *J. Comp. Physiol. A* **186**, 849 (2000).
62. T. Thurm, *Science* **145**, 1063 (1964).
63. H.A. Edwards, *J. Exp. Biol.* **105**, 407 (1983).
64. S. Vogel, *Nature* **385**, 21 (1997).
65. M. Denny and L. Miller, *J. Exp. Biol.* **209**, 4503 (2006).
66. W.A. Linke1, M. Ivemeyer, N. Olivieri, B. Kolmerer, J.C. Ruegg, and S. Labeit, *J. Mol. Biol.* **261**, 62 (1996).
67. L. Tskhovrebova and J. Trinick, *Philos. Trans. Roy. Soc. Lond. B* **357**, 199 (2002).
68. H. Granzier and S. Labeit, *J. Phys.* **541**, 335 (2002).
69. A. Minajeva, C. Neagoe, M. Kulke, and W.A. Linke, *J. Phys.* **540**, 177 (2002).
70. C.A. Opitz, M. Kulke, M.C. Leake, C. Neagoe, H. Hinssen, R.J. Hajjar, and W.A. Linke, *Proc. Natl. Acad. Sci.* **100**, 12688 (2003).
71. W.A. Shear, J.M. Palmer, J.A. Coddington, and P.M. Bonamo, *Science* **246**, 479 (1989).
72. F. Vollrath and D. Porter, *Polymer* **50**, 5623 (2009).
73. F. Vollrath and D. Porter, *Soft Matter* **2**, 377 (2006).
74. L.-Y. Cui, F. Liu, and Z.-C. Ou-Yang, *Thin Solid Films* **518**, 735 (2009).
75. D.H. Hijirida, K.G. Do, C. Michal, S. Wong, D. Zas, and L.W. Jelinski, *Biophys. J.* **71**, 3442 (1996).
76. C. Fu, D. Porter, and Z. Shao, *Macromolecules* **42**, 7877 (2009).
77. E.V. Lugovskoi, P.G. Gritsenko, and S.V. Komisarenko, *Russ. J. Biooganic Chem.* **35**, 393 (2009).
78. J.V. Shah and P.A. Janmey, *Rheol. Acta* **36**, 262 (1997).
79. J.P. Collet, H. Shuman, R.E. Ledger, S. Lee, and J.W. Weisel, *Proc. Natl. Acad. Sci.* **102**, 9133 (2005).
80. J. Shah and P.A. Janmey, *Rheol. Acta.* **36**, 262 (1997).
81. K.M. Weigandt, D.C. Pozzo, and L. Porcar, *Soft Matter* **5**, 4321 (2009).
82. P.A. Evans, K. Hawkins, P.R. Williams, and R.L. Williams, *J. Non-Newt. Fluid Mech.* **148**, 122 (2008).
83. P.A. Evans, K. Hawkins, M. Lawrence, R.L. Williams, M.S. Barrow, N. Thirumalai, and P.R. Williams, *Med. Eng. Phys.* **30**, 671 (2008).
84. A.S. Tatham, L. Hayes, P.R. Shewry, and D.W. Urry, *Biochim. Biophys. Acta* **1548**, 187 (2001).
85. S. Domenek, L. Brendel, M.-H. Morel, and S. Guilbert, *Biomacromolecules* **5**, 1002 (2004).

10
Confinement effects on polymer dynamics

10.1 Spatial confinement

The development of nanotechnology has spurred efforts to understand the properties of materials under physical confinement. Unusually large surface to volume ratios and the intrusion of an external dimension on the cooperativity length scale in condensed matter change the behavior from that in the bulk state. For polymers, which have coil sizes in the tens of nm range, there is the additional factor that the confinement may perturb the shape of the molecules. A polymer chain can lose degrees of freedom under confinement, potentially giving rise to unique properties.

The effect of confinement on the relaxation dynamics of liquids and polymers is complex and there are many anomalies. For example, while generally thin films exhibit lower glass-transition temperatures than the bulk T_g, some experimental methods yield different or negligible dependences on film thickness [1–6]. Many experimentalists employ dielectric spectroscopy to study thin films, but the measured permittivity is not additive in the layer contributions, so that dielectrically inactive regions at the electrode surface can shift the relaxation peak without any actual change in the dynamics [7]. These complications are part of the reason that the question of why confinement sometimes exerts such large effects is far from resolution. Progress requires more complete and consistent experimental characterization of how confinement affects the structure and local dynamics. The effect of confinement on the rheology of polymers is an additional, more challenging problem.

10.1.1 Segmental dynamics

When a material is constrained spatially to a sufficient extent (dimension $h < 100$ nm), its glass-transition temperature often becomes lower than T_g of the bulk state [8,9]. This effect is well established for freestanding films; data for polystyrene measured by different methods are shown in Figure 10.1. Similar reductions in T_g are often observed for films supported on a substrate or material in nanopores (Figure 10.2) [10–13]. These types of spatial confinement tend to yield smaller reductions than seen for freestanding films [14–17], since interaction with the substrate may hinder molecular motions. Phase-segregated block-copolymers (Figure 10.1) [18,19]) sometimes show a T_g reduction similar to that in thin films of the corresponding homopolymer, implying the domains are confined due to the morphology; however, in other experiments this effect is absent [20]. There is an analogy to the size dependence of melting points, although the suppression therein is generally larger than the effect on T_g and has a different origin [8,21,22].

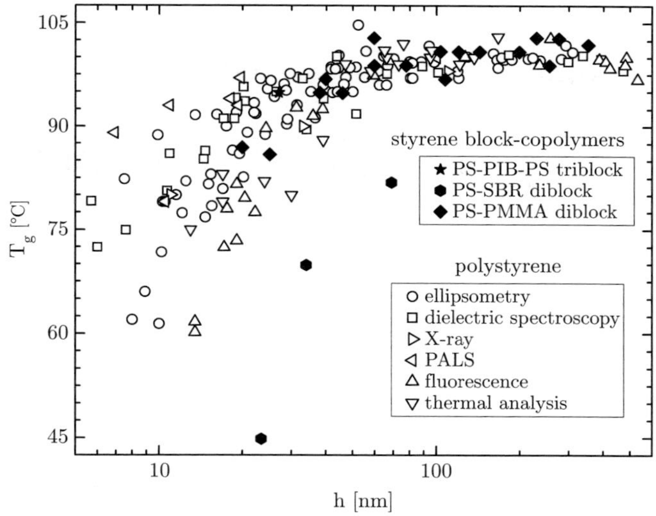

Figure 10.1 Glass-transition temperature of freestanding films of atactic PS measured by the indicated methods (open symbols) and of the styrene block in block-copolymers (filled symbols). For films h is the film thickness and for block copolymers it represents the domain size. The variation of M among samples causes some of the spread of the data, due to the molecular weight dependence of $T_g(h)$. Data from refs. [9,18,19].

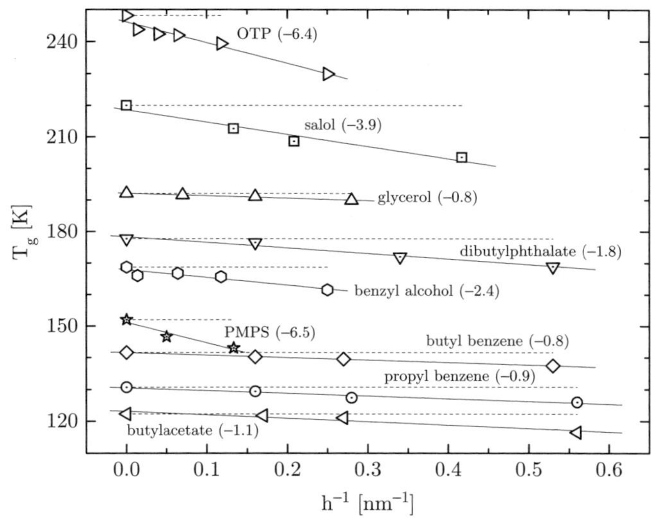

Figure 10.2 Glass-transition temperature for liquids contained in pores as a function of reciprocal pore size. The dotted lines are an extension of the bulk T_g and the solid lines are linear fits, which approximately describe the data. The quantity in parentheses is the change in T_g for confinement in a 10 nm pore. Data from refs. [10–13].

The mechanism for the T_g reduction in confined materials seems to be the enhanced mobility of surface molecules [23–28], a result of fewer constraints from neighbors and/or greater unoccupied volume. Lennard-Jones MD simulations show directly the increase in mobility of particles with increasing distance from a stationary boundary (e.g. wall), with a concomitant narrowing of the relaxation dispersion, consistent with less constrained dynamics (Figure 10.3) [29,30]. The effect can be opposite for fillers with softer surfaces, as seen in the slowing of the dynamics effected by confinement in porous glass versus the faster segmental motion when microemulsions are used to confine the sample (Figure 10.4).

If surface mobility is the origin of the reduction in T_g, it implies that the dynamics in thin films is not homogeneous (in addition to the dynamic heterogeneity intrinsic to condensed matter; Section 2.3.2). Thus, the measured T_g is an average that becomes smaller for thinner films. From this idea, the following expression can be derived for the glass transition of thin films [23]

$$T_g(h) = T_g^{\infty} + \frac{2h_0}{h}\left(T_g^{\infty} - T_g^{surf}\right) \tag{10.1}$$

T_g^{∞} is the bulk value of T_g and h_0 is the thickness of the mobile, surface layer having $T_g = T_g^{surf}$.

This explanation, however, cannot account for many details of the phenomenon. For example, there is a molecular weight dependence, as shown in Figure 10.5 for PS [31] and Figure 10.6 for PMMA [32]. Yet, at least for PS, at lower molecular weights (but still well above the value at which T_g becomes independent of chain length) the confinement effect is invariant to molecular weight (Figure 10.5) [31]. Chemical structure also influences $T_g(h)$. This is seen in the results for pore confinement in Figure 10.2 and comparing the data for PS and PMMA. However, while less sensitive to h than PS, the T_g of PMMA is more affected by

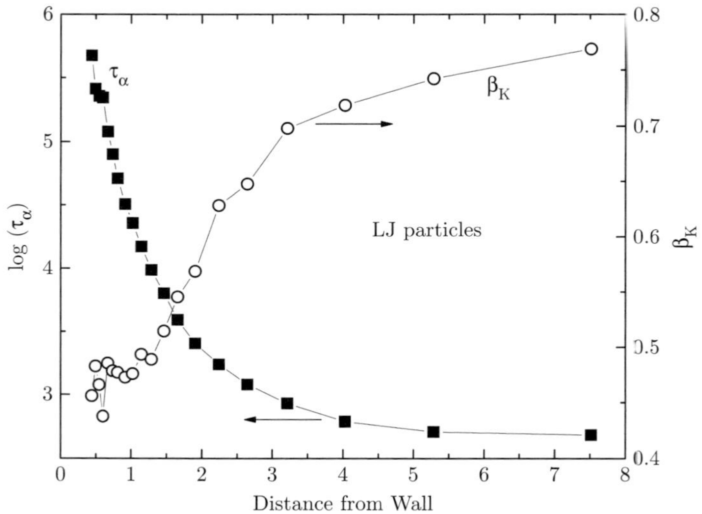

Figure 10.3 Semilogarithmic plots of the local relaxation times from molecular dynamics simulation results for the self-intermediate scattering function, along with the fitted Kohlrausch exponents. Data from refs. [29,30].

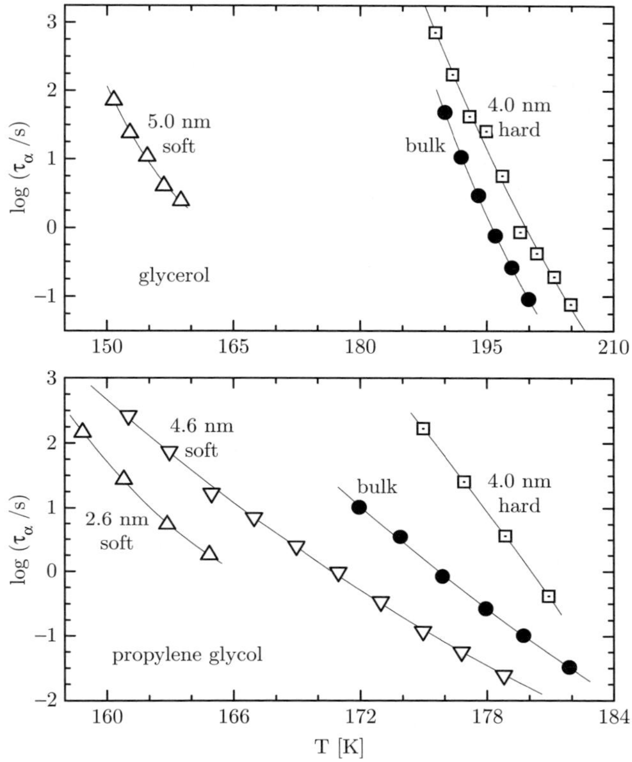

Figure 10.4 Relaxation times for two liquids in the bulk (filled circles) and confined in porous glass (open squares) or liquid droplets (open triangles). Data from ref. [14].

confinement than are other acrylate polymers (Figure 10.7) [33]. A change in tacticity in PMMA can even change the effect from a reduction to an increase in T_g (Figure 10.8) [20,34,35].

In thin films showing lower glass-transition temperatures, quantities presumably associated with greater segmental mobility above T_g, such as the heat capacity (Figure 10.9) [5,36] and thermal expansivity [33,37,38], tend to be smaller under confinement. The magnitude of these effects is reduced when the film is confined by either a bottom or top substrate layer (Figure 10.10 [37]), although the reduction in thermal expansion above T_g seems to be unrelated to the amount by which T_g is reduced (*cf.* data for different polyacrylates in Figure 10.7) [33]. In pore sizes below ~ 1.7 nm, the heat capacity jump at T_g is absent entirely, suggesting a finite length scale requirement for the glass transition [36]. More generally, the properties in the glassy state are essentially unaffected by confinement.

10.1.2 Chain dynamics

Whether the effect of spatial confinement extends to rheological properties depends on the proximity to T_g, since at sufficiently high temperatures the influence of the segmental dynamics on chain motions becomes negligible. Confining geometries for which h is

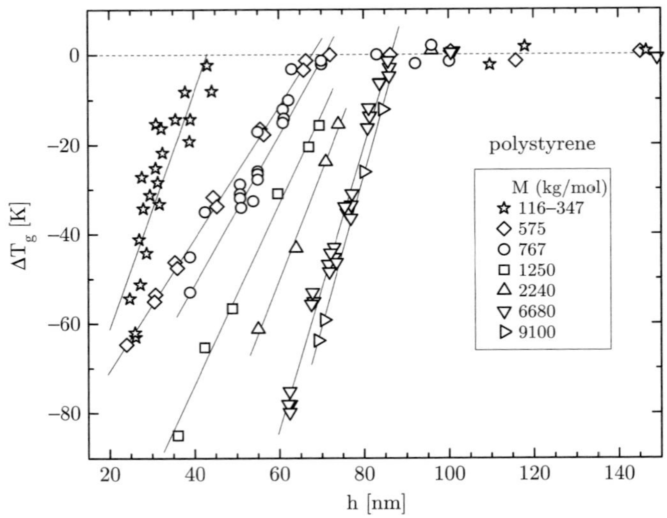

Figure 10.5 Effect of molecular weight on confinement-induced reduction of T_g in freestanding films of PS. Below 347 kg/mol, $T_g(h)$ shows no M dependence. Data from ref. [31].

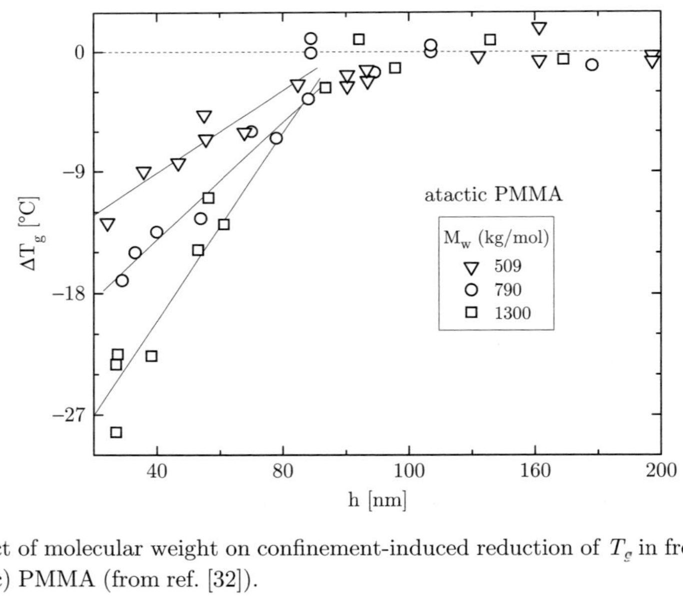

Figure 10.6 Effect of molecular weight on confinement-induced reduction of T_g in freestanding films of (primarily atactic) PMMA (from ref. [32]).

commensurate with or smaller than the coil size may introduce new effects. Thin films of unentangled polystyrene supported on a substrate exhibited viscosities orders of magnitude lower than the bulk viscosity (Figure 10.11) [39]. Using the limiting high M value of the characteristic ratio for PS, $C_\infty = 1.03$ nm, eqn (1.24) gives an upper bound on the coil size, $R_g = \sqrt{\langle r^2 > /6} < 1.4$ nm; thus, even the thinnest films in these experiments were larger than

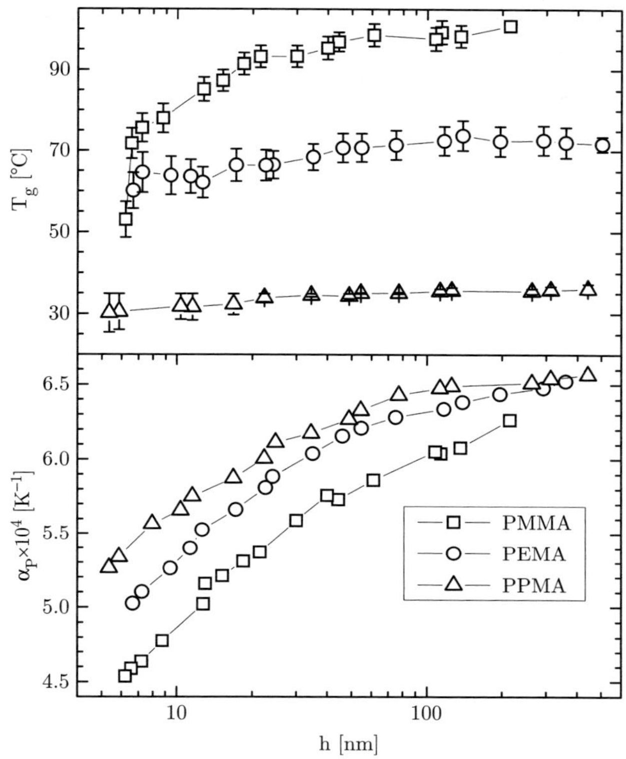

Figure 10.7 Effect of confinement on T_g (upper) and rubbery thermal expansion coefficient of acrylate polymers spin coated onto silicon. Data from ref. [33].

the polymer chains, notwithstanding the large reductions in viscosity. The temperature dependence of η could be described by the VFTH equation (eqn (2.4)), with the Vogel temperature, T_0, having an h dependence similar to that of T_g [39]. In addition to eqn (10.1), the effect of confinement has been parameterized as [40]

$$T_g(h) = \frac{T_g^{\infty}}{1 + \hat{h}/h} \tag{10.2}$$

where $\hat{h}$ is a constant. Confinement effects are absent when $h \gg \hat{h}$; for the data in Figure 10.11, $\hat{h} = 0.6$ nm [39]. Another alternative is [41]

$$T_g(h) = T_g^{\infty}\left(1 - \left(\frac{\hat{h}}{h}\right)^{\xi_h}\right) \tag{10.3}$$

where for polystyrene best-fits to data yield $\hat{h}$=3.2 nm and ξ_h =1.63 [42].

At temperatures 90 K above T_g, for which the influence of the segmental dynamics should be negligible, the flow of entangled polystyrenes ($M > M_e$) through 15-nm diameter cylindrical channels was substantially greater than the flow rate calculated from the bulk viscosity [43].

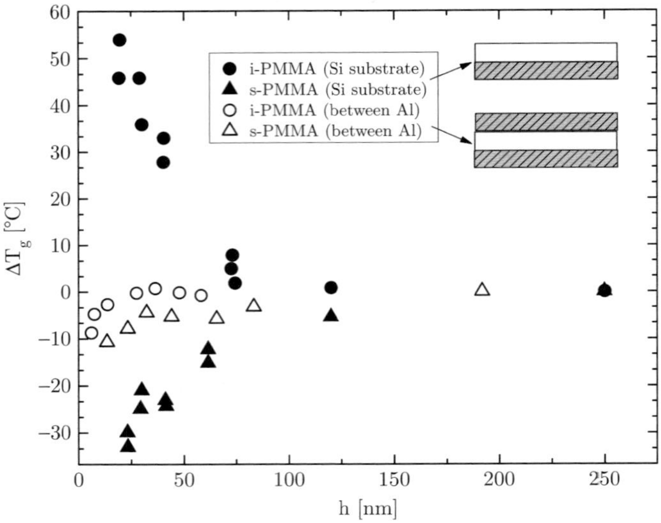

Figure 10.8 Variation of T_g with thickness for isotactic (circles) and syndiotactic (triangles) PMMA films on silicon substrates (solid symbols) and between aluminum plates (open symbols). The effect of confinement and the influence of the substrate both appear to depend on the polymer stereochemistry. Data from refs. [20,34,35].

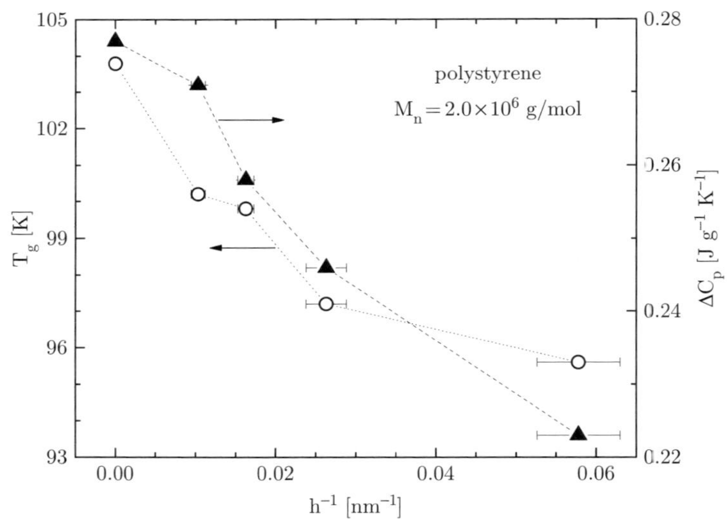

Figure 10.9 Change in T_g (circles) and heat capacity increment at T_g (triangles) as a function of reciprocal film thickness for PS spin coated onto mica. Data from ref. [5].

The channel radius in these experiments was smaller than R_g for even the lowest molecular weight sample ($M_w = 114$ kg/mol), with the flow enhancement greater for larger M (Figure 10.12). This indicates that confinement in the channel extended the chain molecules, thus reducing their degree of entanglement, analogous to the effect of a shear-thinning flow field.

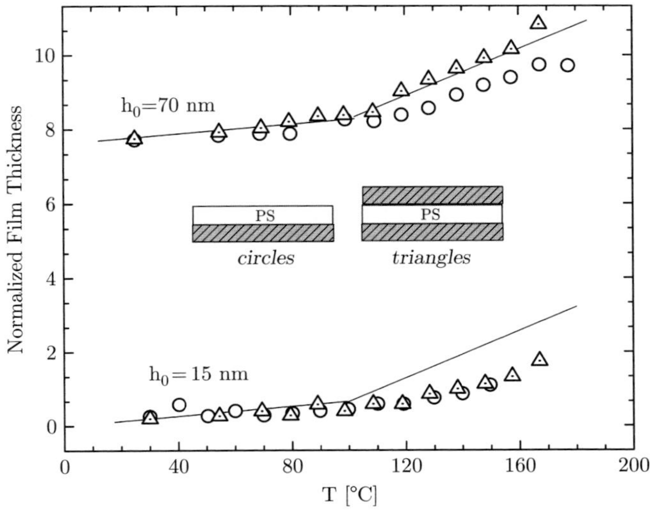

Figure 10.10 Change in thermal expansion above T_g of PS films supported on a rigid substrate and having the indicated initial thickness. Solid represents bulk behavior. The reduction in α_P is larger for thinner films and in the absence of a confining layer on top. Data from ref. [37].

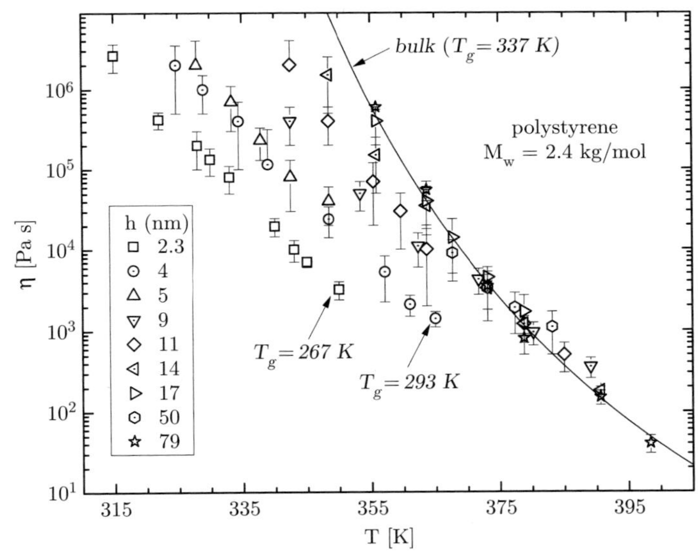

Figure 10.11 Effect of confinement on unentangled PS thin films deposited on silicon. Representative T_g values are indicated; the viscosity reduction under confinement is consistent with the enhancement of the segmental dynamics. Data from ref. [39].

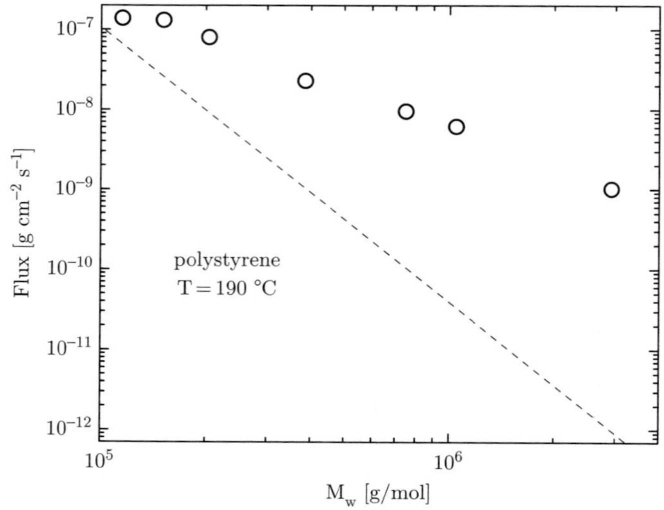

Figure 10.12 Flux of polystyrene through 15-nm cylindrical channels as a function of molecular weight (symbols). The dashed line represents the flow rate calculated from the bulk viscosity. Data from ref. [43].

The stretching of polymer chains confined to nanochannels is of interest in biology, for example to probe the effect of chain flexibility on protein binding [44,45].

Surface-diffusion measurements on plied PS films showed the opposite effect of confinement on mobility [46]. The time for the interfacial region to attain the bulk density, governed by removal of surface asperities and irregularities by chain diffusion, was found to increase, reflecting slower chain motion, with decreasing film thickness (Figure 10.13). For the PS used in the study, $R_g \sim 22$ nm $< h$, implying that the chain dimensions were perturbed by the confinement. It has been suggested that under such circumstance chain molecules adopt a new mode of diffusive motion, a sliding mechanism [47]; however, the behavior in Figure 10.13 is opposite to the enhanced mobility predicted for such motion.

Diffusivity measurements on PS thin films also showed lower mobility, notwithstanding a decrease in T_g (Figure 10.14) [48]. The latter was observed for h as much as six-fold larger than the chain size. However, the mean square displacement, MSD, was very small, corresponding to relaxation at much shorter times than the reorientations and diffusions associated with structural relaxation. Thus, these experiments probed the caged dynamics, not global motion of the chains, with the former evidently slowed down by confinement.

10.2 Filler effects

Polymers often include a significant amount of carbon black, silica, or other reinforcing filler, in order to improve the mechanical properties. Optimal performance requires a significant interfacial region, good dispersion, and interfacial stress transfer [49]. Non-reinforcing fillers such as clay, calcium carbonate, talc, and other mineral fillers are also used. These do not enhance the strength or toughness, but generally lower the cost and may improve selected

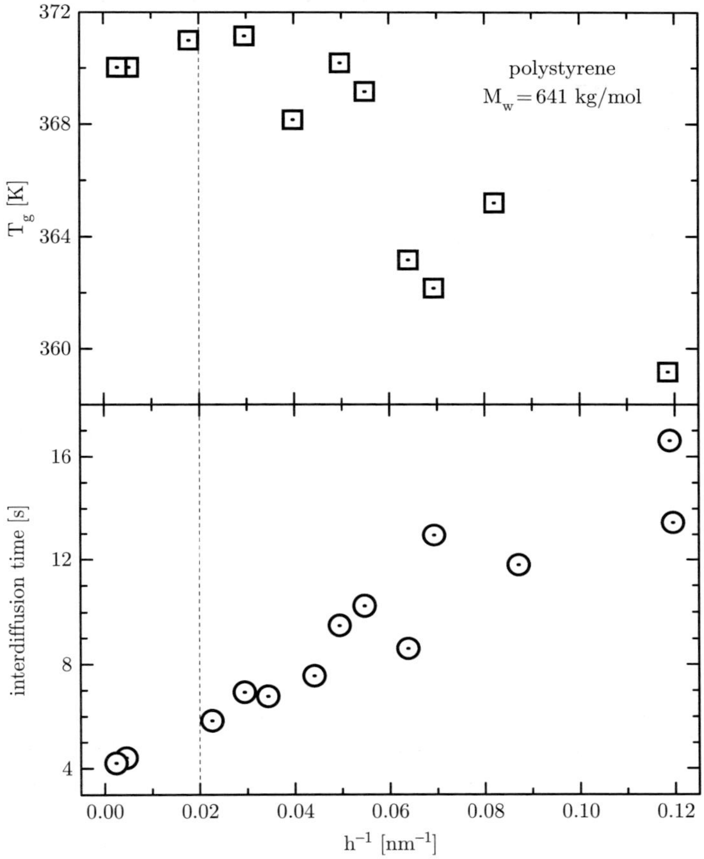

Figure 10.13 Glass-transition temperature and time for interdiffusion between two PS layers having total thickness h with the lower layer supported on a silicon substrate. The dashed line denotes the abscissa value corresponding to $\langle R_g^2 \rangle^{-1/2}$. Data from ref. [46].

properties. A more recent development is the use of nanoparticles, which are smaller than conventional fillers and thus give rise to confinement or other effects specific to nm length scales. Substantial improvements in the mechanical properties can be obtained with a few per cent or less of nanofiller [50–52], although achieving good dispersion can be difficult without chemical modification of the particles [53–55]. Examples of the large changes in stiffness and strength are shown in Figure 10.15 for carbon nanotubes [56] and in Figure 10.16 for nano-layered clay [57]. The focus of this section is the effect of nanofillers on the segmental dynamics. For particles having nm dimensions, unusual effects are seen, similar to those found for thin films and pore confinement. For conventional fillers confinement effects are usually absent, although there is controversy concerning the effect on the dynamics of segments in close proximity to carbon black or silica surfaces.

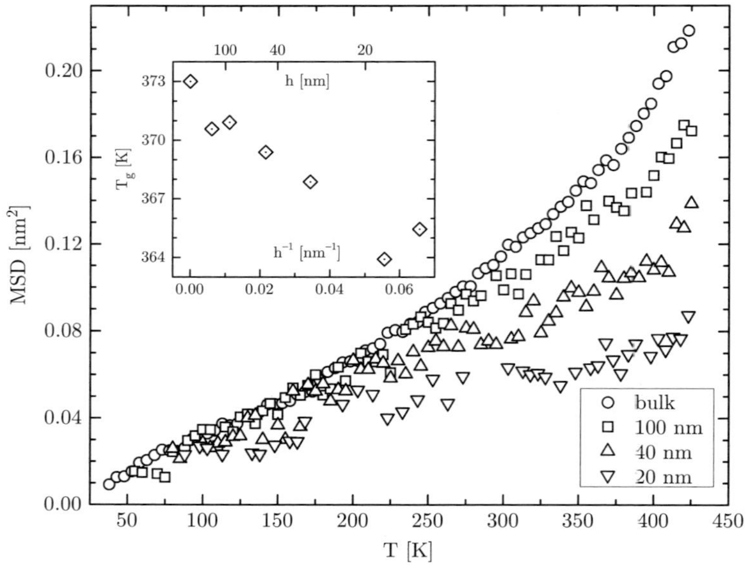

Figure 10.14 Mean squared displacement of PS chains in films having the indicated thickness. The diffusion becomes slower due to confinement, although T_g is reduced (inset). Data from ref. [48].

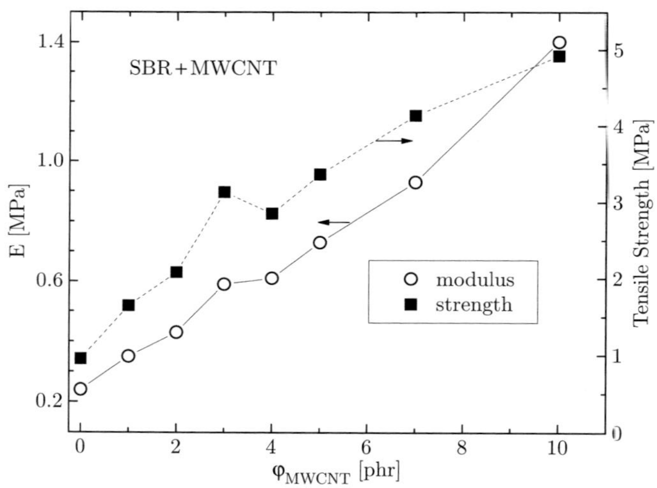

Figure 10.15 Young's modulus (open circles) and tensile stress at break (filled squares) for vulcanized styrene–butadiene copolymer as a function of the concentration of multiwall carbon nanotubes. There is a 45% increase in stiffness and 70% increase in strength for addition of just 1 phr of this nanofiller. Data from ref. [56].

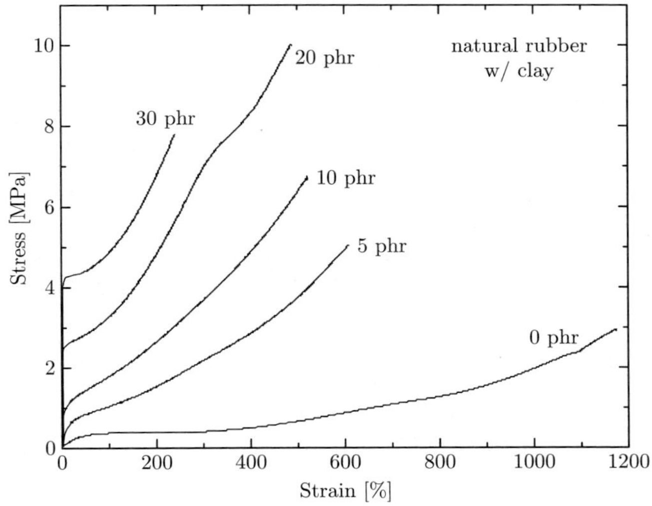

Figure 10.16 Tensile stress–strain curves for uncrosslinked natural rubber having the indicated amount of montmorillonite clay. Data from ref. [57].

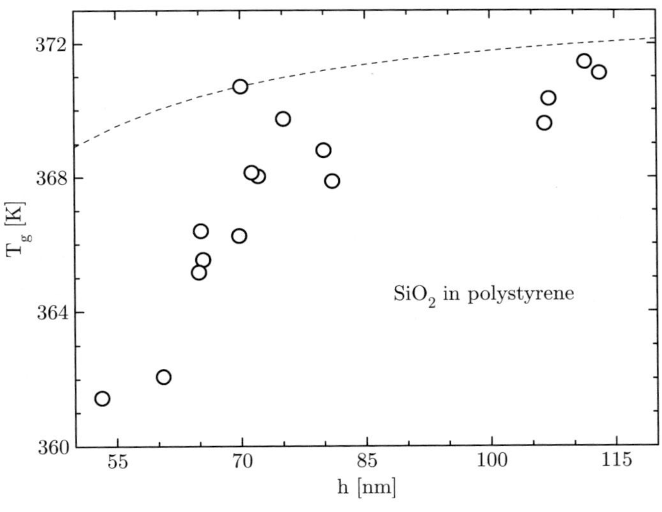

Figure 10.17 T_g of PS homopolymer containing concentrations up to 40 weight % SiO_2 nanoparticles *versus* the distance between particles (open circles). The dashed line is eqn (10.3) fitted to $T_g(h)$ data for thin films of PS (with h now the film thickness). The rough agreement required use of the harmonic average of h obtained using eqn (10.1). Data from ref. [58].

10.2.1 Nanoparticles

Sufficiently small filler particles can enhance segmental mobility [34,58–64], provided there is no adsorption at the filler surface [65,66]. There is a rough correspondence between the effect of particle separation and film thickness on T_g reduction; however, the equivalence is not quantitative [67] and even semiquantitative agreement requires judicious selection of the average value used for h [68].

Results are shown in Figure 10.17 for PS containing various concentrations of 14-nm silica particles, with h now representing the distance between particles [58]. In Figure 10.18 are results for syndiotactic PMMA mixed with Al_2O_3 and nanolayered clay. The effect of the two fillers are comparable and both reduce T_g more than films deposited on silicon substrates at equivalent h. Figure 10.19 [60] shows T_g reductions in PS containing 5 nm gold particles onto which PS chains ($M_w \sim 1$ kg/mol) have been grafted. The change in T_g is much larger than if equivalent chains were added without the Au particles. Interestingly, higher molecular weight ($M_w \sim 50$ kg/mol) grafted chains had the opposite effect on T_g [60]. However, to some extent this is due to the increase in T_g expected from simply adding chains of higher molecular weight, independent of any effect of particles.

Similar to the anomaly seen with confined polymers, the effect of nanoparticles on T_g can differ qualitatively among different polymers. This is illustrated in Figure 10.20 [69], showing data for three polymers. Incorporation of alumina nanoparticles increases, decreases, or has no effect on T_g. Nanolayered clay was found to speed up segmental motion in polypropylene glycol, while the normal mode relaxation times became longer (Figure 10.21) [61]. C_{60} spherical fullerenes ("buckyballs") raised the T_g of PS and PVAc [70].

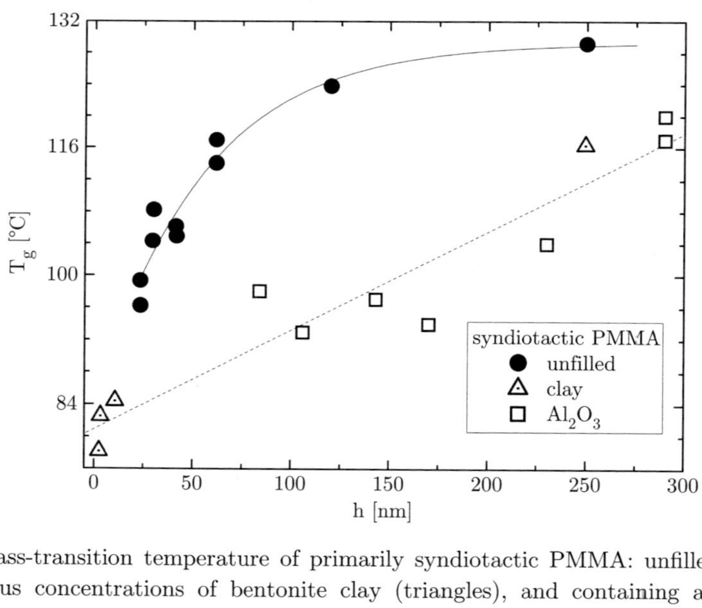

Figure 10.18 Glass-transition temperature of primarily syndiotactic PMMA: unfilled (solid circles), containing various concentrations of bentonite clay (triangles), and containing alumina particles (squares). The abscissa is a measure of the distance between particles, calculated as the ratio of the volume of polymer to the surface area of the filler, or film thickness for the neat material. Data from refs. [35,59].

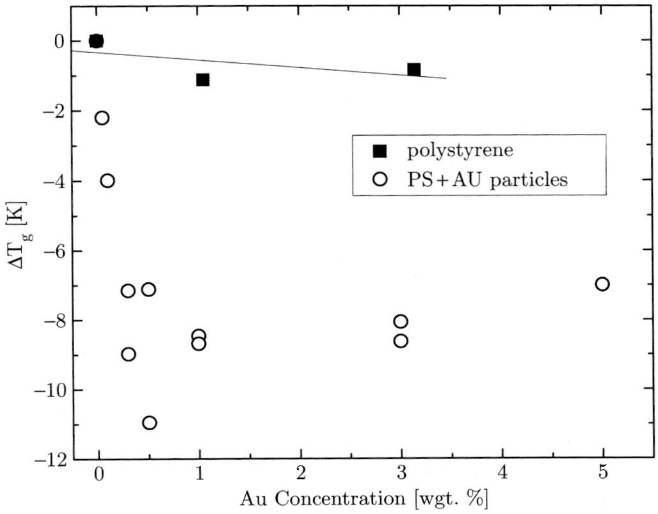

Figure 10.19 Change in calorimetric glass-transition temperature of PS ($M_w = 5.2$ kg/mol) as a function of the concentration of 5-nm gold particles. The particles had PS ($M_w = 1$ kg/mol) grafted to the surface. The solid squares show the change in T_g upon addition of the same molecular weight as the grafted PS but without the Au particles. Data from ref. [60].

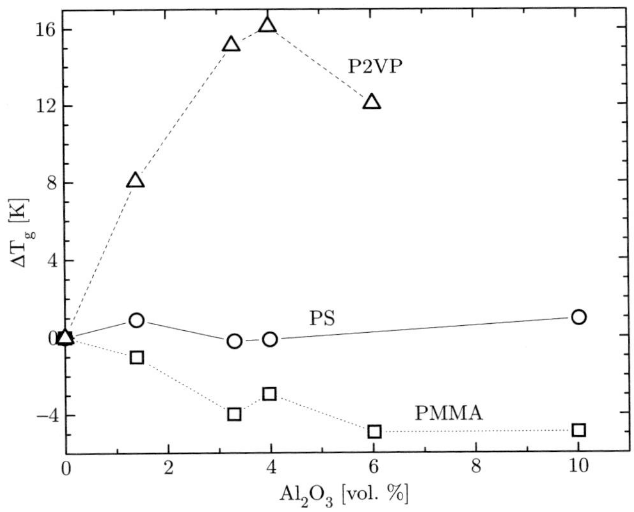

Figure 10.20 Change in T_g from value for neat polymer as a function of concentration of 47 nm alumina spheres for poly-2-vinylpyridine (triangles), polystyrene (circles), and polymethylmethacrylate (squares). Error bars are roughly ±1 K. Data from ref. [69].

In toto, results for nanoparticle reinforcement are complicated, due both to the inherent complexity of filled systems (with variables including particle size, shape, physical and chemical nature of surface, and concentration) and the difficulty of reconciling the disparate effects on the segmental and chain dynamics. Even the same system can give opposite results

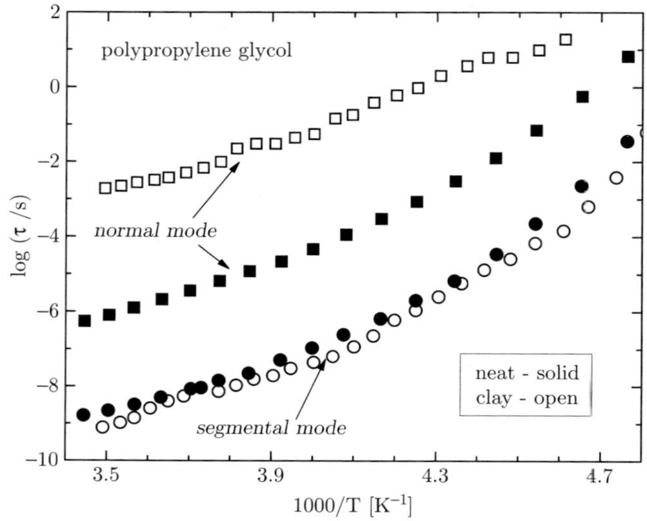

Figure 10.21 Local segmental (circles) and normal mode (squares) relaxation times for PPG (M = 4kg/mol) neat (solid symbols) and containing nano-layered clay (open symbols). Data from ref. [61].

based on the method of sample preparation [69,71]. Nevertheless, while the topic requires further understanding, the breadth of results offers many opportunities for applications of nanocomposites.

10.2.2 Conventional fillers

In general, the particle size of traditional reinforcing fillers is too large to measurably change T_g [65]. However, proximity to a filler surface can affect the mobility of at least some polymer segments, and there is literature reporting the existence of a shell of polymer around filler particles having larger τ_α and higher T_g [72–75]. On the other hand, other studies find no evidence of any change in the polymer dynamics due to filler [76–80]. These differing results reflect to some extent variations in particle size and particle–polymer interaction among different systems. However, some discrepancies in the literature can be ascribed to ambiguous or erroneous methods of data analysis.

NMR has been used frequently to characterize the effect of filler on proton spin–spin relaxation times (T_2) by analyzing spin echoes formed using various pulse sequences. Among these studies there is disagreement on whether an immobilized or higher T_g phase is present in the vicinity of the filler particle (Table 10.1 [81–83, 85–88, 90–94, 96–98]). An immediate complication is that chains attached to filler particles, as well as any chemically crosslinked chains, undergo restricted, anisotropic motion. This suppresses motional narrowing of the NMR spectral lines, complicating efforts to quantify the relaxation processes [86,87]. Experiments cannot always distinguish between chain segments immobilized by their spatial proximity to filler particles versus chain units bonded at specific sites on the filler surface. For example, Dutta *et al.* [88] found that only the olefinic carbons in filled SBR were immobilized, indicating that specific interactions at the carbon black surface were necessary for the effect. The interactions can cause gel formation ("bound rubber") [89], but that is different from a glassy interface.

Table 10.1 NMR results on chain immobilization in carbon-black-filled rubbers.

Polymer	Conclusion	Reference
1,4-polyisoprene	no immobilization	[86, 98]
	immobilization of segments attached to filler	[96]
	reduced segmental mobility	[83,87]
	3–8-nm thick bound layer	[91]
	immobilized layer	[93]
1,4-polybutadiene	immobilization of segments attached to filler	[97]
	interfacial phase with reduced mobility	[83,85]
	immobilized layer	[92]
SBR	no immobilization	[81]
	reduced mobility of olefinic protons bound to filler	[88]
	immobilized layer	[94]
EPDM	increased segmental mobility	[90]
	no immobilization	[82]
	immobilized chains	[92]

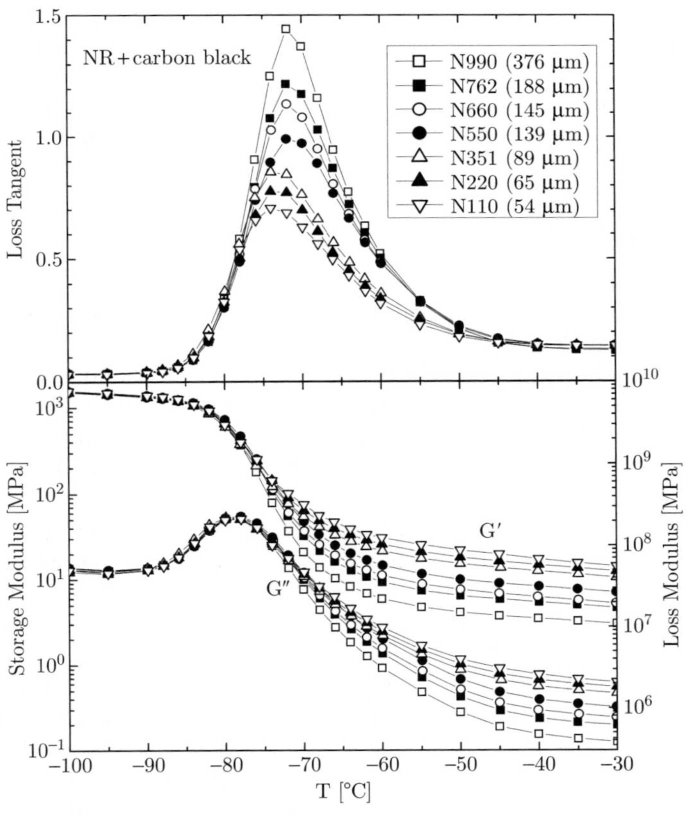

Figure 10.22 (upper) Loss tangent and (lower) dynamic moduli for PBD containing 14% by volume (50 phr) of the indicated carbon blacks. With decreasing aggregate size (indicated in the legend), the plateau modulus increases, causing a shift of the loss tangent towards lower temperature. Data from ref. [101].

In the NMR experiments species can be distinguished according to their mobility by using different pulse sequences [90,91]. However, more often proton-relaxation curves (e.g. free induction decay and/or spin-echo relaxation) are fit to a sum of functions, with the term having the longest time constant identified as the least mobile phase, presumed to be polymer segments residing in the interfacial region. Functions employed include (i) simple exponential decays [88,92,93], which assumes there is no mutual interference among different NMR frequencies (i.e. homogeneous, independent relaxations); (ii) Gaussian decay [94], which is a common shape for dipolar-broadened lines in solids, but neglects the influence of molecular motions; (iii) empirical functions such as the Weibull[1] [91,95,96] and log-normal distributions

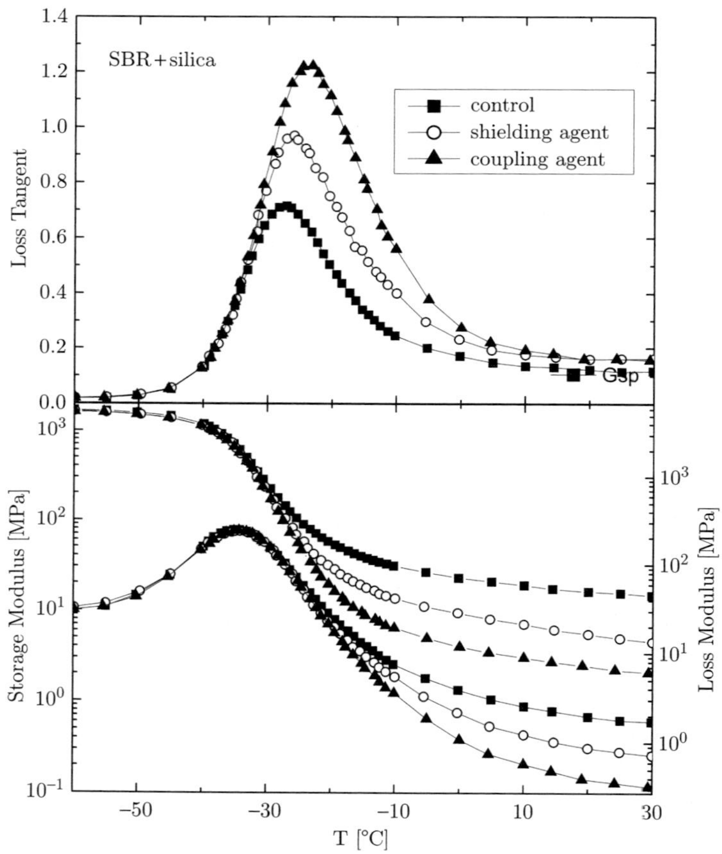

Figure 10.23 (upper) Loss tangent and (lower) dynamic moduli for SBR containing 20% by volume silica (aggregate size 100–1000 nm). Samples had no additive (solid squares), *n*-octyltriethoxysilane as a filler-filler shielding agent (open circles) and 3-mercaptopropyltrimethoxysilane) as a polymer–filler coupling agent (solid triangles). The changes in filler–polymer interaction affect the rubbery modulus, but not the segmental dynamics. Data from ref. [101].

[1] The Weibull function is equivalent to the Kohlrausch function (eqn (1.7)) without restricting the exponent to values between zero and unity. The history of these equations is given in ref. [96].

[90]; and (iv) combinations of different equations to represent different "domains" of polymer [97–100]. Such analyses do not yield unique solutions, since different fitting functions give different results. This accounts in part for the discrepancies in Table 10.1.

A common error in assessments of the effect of filler reinforcement on T_g from mechanical spectroscopy is basing conclusions on the temperature of the maximum in the isochronal loss tangent. This can be misleading because the loss tangent peak in the softening zone of the viscoelastic spectrum is influenced not only by the local segmental dynamics, as seen directly in the loss modulus at lower temperatures, but also by filler-induced changes in the dynamic modulus at higher T. Thus, an increase in rubbery modulus can cause the loss tangent to shift to lower frequency, ostensibly indicating higher T_g, without any actual change in the segmental dynamics. This is illustrated by dynamic-mechanical data in Figure 10.22 [101] for a polybutadiene containing various types of carbon black. With decreasing size of the particles, there is greater reinforcement and hence an increasing rubbery plateau modulus. This causes a shift of the loss tangent toward higher temperatures, ostensibly suggesting a higher T_g; however, there is no change in the loss modulus peak and thus no change in the segmental dynamics. Similar behavior is observed in an SBR with 20% by volume silica particles (Figure 10.23 [101]). Increasing the polymer–filler interaction with a coupling agent or by suppressing filler–filler agglomeration with a shielding agent increases the storage modulus at higher temperatures, but there is no change in the segmental dynamics.

References

1. G.B. McKenna, *J. Phys. IV*, **10**, Pr7–53 (2000).
2. Y. Grohens, L. Hamon, G. Reiter, A. Soldera, and Y. Holl, *Eur. Phys. J. E* **8**, 217 (2002).
3. Z. Fakhraai, J.S. Sharp, and J.A. Forrest, *J. Polym. Sci. Polym. Phys. Ed.* **42**, 4503 (2004).
4. H. Huth, A.A. Minakov, A. Serghei, F. Kremer, and C. Schick, *Eur. Phys. J. Special Top.* **141**, 153 (2007).
5. Y.P Koh and S.L. Simon, *J. Polym. Sci. Polym. Phys. Ed.* **46**, 2741 (2008).
6. A. Serghei, D. Chen, D.H. Lee, and T.P. Russell, *Soft Matter* **6**, 1111 (2010).
7. A. Serghei, M. Tress, and F. Kremer, *J. Chem. Phys.* **131**, 154904 (2009).
8. M. Alcoutlabi and G.B. McKenna, *J. Phys.: Condens. Matter* **17**, R461 (2005).
9. C.B. Roth and J.R. Dutcher, *J. Electroanal. Chem.* **584**, 13 (2005).
10. C.L. Jackson and G.B. McKenna, *J. Non-Cryst. Solids* **131–133**, 221 (1991).
11. A. Schonhals, H. Goering, C. Schick, B. Frick, M. Mayorova, and R. Zorn, *Eur. Phys. J. E* **12**, 173 (2003).
12. J. Zhang, G. Liu, and J. Jonas, *J. Phys. Chem.* **96**, 3478 (1992).
13. J. Koppensteiner, W. Schranz, and M.A. Carpenter, *Phys. Rev. B* **81**, 024202 (2010).
14. F. He, L.-M. Wang, and R. Richert, *Phys. Rev. B* **71**, 144205 (2005).
15. R. Inoue, T. Kanaya, T. Miyazaki, K. Nishida, I. Tsukushi, and K. Shibata, *Mater. Sci. Eng. A* **442**, 367 (2006).
16. C. Le Quellec, G. Dosseh, F. Audonnet, N. Brodie-Linder, C. Alba-Simionesco, W. Haussler, and B. Frick, *Eur. Phys. J. Special Topics* **141**, 11 (2007).

17. K.-L. Akabori, K. Tanaka, A. Takahara, T. Kajiyama, and T. Nagamura, *Eur. Phys. J. Special Topics* **141**, 173 (2007).
18. C.B. Roth and J.M. Torkelson, *Macromolecules* **40**, 3328 (2007).
19. C.G. Robertson, T.E. Hogan, M. Rackaitis, J.E. Puskas, and X. Wang, *J. Chem. Phys.* **132**, 104904 (2010).
20. M. Wubbenhorst, C.A. Murray, and J.R. Dutcher, *Eur. Phys. J. E* **12**, S109 (2003).
21. D.C. Bassett and R. Davitt, *Polymer* **15,** 721 (1974).
22. S.Z.D. Cheng, Z.Q. Wu, and B. Wunderlich, *Macromolecules* **20**, 2802 (1987).
23. J.A. Forrest and J. Mattsson, *Phys. Rev. E* **61**, R53 (2000).
24. C.J. Ellison and J.M. Torkelson, *Nature Mater.* **2**, 695 (2003).
25. R.D. Priestley, C.J. Ellison, L.J. Broadbelt, and J.M. Torkelson, *Science* **309**, 456 (2005).
26. R. Inoue, T. Kanayal, K. Nishida, I. Tsukushi, J. Taylor, S. Levett, and B.J. Gabrys, *Eur. Phys. J. E* **24**, 55 (2007).
27. F. He, L.-M. Wang, and R. Richert, *Eur. Phys. J. Special Topics* **141**, 3 (2007).
28. K.L. Ngai, *J. Polym. Sci. Polym. Phys. Ed.* **44**, 2980 (2006).
29. P. Scheidler, W. Kob, K. Binder, and G. Parisi, *Philos. Mag. B* **82**, 283 (2002).
30. K.L. Ngai, *Philos. Mag. B* **82**, 291 (2002).
31. J.A. Forrest and K. Dalnoki-Veress, *Adv. Coll. Interface Sci.* **94**, 167 (2001).
32. C.B. Roth, A. Pound, S.W. Kamp, C.A. Murray, and J.R. Dutcher, *Eur. Phys. J. E* **20**, 441 (2006).
33. C.G. Campbell and B.D. Vogt, *Polymer* **48**, 7169 (2007).
34. K. Fukao, *Eur. Phys. J. E* **12**, 119 (2003).
35. T.A. Tran, S. Said, and Y. Grohens, *Composites: Part A* **36**, 461 (2005).
36. A. Schonhals, H. Goering, C. Schick, B. Frick, and R. Zorn, *Eur. Phys. J. E* **12**, 173 (2003).
37. D.J. Pochan, E.K. Lin, S.K. Satija, and W.-L. Wu *Macromolecules* **34**, 3041 (2001).
38. C.L. Soles, J.F. Douglas, R.L. Jones, and W.-L. Wu, *Macromolecules* **37**, 2890 (2004).
39. Y. Zhaohui, Y. Fujii, F.K. Lee, C.-H. Lam, and O.K.C. Tsui, *Science*, **328**, 1676 (2010).
40. S. Herminghaus, K. Jacobs, and R. Seemann, *Eur. Phys. J. E* **12**, 101 (2003)
41. J.L. Keddie, R.A.L. Jones, and R.A. Cory, *Europhys. Lett.* **27**, 59 (1994).
42. C.J. Ellison, M.K. Mundra, and J.M. Torkelson, *Macromolecules* **38**, 1767 (2005).
43. K. Shin, S. Obukhov, J.-T. Chen, J. Huh, Y. Hwang, S. Mok, P. Dobriyal, P. Thiyagarajan, and T.P. Russell, *Nature Mater.* **6**, 961 (2007).
44. F. Persson and J.O. Tegenfeldt, *Chem. Soc. Rev.*, **39**, 985 (2010).
45. S.L. Levy and H.G. Craighead, *Chem. Soc. Rev.*, **39**, 1133 (2010).
46. Z. Fakhraai, S. Valadkhan, and J.A. Forrest, *Eur. Phys. J. E* **18**, 143 (2005).
47. K. Dalnoki-Vereess, J.A. Forrest, P.G. deGennes, and J.R. Dutcher, *J. Phys. IV France* **10**, Pr7–221 (2000).
48. R. Inoue, T. Kanaya, K. Nishida, I. Tsukushi, M.T.F. Telling, B.J. Gabrys, M. Tyagi, C. Soles, and W.-L. Wu, *Phys. Rev. E* **80**, 031802 (2009).
49. J.N. Coleman, U. Khan, W.J. Blau, and Y.K Gunko *Carbon* **44**, 1624 (2006).
50. R. Sengupta, S. Chakraborty, S. Bandyopadhyay, S. Dasgupta, R. Mukhopadhyay, K. Auddy, and A.S. Deuri, *Polym. Eng. Sci.* **47**, 1956 (2007).

51. M. Maiti, M. Bhattacharya, and A.K. Bhowmick, *Rubber Chem. Technol.* **81**, 384 (2008).
52. J.E. Mark, *Acct. Chem. Res.* **39**, 881 (2006).
53. Y.T. Vu, J.E. Mark, L.H. Pham, and M. Engelhardt, *J. Appl. Polym. Sci.* **82**, 1391 (2001).
54. M. Galimberti, S. Senatore, L. Conzatti, G. Costa, G. Giuliano, and G. Guerra, *Polym. Adv. Technol.* **20**, 135 (2009).
55. Y.-R. Liang, J. Ma, Y.-L. Lu, Y.-P. Wu, L.-Q. Zhang, and Y.-W. Mai, *J. Polym. Sci. Polym. Phys. Ed.* **43**, 2653 (2005).
56. L. Bokobza, *Polymer* **48**, 4907 (2007).
57. L.F. Valadares, C.A.P. Leite, and F. Galembeck, *Polymer* **47**, 672 (2006).
58. A. Bansal, H. Yang, C. Li, K. Cho, B.C. Benicewicz, S.K. Kumar, and L.S. Schadler, *Nature Mater.* **4**, 693 (2005).
59. B.J. Ash, L.S. Schadler, and R.W. Siegel, *Mater. Lett.* **55**, 83 (2002).
60. H. Oh and P.F. Green, *Nature Mater.* **8**, 139 (2008).
61. G.A. Schwartz, R. Bergman, J. Mattsson, and J. Swenson, *Eur. Phys. J. E* **12**, s13 (2003).
62. M.M. Elmahdy, K. Chrissopoulou, A. Afratis, G. Floudas, and S.H. Anastasiadis, *Macromolecules* **39**, 5170 (2006).
63. J. Mijovic, H.K. Lee, J. Kenny, and J. Mays, *Macromolecules* **39**, 2172 (2006).
64. Y.-H. Lee, A.J. Bur, S.C. Roth, and P.R. Start, *Macromolecules* **38**, 3828 (2005).
65. C.G. Robertson and C.M. Roland, *Rubber Chem. Technol.* **81**, 506 (2008).
66. S.E. Harton, S.K. Kumar, H. Yang, T. Koga, K. Hicks, H.K. Lee, J. Mijovic, M. Liu, R.S. Vallery, and D.W. Gidley, *Macromolecules* **43**, 3415 (2010).
67. J.M. Kropka, V. Pryamitsyn, and V. Ganesan, *Phys. Rev. Lett.* **101,** 075702 (2008).
68. S. Sen, Y. Xie, A. Bansal, H. Yang, K. Cho, L.S. Schadler, and S.K. Kumar, *Eur. Phys. J. Special Top.* **141**, 161 (2007).
69. P. Rittigstein and J.M. Torkelson, *J. Polym. Sci. Polym. Phys. Ed.* **44**, 2935 (2006).
70. J.M. Kropka, V.G. Sakai, and P.F. Green, *Nano Lett.* **8**, 1061 (2008).
71. C. Hub, S.E. Harton, M.A. Hunt, R. Fink, and H. Ade, *J. Polym. Sci. Polym. Phys. Ed.* **45**, 2270 (2007).
72. G. Tsagaropoulos and A. Eisenberg, *Macromolecules* **28**, 396 (1995); **28**, 6067 (1995).
73. P.P.A. Smit, *Rheol. Acta* **5**, 277 (1966).
74. V. Arrighi, I.J. McEwen, H. Qian, and M.B. Serrano Prieto, *Polymer* **44**, 6259 (2003).
75. E.I. Lopez-Martinez, A. Marquez-Lucero, C.A. Hernandez-Escobar, S.G. Flores-Gallardo, R. Ibarra-Gomez, M.J. Yacaman, and E.A. Zaragoza-Contreras, *J. Polym. Sci. Polym. Phys.* **45**, 511 (2007).
76. S. Vieweg, R. Unger, G. Heinrich, and E. Donth, *J. Appl. Polym. Sci.*, **73**, 495 (1999).
77. J.C. Kenny, V.J. McBrierty, Z. Rigbi, and D.C. Douglass, *Macromolecules* **24**, 436 (1991).
78. R.B. Bogoslovov, C.M. Roland, A.R. Ellis, A.M. Randall, and C.G. Robertson, *Macromolecules* **41**, 1289 (2008).
79. V. Arrighi, J.S. Higgins, A.H. Burgess, and G. Floudas, *Polymer* **39**, 6369 (1998).
80. J.G.I. Rodriguez, P. Carreira, A. Garcia-Diez, D. Hui, R. Artiaga, and L.M. Liz-Marzan, *J. Therm. Anal. Calorim.* **87**, 45 (2007).
81. M.A. Waldrop and G. Kraus, *Rubber Chem. Technol.* **42**, 1155 (1969).
82. W. Hu, M.D. Ellus, A.H. Tsou, and S. Datta, *Rubber Chem. Technol.* **80**, 1 (2007).

83. H. Serizawa, T. Nakamura, M. Ito, K. Tanaka, and A. Nomura, *Polymer J.* **15**, 201 (1983).
84. R.J. Roe, D.D. Davis, and T.K. Kwei, *Bull. Am. Phys. Soc. II* **15**, 308 (1970).
85. J.O'Brien, E. Cashell, G.E. Wardell, and V.J. McBrierty, *Macromolecules* **9**, 653 (1976).
86. J. Schaefer, S.H. Chin, and S.I. Weissman, *Macromolecules* **5**, 798 (1972).
87. C.R. Dybowski and R.W. Vaughan, *Macromolecules* **8**, 50 (1975).
88. N.K. Dutta, N.R. Choudhury, B. Haidar, A. Vidal, J.B. Donnet, L. Delmotte, and J.M. Chezeau, *Polymer* **35**, 4293 (1994).
89. B. Meissner, *J. Appl. Polym. Sci.* **50**, 285 (1993).
90. V.M. Litvinov and P.A.M. Steeman, *Macromolecules* **32,** 8476 (1999).
91. J.C. Kenny, V.J. McBrierty, Z. Rigbi, and D.C. Douglass, *Macromolecules* **24**, 436 (1991).
92. S. Kaufman, W.P. Slichter, and D.D. Davis, *J. Polym. Sci. Polym. Phys.* **9**, 829 (1971).
93. T. Nishi, *J. Polym. Sci. Polym. Phys. Ed.* **12**, 685 (1974).
94. H. Luchow, E. Breier, and W. Gronski, *Rubber. Chem. Technol.* **70**, 747 (1997).
95. N. Kida, M. Ito, F. Yatsuyanygi, and H. Kaido, *J. Appl. Polym. Sci.* **61**, 1345 (1996).
96. M. Cardona, R.V. Chamberlin, and W. Marx, *Annal. Phys.* **16**, 842 (2007).
97. J.P. Cohen Addad and P. Frebourg, *Polymer* **37**, 4235 (1996).
98. J. Leisen, J. Breidt, and J. Kelm, *Rubber Chem. Technol.* **72**, 1 (1999).
99. J. Berriot, F. Lequeux, L. Monnerie, H. Montes, D. Long, and P. Sotta, *J. Non-Cryst. Solids* **307–310**, 719 (2002).
100. G. Simon, K. Baumann, and W. Gronsky, *Macromolecules* **25**, 3624 (1992).
101. C.G. Robertson, C.J. Lin, M. Rackaitis, and C.M. Roland, *Macromolecules* **41**, 2727 (2008).

Glossary of Symbols

a_i	Ogden coefficient
a_T	time-temperature shift factor
$\hat{a}$	Mark-Houwink exponent
A	VFTH parameter
A_D	prefactor in Doolittle equation (eqn 6.1)
A_e	prefactor in relation between entanglement density and packing length
A_H	Helmholtz energy
A_η	coefficient in transient viscosity equation (eqn 3.23)
A_Ψ	coefficient in equation for normal stress overshoot (eqn 3.24)
$\hat{A}$	material constant in Maier-Saupe potential equation
$A(t)$ and $B(t)$	microscopic quantities used to define $C(t)$
B	VFTH parameter
b_i	Ogden coefficient
B_D	coefficient in Doolittle equation (eqn 6.1)
$B_{1,1}$,$B_{2,2}$	functions in constrain-junction elasticity model
c	concentration
c_1, c_2	WLF constants
$\hat{c}$	transverse sound speed
Δc_P	isobaric heat capacity increment at T_g
Δc_V	isochoric heat capacity increment at T_g
C	stress-optical coefficient
C_G	glassy component of stress-optical coefficient
C_R	rubbery component of stress-optical coefficient
$C(t)$	time correlation function
C_{ij}	coefficients in phenomenological strain energy equations
C_n	characteristic ratio
C_p	isobaric heat capacity
$C_{ph}(t)$	time correlation function for phantom network
C_V	isochoric heat capacity
C_∞	high polymer limiting value of the characteristic ratio
D	diffusion constant
D^*	dynamic compliance for uniaxial strain
D_{rot}	rotational diffusion constants
D_R	Rouse diffusion constant
D_t	translational diffusion constant
e	Euler's number ($= 2.718\ldots$)

e	charge of an electron (= 1.60×10^{-19} coulombs)
E	energy
$\hat{E}$	electric field strength
E_a	isobaric activation energy
E_V	isochoric activation energy
E^*	complex dynamic tensile modulus
E^0_a	non-cooperative activation energy of the coupling model
E^*_a	observed (apparent) activation energy in the presence of intermolecular coupling
E^{lin}_a	activation energy for terminal relaxation of linear chain
E^{Br}_a	activation energy for terminal relaxation of branched chain
f_α	relative strength of the α-process in eqn (2.17)
f_{cnstr}	function describing suppression of chain fluctuations in constraint models of rubber
f_e	chain elastic force
f_{gel}	gel fraction of polymer
f_N	trapping factor in eqn (4.55)
f_s	instantaneous value of self-intermediate scattering function
f_V	fractional free volume
$f(\lambda)$	general constraint function in elasticity theories
$f(\theta)$	orientational distribution function
f_2	Hermans orientation function
$F(q,t)$	*intermediate scattering function*
$F_s(q,t)$	*self-intermediate scattering function*
g_p	mode weighting factor in Rouse normal mode spectrum
$g(\epsilon)$, $g(\gamma)$	damping function in constitutive equations
g_{ij}	deformation gradient matrix ($=\frac{\partial x_i}{\partial X_j}$)
g_{11}	radial pair distribution function
G_N	plateau modulus
$G(t)$	stress relaxation modulus
G_∞	high frequency limiting value of shear modulus
$G''(\omega)$	dynamic shear modulus
h	film thickness or distance between particles
$\hat{h}$	parameter in eqns (10.2) and (10.3) for effect of confinement on T_g
h_0	surface thickness (equation 10.1)
$H(\tau)$	distribution of relaxation times
$H_K(\tau)$	Kohlrausch distribution of relaxation times
I_1, I_2, I_3	strain invariants
I_m	maximum value of first strain invariant in Gent model
$J(t)$	compliance function

J_r	recoverable compliance
$J^0{}_S$	steady state recoverable compliance
k	summation index in eqn 1.9
k_B	Boltzmann constant
k_{DSE}	fractional Debye-Stoke-Einstein exponent
k_{FF}	parameter in eqn 1.16 for $T_g(M_n)$
k_{UK}	parameter in eqn 1.18 for $T_g(M_n)$
$k(t)$	time dependent rate constant
k_0	rate constant for exponential decay
k	wave vector
$\lvert\mathrm{k}\rvert$	wave number
K	phonon force constant
$K(x)$	function in constrained-junction model
$\hat{K}$	Mark-Houwink coefficient
l	backbone bond length
L_c	chain contour length
l_K	Kuhn step length
l_p	packing length (length scale at which polymers begin to interpenetrate)
m	fragility
m_{LJ}	repulsive exponent in LJ potential
m_V	isochoric fragilty
m_0	repeat unit molecular weight
M	molecular weight
M_{Br}	branch molecular weight
M_c	molecular weight between crosslinks
M_e	molecular weight between entanglements
M_n	number average molecular weight
M_{sol}	soluble fraction of polymer
M_w	weight average molecular weight
M_V	viscosity average molecular weight
M_O	number of terms in Mooney and Ogden equations
m_0	monomer (repeat unit) molecular weight
n_i	refractive index
n	coupling parameter of coupling model
n_b	number of backbone bonds
n_K	number of Kuhn steps
n_x, n_y, n_z	principal components of refractive index
Δn	birefringence
N	network chains per unit volume
$\hat{n}$	number of chains (used in definition of number avr. molecular weight)

N_b	number of Gaussian submolecules in chain
N	degree of polymerization
N_b	number of Gaussian submolecules in eqn (1.31)
$\hat{N}$	= parameter in Arruda-Boyce model characterizing limited chain extensibility
N_A	Avogadro's number (or Avogadro constant)
N_c	number of monomer units between crosslink junctions
Δn_{cc}	birefringence for constrained chain model
$\Delta n_{phantom}$	birefringence conforming to stress optical law
N_c	number of dynamically correlated molecules or chain segments
N_1	1st normal stress
O^*	complex strain-optical coefficient
p	mode index of Rouse model
P	polarization
$\hat{p}_A$	probability of filling a given site with segment A
P_i	polydispersity index
P_j	Legendre polynomial
p_0	initial pressure
$P(r)$	end-to-end probability distribution function
p_v	virial pressure
Q	momentum transfer
r	spatial distance
r_i	spatial coordinate
$\bar{r}$	spherical radius in eqn (2.20)
$\langle r^2 \rangle$	mean square end-to-end distance
r_{LJ}	molecular size in Lennard-Jones potential
r_0	amplitude in equation of motion of Rouse
R	gas constant
R_g	radius of gyration
$\langle R_g^2 \rangle$	mean square radius of gyration
$\hat{R}$	= cycle rank
S	entropy
S_{ref}	reference entropy
S_f	static structure factor
t_c	crossover time of coupling model
T_A	temperature at the onset of Arrhenius behavior
T_B	temperature at the dynamic crossover
T_c	liquid crystal phase transition temperature
T_f	fictive temperature
T_g	glass transition temperature

$T_{g,\infty}$	high polymer limiting value of glass transition temperature
T_g^{surf}	glass transition temperature of surface layer of film
T_g^{∞}	glass transition temperature of film in limit of large thickness
T_0	Vogel temperature (VFTH equation)
δT	temperature interval glass transition
u	exponent for molecular weight dependence of D in Table 1.2
u	is a unit vector parallel to the bond
$U(r)$	intermolecular potential
V	specific volume
V_c	volume at liquid crystal phase transition
V_{chain}	volume pervaded by a chain
V_X	volume of network during crosslinking
w	Valanis-Landel strain energy function
W	strain energy
x	distance
z	ion charge
$\hat{\alpha}$	sound attenuation coefficient
α_{CC}	Cole-Cole function shape parameter
α_{HN}	Havriliak-Negami shape parameter
$\Delta\alpha_P$	step changes in isobaric thermal expansion
β_{HN}	Havriliak-Negami shape parameter
β_K	Kohlrausch exponent
γ	shear strain
$\hat{\gamma}$	scaling exponent
γ_G	Grüneisen parameter
γ_0	strain amplitude
γ_r	recoverable shear strain
γ	shear rate
Γ	thermodynamic ratio
$tan\ \delta$	loss tangent
ϵ	uniaxial strain
ϵ_{LJ}	Lennard-Jones energy
ϵ_0	static dielectric constant
$\epsilon(t)$	time dependent dielectric permittivity
ϵ’	real part of dielectric permittivity
ϵ”	dielectric loss
$\Delta\epsilon$	dielectric strength
$\Delta\epsilon$,	dielectric strength
ζ	friction factor
ζ_{chain}	friction factor of chain (= n times ζ in chapter 1)

ξ_h	exponent in eqn 10.2 for effect of h on T_g
η_{ss}	steady state shear viscosity
η^+	transient shear viscosity
η_{max}	maximum in transient shear viscosity
η_0	zero-shear-rate viscosity
θ	angle (of scattering or orientation)
κ	constraint severity parameter of elasticity models
κ_T	isothermal compressibility
κ_r	temperature coefficient of chain expansion
$\Delta\kappa_t$	step change in isothermal compressibility
K	function in Flory elasticity model
λ	stretch ratio
$\hat{\lambda}$	wavelength
μ	number density of network junctions
$\hat{\mu}$	ion mobility
ν	frequency in Hertz
ξ	dynamic correlation length
ξ_F	parameter in Flory elasticity model to describe no affine distortion
ξ	dynamic correlation length
Π	Prigogine-Defay ratio
ρ	mass density
ρ_k	wave vector dependent local density
$\hat{\sigma}$	Cauchy (true) stress
σ	engineering stress
σ_{CJ}	elastic stress of constrained-junction model
σ_{DC}	dc-conductivity
$\sigma_{phantom}$	elastic stress of phantom network
σ_V^2	variance of chain length distribution
ς	infinitesimal perturbation (eqn 1.51)
τ_α	local segmental relaxation time
τ_{JG}	Johari-Goldstein secondary relaxation time
τ_{CC}	Cole-Cole relaxation time
τ_{max}	relaxation time determined from peak frequency
$\langle\tau\rangle$	mean relaxation time
τ_K	Kohlrausch relaxation time
τ_p	relaxation time of pth Rouse mode
τ_d	tube disengagement time (longest relaxation time of tube model)
τ_{ph}	correlation time of phantom network
τ_R	longest Rouse relaxation time
τ_{rec}	time for recovery after cessation of shear flow

τ_0	non-cooperative rel. time of coupling model
τ_∞	prefactor in VFTH equation
τ_η	terminal relaxation time
τ_N	normal mode relaxation time
υ	used again as exponent on M (like in chapter one)
υ_B	Boson peak frequency
ϕ	network junction functionality
φ_A	volume fraction of component A
φ_A^{eff}	local concentration of segment A in Lodge-McLeish model
$\langle \delta\varphi_{PS}^2 \rangle$	mean square concentration fluctutation of component A
φ_A^{self}	local concentration of segment A due to chain connectivity
χ_i ith	order of electric susceptibility tensor
$\chi(t)$	general linear susceptibility
$\chi_4(t)$	four-point dynamic susceptibility
$\chi_T(t)$	two-point dynamic correlation function (approximation to $\chi_4(t)$)
$\chi_T^{\max}$	maximum value of $\chi_T(t)$
$\phi_T =$	Stickel temperature-derivative function
$\phi_P =$	Stickel pressure-derivative function
ψ	expression in Havriliak-Negami equation
ψ_1	first normal stress coefficient
Ψ_{ss}	steady state first normal stress coefficient
Ψ_{max}	overshoot maximum in first normal stress coefficient
ψ_1^0	equilibrium first normal stress coefficient
ω	circular frequency (rad/s)

Index